AF581685

Emerging Power Electronics Technologies for Sustainable Energy Conversion

Emerging Power Electronics Technologies for Sustainable Energy Conversion

Editor

Francisco Perez-Pinal

MDPI • Basel • Beijing • Wuhan • Barcelona • Belgrade • Manchester • Tokyo • Cluj • Tianjin

Editor
Francisco Perez-Pinal
Departamento de Ingeniería
Electrónica
TecNM-Instituto Tecnológico
de Celaya
Celaya, Guanajuato
Mexico

Editorial Office
MDPI
St. Alban-Anlage 66
4052 Basel, Switzerland

This is a reprint of articles from the Special Issue published online in the open access journal *Micromachines* (ISSN 2072-666X) (available at: www.mdpi.com/journal/micromachines/special_issues/Power_Electronics_Converters).

For citation purposes, cite each article independently as indicated on the article page online and as indicated below:

LastName, A.A.; LastName, B.B.; LastName, C.C. Article Title. *Journal Name* **Year**, *Volume Number*, Page Range.

ISBN 978-3-0365-6306-0 (Hbk)
ISBN 978-3-0365-6305-3 (PDF)

Contents

About the Editor

Francisco Perez-Pinal

Francisco J. Perez-Pinal (IEEE M'02-SM'16) received his PhD degree (2008) in electrical engineering from the Universidad Autonoma de San Luis Potosi, Mexico.

He was a research associate at the Manchester University, United Kingdom (2006), and the Illinois Institute of Technology, USA (2007). He was also a CL assistant professor (2010-2011) at McMaster University (Canada), and professor (2011-2013) at Mohawk College of Applied Arts and Technology (Canada). Dr. Perez-Pinal is currently professor and director of the Sustainable Transportation Laboratory (STL) at the Electrical and Electronic Engineering Department at Tecnologico Nacional de Mexico, Instituto Tecnologico de Celaya, Mexico. He is also adjunct associate professor at McMaster University (Canada), in the Mechanical Engineering Department.

Dr. Perez-Pinal is the Head of the Electromobility Research Network, Tecnologico Nacional de Mexico, México. Dr. Perez-Pinal is the recipient of the IETE J C Bose Memorial Award 2020, Best engineering-oriented paper, India. He is also the recipient of the first prize, National best PhD dissertation, area of energy efficiency, IIE, CFE, FIDE, Mexico. He is also a recipient of the National Prize of Renewable Energy, area of innovation, Secretary of Energy, Mexico. He is also a recipient of the COMEXUS-Fulbright Garcia-Robles, USA.

Dr. Perez-Pinal is a member of the IEEE Power Electronics Society (PELS), IEEE Industrial Electronics Society (IES), IEEE Transportation Electrification Community and Technical Committee on Electrical Transportation Systems, IEEE Power Electronics Society. He is the principal author/coauthor of 100 journal and conference papers. His major research interests include power electronics, electric motor drives, transportation electrification, e-mobility, renewable energy systems, and energy conversion systems. Dr. Perez-Pinal serves as Associate Editor for IEEE Access, and Topic Editor for Micromachines, MDPI.

Preface to "Emerging Power Electronics Technologies for Sustainable Energy Conversion"

This Special Issue was edited by Dr. Francisco J. Perez-Pinal to help you explore the latest research on power electronics (PE) at sustainable energy conversion (SEC).

In broad terms, PE is an electronics'branch that links the best effective way for energy suppliers and loads, the applications of which can vary from a milliwatts to megawatts. At the beginning of the last decade, PE was believed to be a mature technology; however, in the last few years, emerging PE´s concepts and components have recently appeared, i.e., wide-bandgap semiconductors, advanced modelling methods, complex control techniques, and innovative ferromagnetic materials. As a result, PE is experiencing a transformation, and some examples are SEC, renewable energy (RE), electrified mobility (E-mobility), industry 4.0, digitalization, and digital twins. Therefore, this Special Issue summarizes in a single reference some stylish and notable solutions that deal with this PE´s revolution on SEC.

Unlike another Special Issues on PE at SEC, basic experience on converter design, modelling tools, classical control techniques, and numerical analysis on AC-DC, DC-DC, DC-AC and AC-AC is assumed. The manuscripts presented in this Special Issue cover the fundamentals and applications of silicon-carbide and gallium trioxide semiconductors, modelling methods, advanced control techniques, novel converter configurations, and up-to-date ferromagnetic materials.

It is important to mention that you do not need to read the content in sequence, as each manuscript is self-explained; and those contain a section of motivation, detailed description, numerical and experimental validation results, a comparison with other techniques reported at the state of the art and concluding remarks, this last with the main aim to enhance the development of new solutions to current and coming demands.

There is not room in this Special Issue to cover another emerging area of PE in topics such as: renewable energy (RE), electrified mobility (E-mobility), industry 4.0, digitalization, and digital twins. For more details on those hot topics, interested readers are referred to other Special Issues of MDPI´s journals and books.

I would like to take this occasion to acknowledge all the authors for submitting their work to this Special Issue. I would also like to recognize all the anonymous reviewers for contributing their efforts to providing precise and timely reviews to ensure the quality of this Special Issue. Last but not least, I would like to thank *Micromachines*'editorial staff for providing me guidance during the overall steps related to the editorial process.

Francisco Perez-Pinal
Editor

MDPI

Editorial

Editorial for the Special Issue on Emerging Power Electronics Technologies for Sustainable Energy Conversion †

Francisco J. Perez-Pinal 

Tecnológico Nacional de México, Instituto Tecnológico de Celaya, Antonio García Cubas Pte. 600, Celaya 38010, Mexico; francisco.perez@itcelaya.edu.mx

† This article belongs to the Special Issue Emerging Power Electronics Technologies for Sustainable Energy Conversion.

Citation: Perez-Pinal, F.J. Editorial for the Special Issue on Emerging Power Electronics Technologies for Sustainable Energy Conversion. *Micromachines* **2022**, *13*, 539. https://doi.org/10.3390/mi13040539

Received: 28 March 2022
Accepted: 29 March 2022
Published: 30 March 2022

Power electronic (PE) technology became considered a mature technology over the last century. Widespread silicon-based components and the emergence of modern modelling and control techniques have been the basis of the latest technological era.

Nevertheless, today we are living through a PE revolution, essentially motivated by three components: the arrival and application of wide-bandgap semiconductor technologies such as silicon-carbide, gallium-nitride and others in advanced power converter configurations; the appearance of improved modelling and control techniques such as non-integer philosophy to design fractional controllers; and the entrance of advanced ferromagnetic materials such as amorphous materials, nanocrystalline, powder cores, and Fe-Si alloys, to design magnetic components. The addition and combination of those three basic factors is predicted to be a feasible proposal to solve challenges related to cost, power consumption, power density and efficiency towards the development of small size and weight PE applications.

Accordingly, this Special Issue summaries nine research papers and one review on a wide range of power electronics technologies for sustainable energy conversion. Half of the manuscripts of this Special Issue [1–5] report applications of wide-bandgap and novel power electronic converters. Four of the papers [6–9] describe topics related to the modelling and control techniques of PE, and the remaining paper covers various aspects related to ferromagnetic materials.

In particular, Ma et al. [1] propose a distributed low voltage ride-through compensator (LVRTC) to be used in every wind turbine generator (WTG) for wind-farm applications. A step-by-step procedure to model and design the proposed controller using a dq-axis current decoupling algorithm and quantitative design of the three-phase inverter controller is given. Numerical results and hardware tests using a 2 kVA, six Silicon-Carbide switching device and a digital signal processor (DSP) as control core are reported. It is stated that the application of distributed LVRTCs increases reliability and flexibility, but decreases the required capacity and total price.

Following the same trend, the performance of an active neutral-point-clamped (ANPC) inverter operating with silicon (Si) and gallium trioxide (Ga_2O_3) devices is evaluated [2]. In this proposal, both devices are switched at different rates. The Si is operated at low frequency and the Ga_2O_3 is operated at high frequency. This switching configuration results in a homogeneously distributed power loss. Another interesting characteristic of the proposed arrangement is that the common mode current stress on the switches is reduced, a low common-mode voltage on the output is achieved and the fault current in the switching devices is mitigated. It is necessary to mention that the hybridization of ultra-wide bandgap (UWBG) semiconductor devices is still in development and its early stages. However, these promising results open the door to soon-achievable advances in this research direction.

On the other hand, a novel three-phase six-level multilevel inverter (MLI) controlled by a nearest-vector modulation strategy is reported in [3]. Numerical and practical results on a low-power test bed were reported. Advantages of the proposed MLI include a reduced number of devices, low total harmonic distortion (THD), and high efficiency (near 98%). The addition of the previous characteristic makes it suitable for high-voltage operations, utilization as an interconnected device in grid systems, and to be applied as a current and voltage compensation system, i.e., shunt active power filter (SAPF), dynamic voltage restorer (DVR) and unified power quality conditioner (UPQC).

Another interesting converter proposal was reported in [4]. In this manuscript, Loera-Palomo et al. proposed a novel non-cascade quadratic structure based on the interconnection of basic step-up and step-down configurations. The proposed converter has a noninverting output voltage and a minimum number of devices, and its quadratic property is valid only in continuous inductor conduction mode. A steady state (voltage and current source) boundary between continuous and discontinuous conduction mode, semiconductor stress analysis, dynamic model, numerical and experimental results are reported. In addition, a perturb and observe (P&O) maximum power point tracking was used to interconnect the reported system to a PV module. It is important to remark that the converter allows for a wide input voltage variation with a fixed output voltage, which is translated into a high-voltage conversion ratio property. This is a desirable characteristic in renewable energy applications where the energy source is variable.

Following this trend, in [5] a different converter with a quadratic voltage gain, common ground, and non-series power transfer is proposed. It consists of two traditional boost converters with a capacitor between both, which allows for parallel energy transmission between converters and output. The operation characteristics considering volt-second steady-state analysis, continuous conduction mode, ideal components, and pre-defined switching commutation, are reported. Additionally, a qualitative and practical quantitative comparison with other quadratic converters is also given, and bilinear and linear models of the proposed converter are also developed. The main advantage of this converter is that it increases the converter's power efficiency due to the non-series power transfer, which was verified by a 500 W laboratory prototype. It is worth noting that the quadratic gain and practical high efficiency reported, above 95%, make this converter suitable for energy-harvesting systems.

Alternatively, a new resistive-inductance (RL) model for a proton-exchange membrane fuel cell (PEMFC) is reported in [6], which is compared with three well-known counterparts, i.e., nonlinear state-space, generic MATLAB™, and a resistive capacitance (RC). A step-by-step example to determine the proposed RL model's parameters is also given. The four models were compared with static and dynamic responses of a practical 1.2 kW fuel-cell module; current–voltage graph, and time/voltage, respectively. Root-mean-square error (RMSE) was selected as the criterion during the static performance. On the contrary, for the dynamic performance, a load change of almost three times was applied and the different models' performances were recorded. It was noticed that the non-linear model achieves the best functioning compared with the practical results, but it has the highest computational effort. Indeed, the proposed RL model achieves the second-best performance in both tests, but it is simple and accurate, which makes it a good candidate for real-time applications.

Following this scope, an all-inclusive conduction-mode independent (CMI) modelling for a boost converter is reported in [7]. This model is valid for the following cases: switched constant power (CPL), non-switched CPL, switched resistive load, and unswitched resistive load. A nonlinear controller, stability analysis based on Lyapunov, and numerical and experimental validation are also described for the CMI model with CPL. It is important to remark that, in contrast to traditional alternatives, the proposed controller does not rest on frequency control, and it is able to stabilize the boost converter with a CPL regardless of the operating conduction mode.

Another interesting manuscript is presented in [8]. Three pulse-width modulation methods derived from the traditional Space Vector PWM (SVPWM) technique were reported and applied to the transformerless photovoltaic DCM-232. This consists of a six-switch three-phase inverter plus four switches and two diodes that decouple photovoltaic supplies to generate two independent DC sources. The main aim of the proposed modulation methods was to control the decoupling switches to keep the common mode voltage constant, thereby achieving a leakage ground-current reduction. Numerical and practical results and a comparative efficiency analysis between the proposed PWM techniques and selected alternatives available in the literature were performed. It was concluded that all proposed modulation techniques were able to reduce the common mode current. As a result, the power inverter was able to fulfill international connection standards.

An advanced non-integer control strategy for a buck/boost converter was reported in [9]. To develop such kind of controller, the following steps were followed: an approximation, synthesis, and an implementation pathway. Here, the proposed non-linear synthesis is achieved through a biquadratic module that exhibits a flat phase response. Numerical and experimental results are provided to validate the usefulness of this approach. It is important to mention that implementation was carried by using analog operational amplifiers, which helped to achieve a fast practical response. However, the proposed non-integer approximated controller can also be extended to a digital version. To this end, traditional direct or indirect digital approximation can be used by using any sampled transformation.

Finally, a review of representative ferromagnetic core loss models is given in [10]. It includes amorphous, nanocrystalline powder cores and Fe-Si alloys. Characteristics, advantages, limitations, and applications of the more popular developed models reported in the literature are given. It is concluded that due to the non-linear features of ferromagnetic materials, the complexity of developing a unique power core loss model, which includes different ferromagnetic materials, and its respective validation, it is still incipient. Therefore, magnetic, and thermal loss calculus, modelling and ferromagnetic validation are still open research areas. The development of new ferromagnetic materials with specific features (core material, waveform, density flux, application, frequency range, among others) is still in the research and validation stages.

I would like to take this opportunity to thank all authors for submitting their manuscripts to this Special Issue. I would also like to acknowledge all reviewers for dedicating their time to providing careful and timely reviews to ensure the quality of this Special Issue.

Funding: This research received no external funding.

Acknowledgments: The author would like to thank CONACyT México, and TecNM.

Conflicts of Interest: The author declares no conflict of interest.

References

1. Ma, C.-T.; Shi, Z.-H. A Distributed Control Scheme Using SiC-Based Low Voltage Ride-through Compensator for Wind Turbine Generators. *Micromachines* **2022**, *13*, 39. [CrossRef] [PubMed]
2. Meraj, S.T.; Yahaya, N.Z.; Hossain Lipu, M.S.; Islam, J.; Haw, L.K.; Hasan, K.; Miah, M.S.; Ansari, S.; Hussain, A. A Hybrid Active Neutral Point Clamped Inverter Utilizing Si and Ga_2O_3 Semiconductors: Modelling and Performance Analysis. *Micromachines* **2021**, *12*, 1466. [CrossRef] [PubMed]
3. Meraj, S.T.; Yahaya, N.Z.; Hasan, K.; Hossain Lipu, M.S.; Masaoud, A.; Ali, S.H.M.; Hussain, A.; Othman, M.M.; Mumtaz, F. Three-Phase Six-Level Multilevel Voltage Source Inverter: Modeling and Experimental Validation. *Micromachines* **2021**, *12*, 1133. [CrossRef] [PubMed]
4. Loera-Palomo, R.; Morales-Saldaña, J.A.; Rivero, M.; Álvarez-Macías, C.; Hernández-Jacobo, C.A. Noncascading Quadratic Buck-Boost Converter for Photovoltaic Applications. *Micromachines* **2021**, *12*, 984. [CrossRef] [PubMed]
5. Diaz-Saldierna, L.H.; Leyva-Ramos, J. High Step-Up Converter Based on Non-Series Energy Transfer Structure for Renewable Power Applications. *Micromachines* **2021**, *12*, 689. [CrossRef] [PubMed]
6. Belhaj, F.Z.; El Fadil, H.; El Idrissi, Z.; Intidam, A.; Koundi, M.; Giri, F. New Equivalent Electrical Model of a Fuel Cell and Comparative Study of Several Existing Models with Experimental Data from the PEMFC Nexa 1200 W. *Micromachines* **2021**, *12*, 1047. [CrossRef] [PubMed]
7. Parada Salado, J.G.; Herrera Ramírez, C.A.; Soriano Sánchez, A.G.; Rodríguez Licea, M.A. Nonlinear Stabilization Controller for the Boost Converter with a Constant Power Load in Both Continuous and Discontinuous Conduction Modes. *Micromachines* **2021**, *12*, 522. [CrossRef] [PubMed]

8. Vazquez-Guzman, G.; Martinez-Rodriguez, P.R.; Sosa-Zuñiga, J.M.; Aztatzi-Pluma, D.; Langarica-Cordoba, D.; Saldivar, B.; Martínez-Méndez, R. Hybrid PWM Techniques for a DCM-232 Three-Phase Transformerless Inverter with Reduced Leakage Ground Current. *Micromachines* **2022**, *13*, 36. [CrossRef] [PubMed]
9. Sánchez, A.G.S.; Soto-Vega, J.; Tlelo-Cuautle, E.; Rodríguez-Licea, M.A. Fractional-Order Approximation of PID Controller for Buck–Boost Converters. *Micromachines* **2021**, *12*, 591. [CrossRef] [PubMed]
10. Rodriguez-Sotelo, D.; Rodriguez-Licea, M.A.; Araujo-Vargas, I.; Prado-Olivarez, J.; Barranco-Gutiérrez, A.-I.; Perez-Pinal, F.J. Power Losses Models for Magnetic Cores: A Review. *Micromachines* **2022**, *13*, 418. [CrossRef] [PubMed]

Article

A Distributed Control Scheme Using SiC-Based Low Voltage Ride-Through Compensator for Wind Turbine Generators

Chao-Tsung Ma * and Zong-Hann Shi

Applied Power Electronics Systems Research Group, Department of EE, CEECS, National United University, Miaoli City 36063, Taiwan; M0921008@o365.nuu.edu.tw
* Correspondence: ctma@nuu.edu.tw

Abstract: As the penetration of renewable energy power generation, such as wind power generation, increases low voltage ride-through (LVRT), control is necessary during grid faults to support wind turbine generators (WTGs) in compensating reactive current to restore nominal grid voltages, and maintain a desired system stability. In contrast to the commonly used centralized LVRT controller, this study proposes a distributed control scheme using a LVRT compensator (LVRTC) capable of simultaneously performing reactive current compensation for doubly-fed induction generator (DFIG)-, or permanent magnet synchronous generator (PMSG)-based WTGs. The proposed LVRTC using silicon carbide (SiC)-based inverters can achieve better system efficiency, and increase system reliability. The proposed LVRTC adopts a digital control scheme and dq-axis current decoupling algorithm to realize simultaneous active/reactive power control features. Theoretical analysis, derivation of mathematical models, and design of the control scheme are initially conducted, and simulation is then performed in a computer software environment to validate the feasibility of the system. Finally, a 2 kVA small-scale hardware system with TI's digital signal processor (DSP) as the control core is implemented for experimental verification. Results from simulation and implementation are in close agreement, and validate the feasibility and effectiveness of the proposed control scheme.

Keywords: wind turbine generator (WTG); permanent magnet synchronous generator (PMSG); low voltage ride-through (LVRT); silicon carbide (SiC)-based inverter

Citation: Ma, C.-T.; Shi, Z.-H. A Distributed Control Scheme Using SiC-Based Low Voltage Ride-Through Compensator for Wind Turbine Generators. *Micromachines* **2022**, *13*, 39. https://doi.org/10.3390/mi13010039

Academic Editor: Francisco J. Perez-Pinal

Received: 25 November 2021
Accepted: 27 December 2021
Published: 28 December 2021

1. Introduction

In recent years, the worldwide demand for energy has been rapidly growing to support the development of various modern technologies; thus, the vast use of fossil fuels has negatively affected the environment. Therefore, renewable energy (RE)-based distributed power generation (DG) has been intensively researched in the past decade. In the near future, the penetration of various grid-connected DG systems is expected to keep increasing. As a result, utilizing advanced power converter-based compensators to work with RE-based generation systems and existing grids are one of the most important tasks in ensuring the voltage stability and power quality (PQ) of distribution systems regardless of the naturally intermittent characteristic of RE-based power generation. Wind power generation is currently one of the most promising RE sources because it can generate a considerable amount of electrical energy in a short period of time. Commonly used generators for wind turbine generators (WTGs) include permanent magnet synchronous generators (PMSGs) [1], doubly-fed induction generators (DFIGs) [2], and brushless wound-rotor doubly-fed generators [3]. Currently, most research papers on grid-connected PMSG-based WTGs (PMSG-WTGs) found in open literature have focused on advanced control schemes, low voltage ride-through (LVRT) capacity, and operating modes. M. Jamil et al. [4] suggested that, among commonly used generators in WTGs, PMSGs have more advantages compared with conventional DC and induction generators. The topologies of PMSG-WTG power converters were reviewed, including a thyristor-based inverter, hard-switched pulse with modulation (PWM) converter, multilevel converter, matrix converter, and Z-source inverter. For conventional small- and medium-size WTGs, power interfaces usually adopt the configuration that consists of a diode bridge rectifier and a buck boost converter. Although this configuration has the advantages of a simple circuit, low cost, and easy control, it causes a large amount of current harmonic and lowered power factor (PF) in the three-phase AC side, which leads to voltage pulses that lower PQ and efficiency, and shorten the life of the generator. A good WTG power interface must offer high conversion efficiency, low power consumption, PF correction, and low current harmonic. When a WTG is connected to a grid, appropriate control strategy is necessary to meet the power company's LVRT specifications. On this basis, this study proposes a distributed control scheme to enhance LVRT capabilities of small- and medium-size PMSG-WTGs using silicon carbide (SiC)-based inverters.

Generally, PQ improvement is necessary for distribution systems embedded with DG systems because of the uncertainty and intermittency of RE sources, which can easily result in various undesirable effects, such as voltage sag. In [5], a grid-connected direct-drive PMSG-WTG model was simulated for various wind speeds and load conditions using PI-based controller. The results showed a properly regulated DC link voltage and output power. M. H. Qais et al. [6] reviewed LVRT capability enhancement methods for a grid-connected PMSG driven directly by a variable speed WT (VSWT). Later, the authors used a gray wolf optimizer (GWO) to tune eight PI controllers to improve LVRT capability and MPPT performance. In comparison with the genetic algorithm (GA) and simplex method, GWO algorithm yielded the best convergence, MPPT capability, and LVRT performance during symmetrical and asymmetrical faults [7]. In [8], a resistor-type superconducting fault current limiter (R-SFCL) was combined with superconducting magnetic energy storage (SMES) to improve the LVRT capability of a 2.5 MW PMSG-WTG. Improved transient stability and reduced cost for superconducting devices were achieved. P. Xing et al. [9] proposed a fast compositive control of LVRT for a 2 MW PMSG-WTG by consuming initial excessive energy with a crowbar circuit, and then converting excessive energy into rotor kinetic energy. In [10], a perturbation-observing nonlinear adaptive control was proposed for a dynamic voltage restorer (DVR) embedded with an energy storage system (ESS). This simple and robust control scheme does not require an accurate system model and full-state feedback. The results showed better performance than conventional fixed-gain vector control (VC) and feedback linearizing control based on an accurate system model. The authors used 10 MW PMSG- and DFIG-based WTGs to verify the control. M. Jahanpour-Dehkordi et al. [11] combined VC and direct torque control (DTC) using two hysteresis current controllers to achieve fast and smooth LVRT for a 2 MW PMSG-WTG. This method is better than DTC control and VC control in terms of smoothness and speed, respectively. In [12], the LVRT capability of a 2.5MVA full-scale PMSG-WTG was enhanced with active damping control and DC-link voltage bandwidth retuning. M. Nasiri and R. Mohammadi [13] enhanced the LVRT capability of a 1.5 MW PMSG-WTG with proposed back-to-back converter controllers and active power limiter without the need for external devices.

Some advanced control strategies for DFIG and PSMG wind turbines can be found in [14–18]. In [14], advanced control strategies were proposed to control the pitch angle, and the rotor and grid side converters of the doubly fed induction generator (DFIG)-based wind turbine (WT) for enhancing the LVRT capability. The converter systems used in the study include a back-to-back converter and a DC chopper. Simulation results showed that with the proposed control means, the LVRT capability of WT generators can be enhanced, and the oscillations in the stator and rotor currents can be effectively reduced. Both low-wind and high-wind speeds cases were investigated to verify the proposed control algorithm. With the same hardware system used in [14], the same group of authors of [15] proposed a LVRT control method to achieve the real-time regulation of the rotor's excessive inertia energy. The reactive power capacity was simultaneously handled to satisfy the grid code.

It was demonstrated with simulation results that the control topology or parameters can be changed at different stages of voltage fault period. In [16], a LVRT control scheme for the permanent-magnet synchronous generator (PMSG) with a back-to-back neutral-point-clamped, three-phase converter was proposed. In the studied grid voltage dip scenario, the proposed controllers designed for generator-side and grid-side converters can work concurrently to meet the LVRT requirement by storing the active power in the mechanical system of WT, and regulating the dc-link voltage at a constant value. Results from both simulation and experimental tests verified the effectiveness of the proposed method. The system stability issues of weak AC grid-connected DFIG-based wind turbines during LVRT were comprehensively investigated in [17]. To study the instability mechanism of a DFIG with back-to-back converter systems during a weak grid fault, a small signal state-space model was established. The effectiveness of the proposed LVRT control methodologies were validated with simulation and experiments. In [18], a wind speed combination model (WSCM) was proposed using the field-measured data, which was then used as equivalent wind speeds for each WTG when modeling the entire wind farm. Results from a case study verified the correctness of WSCM and the performance of the proposed algorithm in predicting the LVRT entering status of WT generators in the wind farm.

The use of a static synchronous compensator (STATCOM) for LVRT capability enhancement at the point of common coupling (PCC) was discussed in [19–28]. P. Dey et al. [19,20] developed a controller based on a pitch angle control and a flux weakening control and a STATCOM to enhance the LVRT capability of a PMSG-WTG. As a result, enhanced LVRT performance, reduced DC-link capacitor overvoltage, and the connection between the WTG and the grid during faults were achieved compared with conventional braking chopper-based LVRT strategy. In [21], a coordinated current control scheme was proposed for a 100 MW PMSG-WTG with a 20 MW STATCOM for severe LVRT by exchanging control roles between the two converters, such that the DC link voltage was controlled by the machine-side converter (MSC), and the grid active power was controlled by the grid-side converter (GSC). Synchronization was maintained, and reactive power requirement was satisfied. J. Yao et al. [22] studied a cost-effective capacity configuration strategy for the improvement of LVRT capability of a hybrid wind farm consisting of a 30 MW fixed-speed induction generator (FSIG) in conjunction with a 30 MW PMSG by using the PMSG as a STATCOM during grid fault. For LVRT capability improvement of DFIG-WTGs using STATCOMs, a PI controller with robust control technique was proposed in [23], where the reliability of the WTG was ultimately increased, and a fuzzy adaptive proportional-integral-derivative (PID) controller was proposed in [24], with enhanced control accuracy and performance. For self-excited induction generator (SEIG)-WTGs, M. I. Mosaad et al. [25] proposed a model reference adaptive control (MRAC) of STATCOM, yielding more efficient and more robust performance than that of a GA-tuned PI controller. In [26], the LVRT capability enhancement of STATCOM for the terminal voltage of a squirrel cage induction generator (SCIG)-WTG was studied in a wind farm in Bizerte, Tunisia. T. Tanaka et al. [27,28] used an 80MVar/33kV STATCOM based on multilevel converters for LVRT capability enhancement of non-specific-type offshore WTGs.

Most of the above reviewed applications of STATOM for LVRT capability enhancement adopted the configuration of centralized compensation, where a single STATCOM module was employed at the PCC to take charge of the control goal. However, this configuration is expensive, and the required system capacity is normally high. Moreover, if a single malfunction occurs in the STATCOM, then the LVRT capability of the system may deteriorate drastically. In this regard, this study aims to propose a distributed LVRT compensator (LVRTC) for every separate WTG. The use of distributed LVRTCs increases reliability and flexibility, while reducing the required capacity and overall costs. In the next section, LVRT specifications will be explained followed by the proposed distributed LVRTC and WTG system configuration. The mathematical models required for the controller design using a dq-axis current decoupling algorithm and quantitative design of the three-phase inverter controller are explained in Section 3. In Section 4, typical results of simulation in a computer software environment and hardware tests using TI's DSP as control core are presented. Finally, the conclusion of this work is summarized in the last section.

2. LVRT Specifications and WTG System Configuration

2.1. LVRT Specifications

In a wind farm, the stability and reliability of the system are two of the most important factors for operation. When disturbances occur, it is required that a grid-connected WTG doesn't simply disconnect from the grid, such that the power system has a chance to maintain its voltage stability. As a result, a WTG must be able to feed appropriate active power and reactive power to restore the system voltage and frequency back to nominal values. In other words, modern grid-connected WTGs must be equipped with reactive power controls and the ability to eliminate faults, especially for cases where the penetration of wind power generation is high. Specifications of grid-connected WTGs include frequency range, voltage tolerance, and LVRT capability. Among these specifications, LVRT capability is the most important criterion in a WTG embedded system.

To regulate and facilitate the development of various RE generations, German electric utility company, E.ON, has set tolerance requirements for newly installed WTGs, as shown in Figure 1 [29]. In Figure 1, when the grid voltage drops to zone 1 or zone 2, the WTG has to maintain the connection with the grid. For voltage restoration, if the grid voltage drops to zone 2, and the fault is later removed, the WTG must restore the power with a rate of 10% per second. In zone 3, the WTG is allowed to be disconnected from the grid to protect the generator, and when the fault is removed, the WTG must be reconnected with the grid, and restore the power with a rate of 20% per second. If the failure continues for more than 1.5 s, the WTG can be protected from the network through disconnection. Moreover, the WTG should also provide reactive power, as conventional power systems do, to help stabilize and restore the voltage, as shown in Figure 2 [29]. As can be seen in Figure 2, when the voltage drops out of the non-action zone (±10%), the required percentage of reactive current from the WTG is two times the percentage of the voltage sag. For example, when the voltage drop is 50%, the WTG must provide reactive current of 100% system capacity within 20ms, as shown in Equation (1). In Equation (1), i_{fq}^* denotes the per unit value of the reactive current command in q-axes, V_{dN} is the d-axes system nominal voltage, and V_d represents the d-axes monitored system voltage.

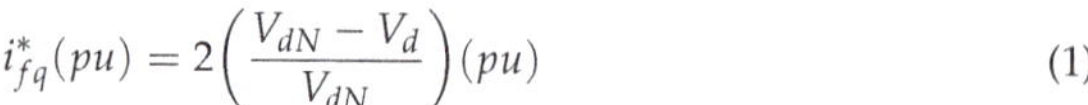

$$i_{fq}^*(pu) = 2\left(\frac{V_{dN} - V_d}{V_{dN}}\right)(pu) \quad (1)$$

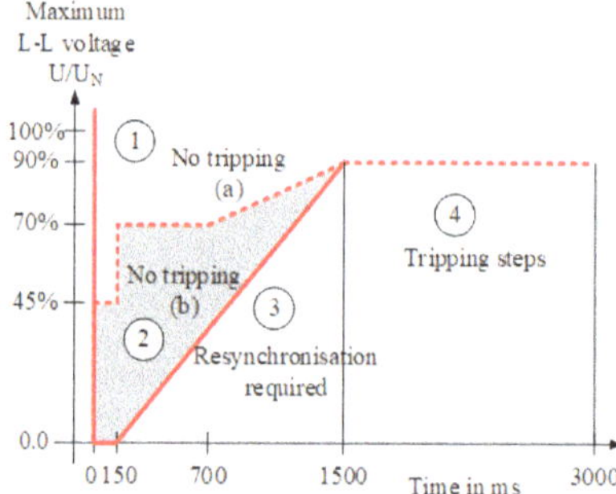

Figure 1. Requirements of low-voltage tolerance for WTGs by E.ON. (U denotes the system voltage, U_N denotes the system nominal voltage).

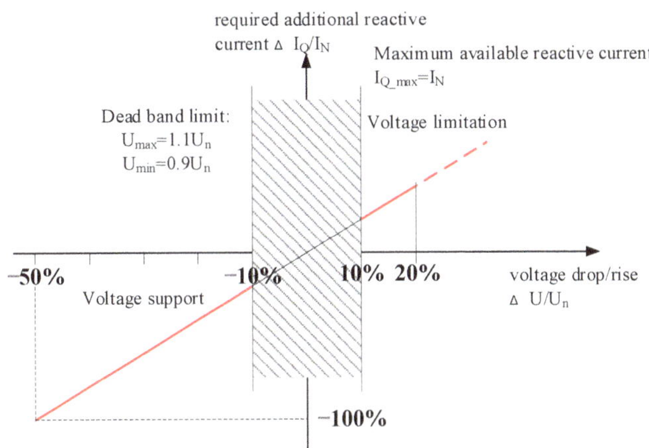

Figure 2. Requirements of reactive current compensation for WTGs by E.ON. (U denotes the system voltage, U_N denotes the nominal voltage of the system; I_Q denotes the reactive current; I_N denotes the nominal current of the LVRTC in this study).

2.2. *WTG Interface*

Figure 3 shows schematic system architecture of the proposed distributed LVRTC system investigated in this study.

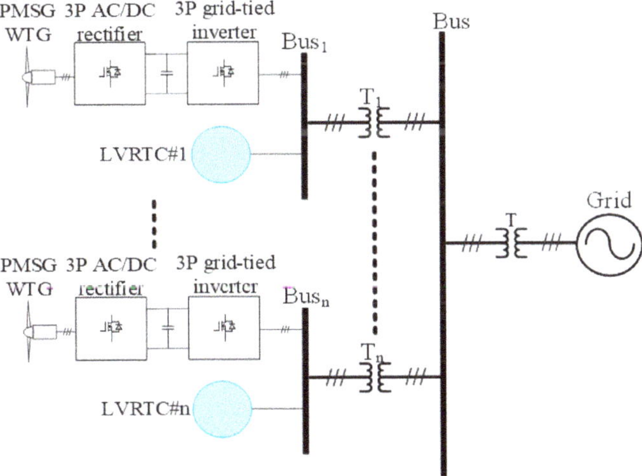

Figure 3. Conceptual WTG schematic architecture and allocation of the proposed LVRTC.

Mathematical Model Derivation of Grid-connected Three-phase Inverter.

The control goal of the proposed LVRTC using a three-phase grid-connected inverter system shown in Figure 4 is to maintain the DC link voltage, and output the reactive power as commanded. In Figure 4, L_f is the filter inductance; i_{fa}, i_{fb}, i_{fc} are, respectively, the three currents of the output filter; r_l denotes the equivalent resistance of the inductor; v_{an}, v_{bn}, v_{cn} are, respectively, the three phase-voltages of the system; C_d is the dc capacitor of the inverter. The inverter model for the related controller design can be established according to Kirchhoff's laws. The three-phase coordinate system (here assumed to be positive phase sequence) is converted into a synchronous rotating coordinate system.

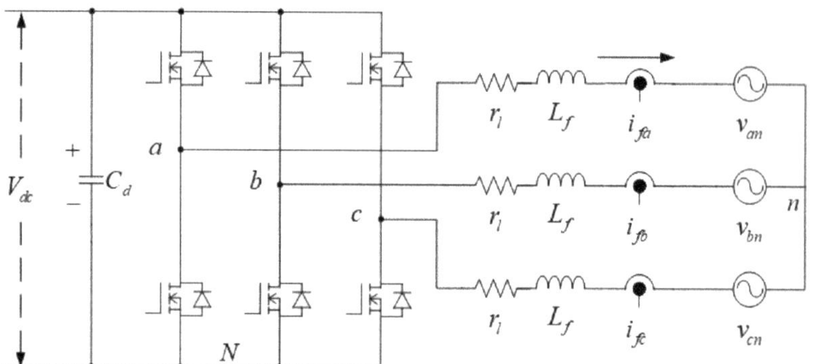

Figure 4. Schematic architecture of the proposed LVRTC on a three-phase grid-connected inverter.

In this study, pulse width modulation (PWM) is adopted for power switching, where three-phase sine waves are used. The switching equation can be expressed as follows:

$$e_{iN} = \left(\frac{1}{2} + k_{pwm}\right) V_{dc}\bigg|_{i=a,b,c}, \tag{2}$$

where k_{pwm} is constant and defined as $V_{dc}/2V_{tri}$, and V_{tri} represents the triangular carrier wave. Assuming that the three-phase power supply is balanced, dq-axis voltages are as follows: $v_d = V_m$, and $v_q = 0$, where V_m represents the grid peak phase voltage. As a result, active and reactive powers are as follows: $P_{ac} = 3V_m i_{fd}/2$, and $Q_{ac} = 3V_m i_{fq}/2$.

3. Controller Design for LVRTC

To achieve a better dynamic performance, the proposed LVRTC on a three-phase inverter adopts a dual-loop control scheme, where the inner loop controls the output currents, and the outer loop controls the DC voltage and reactive power. When the WTG outputs power, the DC voltage controller regulates the DC link voltage to follow the voltage command, such that the link stays unaffected by the control of reactive power. Under normal circumstances, higher power factor and low-harmonic currents are required to reduce losses, and improve system stability. With a proper mathematical derivation, system voltage and current equations can be decoupled into d and q axes, as shown as follows:

$$u_d = -v_d + \omega_e L_f i_{fq} + e_d; \tag{3}$$

$$u_q = -v_q + \omega_e L_f i_{fd} + e_q, \tag{4}$$

where $u_{d,q}$ represents the inverter switching node dq-axis voltage command, $v_{d,q}$ represents grid dq-axis voltage, L_f is the filter inductance, r_l denotes the equivalent resistance of the inductor. $I_{fd,q}$ represents dq-axis filter current, and $e_{d,q}$ represents inverter switching node dq-axis voltages. After derivation, we obtain the plant transfer function and dq-axis current controller transfer function as follows:

$$H_1 = \frac{1}{sL_f + r_1}; \tag{5}$$

$$G_{iPI} = \frac{k_{iP}s + k_{iI}}{s} = k_{iP}\left(1 + \frac{1}{T_i s}\right)\bigg|_{i=d,q}, \tag{6}$$

where k_{iP} represents the proportional gain, k_{iI} represents the integral gain, and T_i represents the integral constant and is defined as k_{iP}/k_{iI}. The inner loop current control structure is shown in Figure 5.

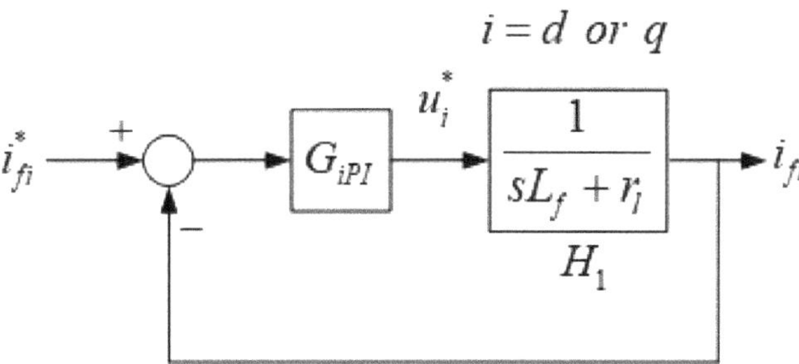

Figure 5. Schematic architecture of the current controller.

As shown in Figure 5, the close-loop transfer function of the current controller can be derived as follows:

$$\frac{i_{fi}}{i_{fi}^*} = \left. \frac{\frac{k_{iP}(T_i s+1)}{T_i s} \frac{1}{r_l(T_j s+1)}}{1 + \frac{k_{iP}(T_i s+1)}{T_i s} \frac{1}{r_l(T_j s+1)}} \right|_{i=d,q}, \tag{7}$$

where T_j represents the time constant, and is defined as L_f/r_l. Letting $T_d = T_q = T_j$, we obtain the following:

$$\frac{i_{fd}}{i_{fd}^*} = \frac{1}{\left(L_f/k_{dp}\right)s + 1}; \tag{8}$$

$$\frac{i_{fq}}{i_{fq}^*} = \frac{1}{\left(L_f/k_{qp}\right)s + 1}. \tag{9}$$

The DC voltage controller is constructed as shown in Figure 6, where the DC voltage error is used to obtain the current command through controller G_{vPI}. Controller and plant transfer functions are expressed as follows:

$$H_2 = \frac{1}{sC_d}; \tag{10}$$

$$G_{vPI} = \frac{k_{vP}s + k_{vI}}{s} = k_{vP}\left(1 + \frac{1}{T_v s}\right), \tag{11}$$

where C_d is the dc capacitor of the inverter. k_{vP} represents proportional gain, k_{vI} represents integral gain, and T_v represents integral constant, and is defined as k_{vP}/k_{vI}. The control structure is shown in Figure 6.

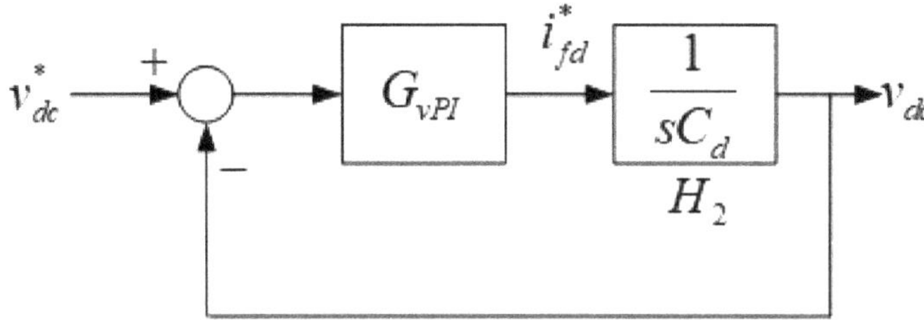

Figure 6. Schematic architecture of the DC voltage controller.

Based on Figure 6, the close-loop transfer function of the DC voltage controller can be expressed as shown in Equation (12). As a result, the complete inverter controller can be obtained as shown in Figure 7.

$$\frac{v_{dc}}{v_{dc}^*} = \frac{\frac{k_{vP}(T_v s+1)}{T_v s}\frac{1}{C_d s}}{1+\frac{k_{vP}(T_v s+1)}{T_v s}\frac{1}{C_d s}}. \tag{12}$$

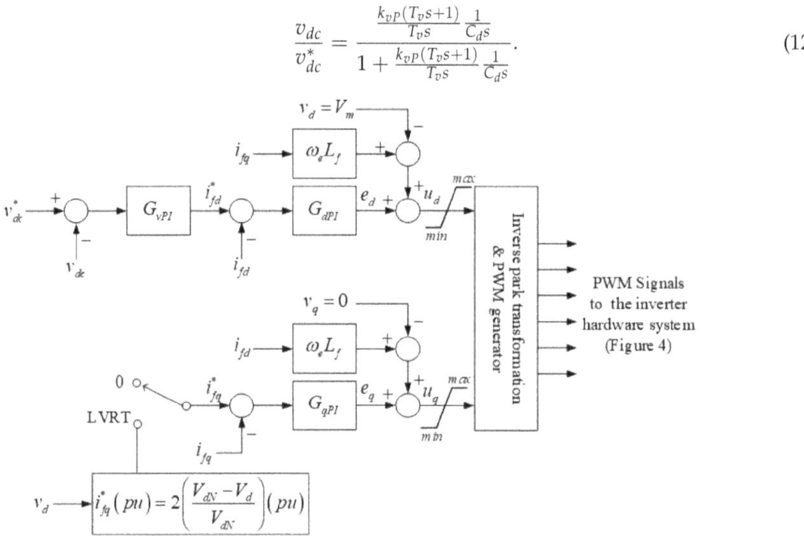

Figure 7. Schematic architecture of the complete LVRTC controller (ω_e denotes the angular frequency of the system).

As shown in Figures 2 and 7, when the grid voltage falls below 90% of the nominal value, the reactive current control command will be switched to LVRT mode, and perform reactive power command tracking control. In this study case, the system specifications of the inverter are as follows: system rating: 2 kVA; grid voltage $V_{L\text{-}L}$: 220 V/60 Hz; DC voltage V_{dc}: 400 V; filter inductance L_f: 3 mH; filter inductor resistance r_l: 3.5 Ω; DC link regulator capacitance C_d: 560 μF; and switching frequency f_{SW}: 18 kHz. According to these specifications, the plant transfer functions of the DC voltage loop and current loop are as follows:

$$H_1 = \frac{1}{0.003s + 3.5}; \tag{13}$$

$$H_2 = \frac{1}{5.6 \times 10^{-4} s}. \tag{14}$$

In this application case, a dual-loop control scheme is used to achieve a better voltage regulation on the DC bus. Theoretically, the outer loop (the DC voltage controller) should have a smaller bandwidth than the inner loop (the current controller). In practice, these parameters are properly chosen to achieve the desired dynamic performance [30]. Here, the close-loop bandwidth of the DC voltage controller is set at 500 Hz, and the bandwidth of the current response is set at 1/10 times the switching frequency, which is 1.8 kHz. This approach yields proportional and integral gains of the current control loop: $k_{dP} = k_{qP} = 33.93$, and $k_{dI} = k_{qI} = 39584$. Next, Equation (12) can be expressed in the form of a second order standard equation, with damping ratio ξ, and undamped natural frequency ω_0, as shown as follows:

$$s^2 + 2\xi\omega_0 s + \omega_0^2 = 0, \tag{15}$$

where $k_{vP} = 2\xi\omega_0 C_d$, $T_v = 2\xi/\omega_0$, and, $k_{vP} = k_{vP}/T_v$. ξ and ω_0 can be chosen according to the desired dynamic performance. Letting $\xi = 1$ and $\omega_0 = 500$ yields the following results: $k_{vP} = 0.56$, and $k_{vI} = 140$. Figure 8 shows the Bode plots of the DC voltage loops.

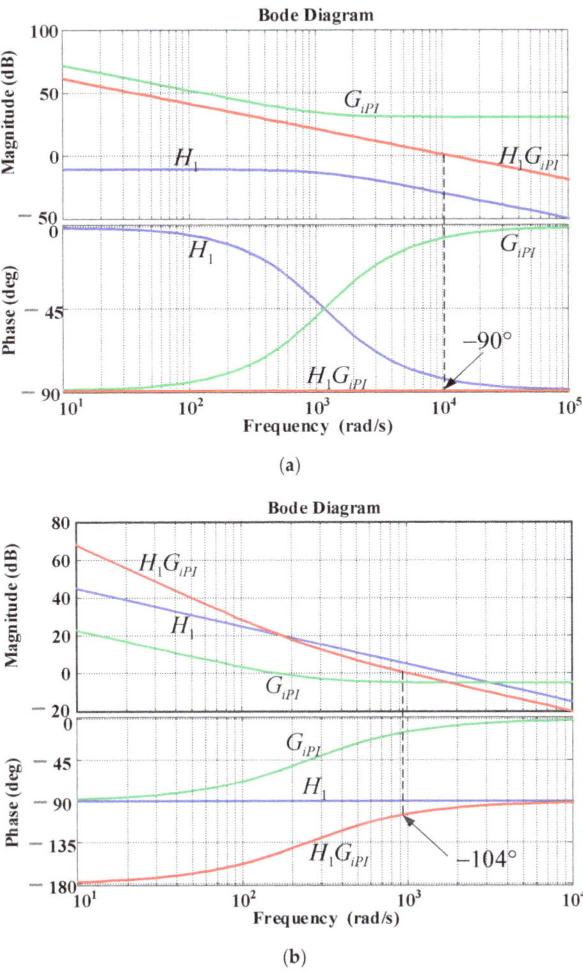

Figure 8. Bode plots of designed controller: (**a**) inner loop current controller; (**b**) DC voltage controller.

4. Simulation and Implementation

In this study, the POWERSIM simulation software is used in both simulation and hardware implementation, and a personal computer equipped with an i7-9700 CPU, @ 3.00 GHz is used as a control desk. The sampling frequency used in the hardware design of the proposed LVRTC is 50 kHz. It is the same as that of the switching frequency used. Figure 9a shows the schematic architecture of the simulated LVRTC system connected to the power grid. For the simulation scenario, the grid voltage variation sequence is planned as follows: 0.5 pu, 0.6 pu, 0.7 pu, 0.8 pu, 0.9 pu, 1.0 pu, 1.2 pu, and then 1.0 pu. The corresponding reactive power commands for the LVRTC are generated according to the grid code. The simulation results are shown in Figure 10. Figure 11 shows the detailed steady-state waveforms of Figure 10. It can be observed that the LVRTC successfully feeds appropriate reactive current to the grid during faults. Figure 12 shows the results of the experimental tests using the constructed 2 kVA SiC-based three-phase inverter as shown in Figure 9b, and a programmable AC power supply emulating the grid. In this study, the TI's DSP (TMS320F28335) is used as the control core for the proposed LVRTC hardware, in which six SiC switching devices using ON Semiconductor's NTHL060N090SC1 with the driver integrated circuits of Si8271 and the sensor devices for sensing currents and voltages are used to facilitate a flexible experimental system. Regarding the task of DSP's programming, the SimCoder tool of POWERSIM software (2021) is used to convert the AD modules, various functional blocks, and controllers into the required DSP control program. Then, the DSP control program can be burned into the DSP chip. In the hardware implementation stage, the control desk (PC) is used to communicate with the DSP, and the controller parameters can be adjusted online to achieve the best control performance and accuracy. Figure 9c shows the DSP programming steps in hardware implementation of LVRTC. In this study, the identical system condition used in the simulation case is implemented here in the hardware test. Figure 13 shows the detailed steady state waveforms of Figure 12. It can be clearly observed that the implementation results are in good agreement with the simulation results.

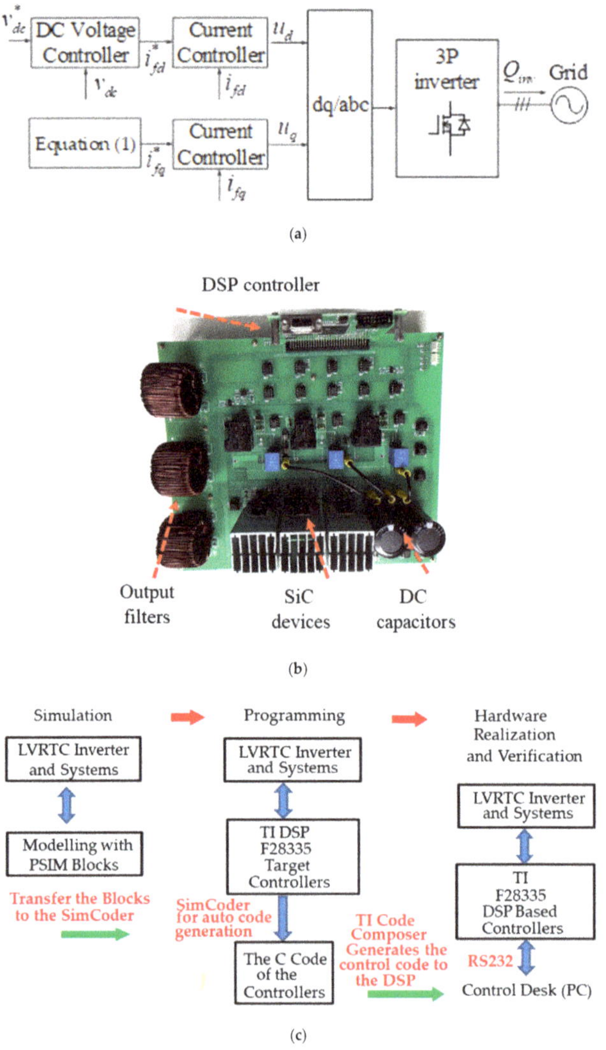

Figure 9. (**a**) Schematic of simulated grid-connected LVRTC; (**b**) the photograph of the LVRTC SiC-based 3P inverter circuit prototype; (**c**) the DSP programming steps in hardware implementation of LVRTC.

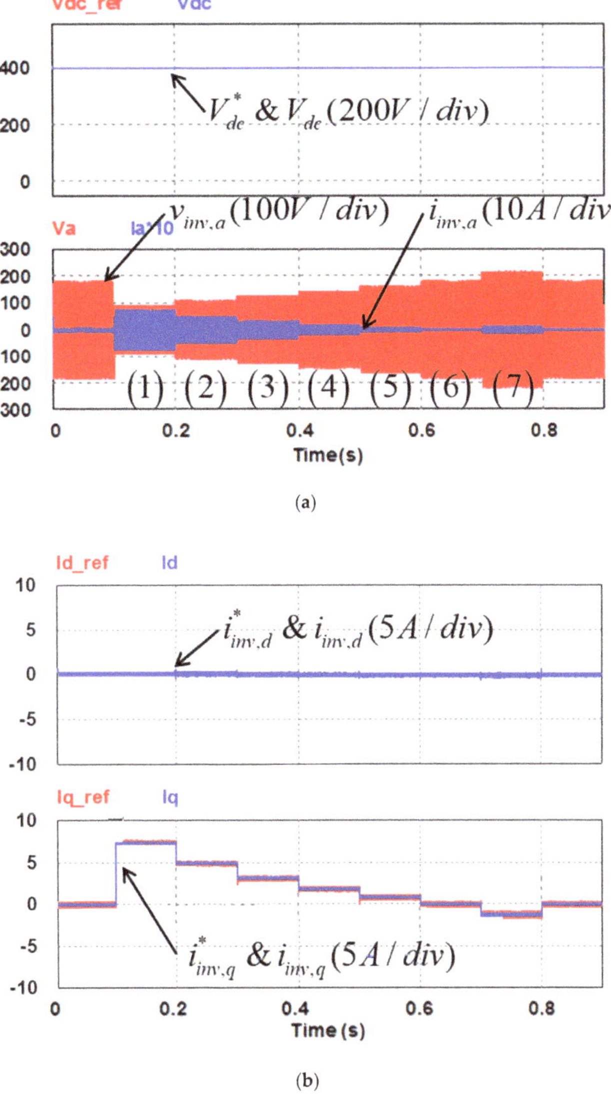

Figure 10. Simulation results of LVRT capability: (**a**) DC voltage and its command signal/grid phase A voltage and current (20 times); (**b**) inverter d-axis current and its command/q-axis current and its command.

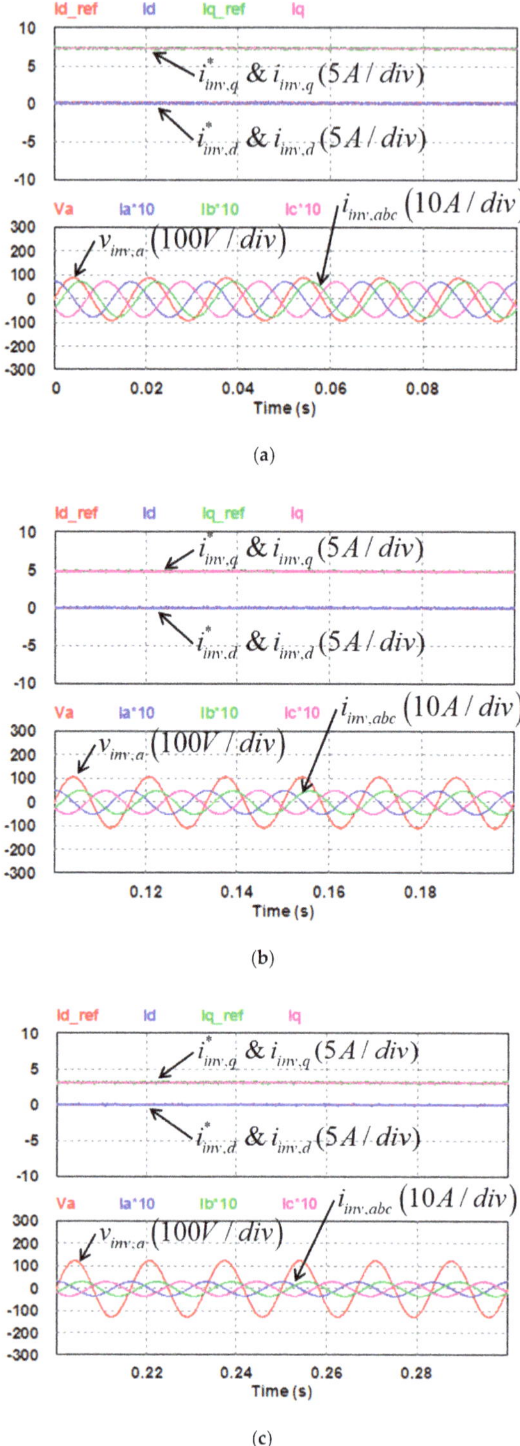

Figure 11. *Cont.*

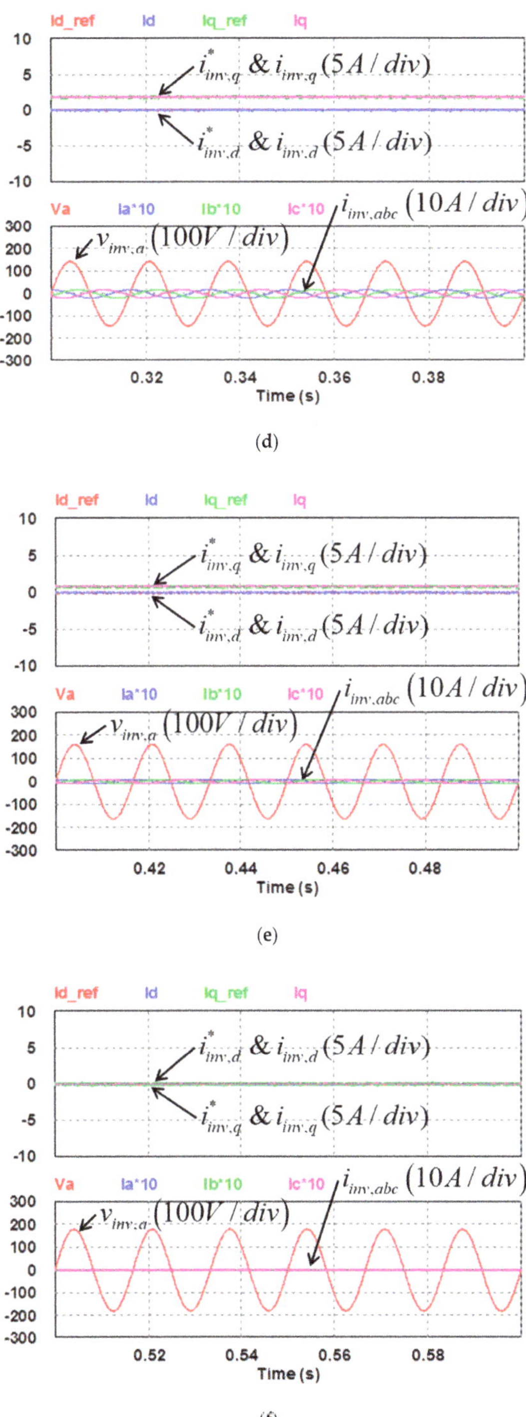

Figure 11. *Cont.*

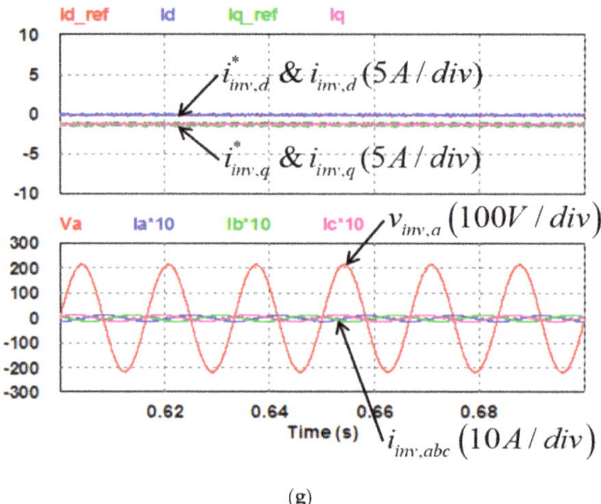

(g)

Figure 11. Detailed view of inverter dq-axis currents and their commands/phase A voltage (V) and three-phase currents (A) in Figure 10: (**a**) stage (1); (**b**) stage (2); (**c**) stage (3); (**d**) stage (4); (**e**) stage (5); (**f**) stage (6); (**g**) stage (7).

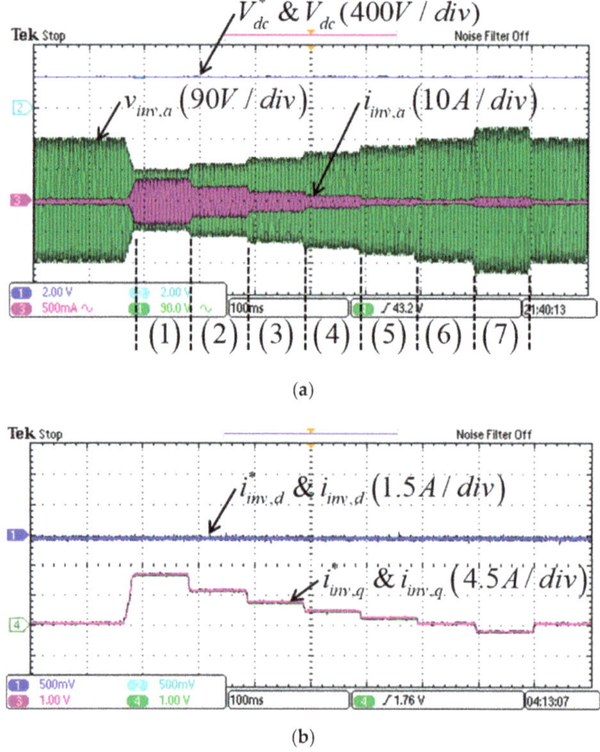

(a)

(b)

Figure 12. Implementation results of LVRT capability: (**a**) DC voltage and its command signal (400 V/div)/grid phase A voltage (90 V/div) and current (5 A/div); (**b**) inverter d-axis current and its command (1.5 A/div)/q-axis current and its command (4.5 A/div).

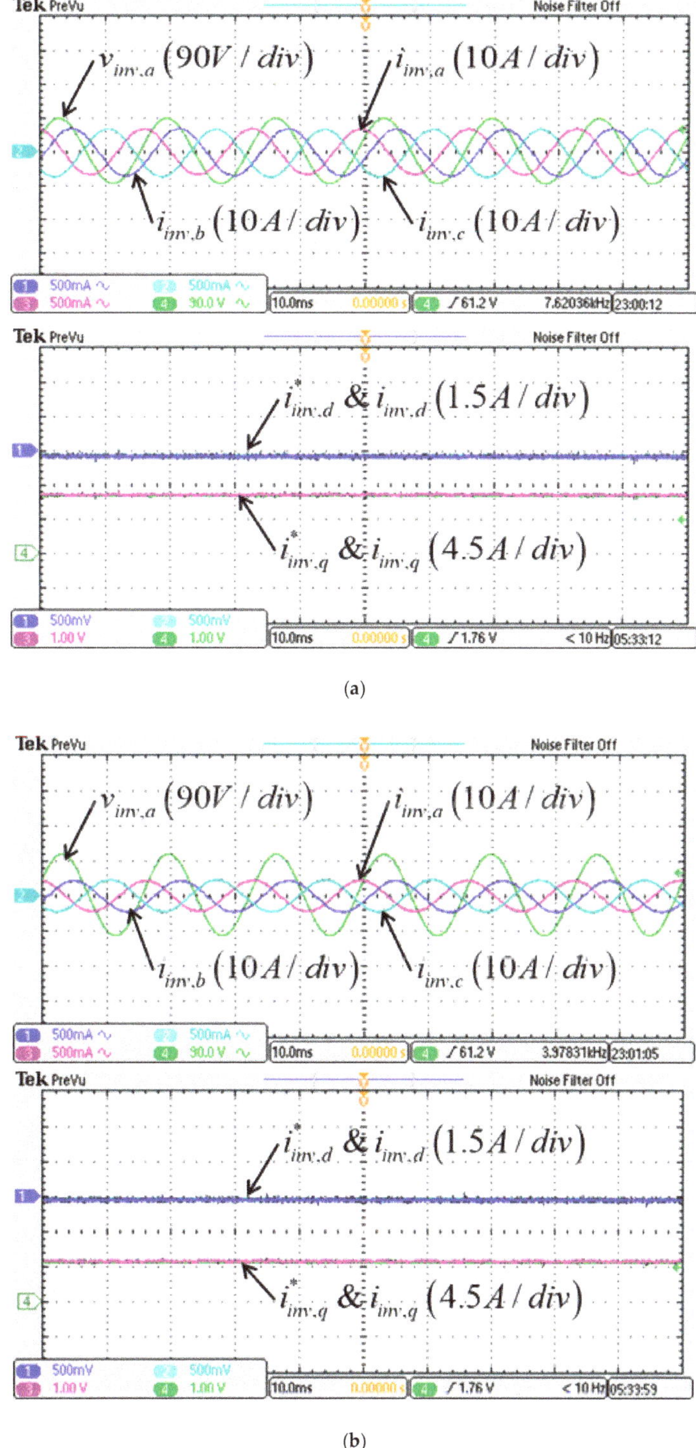

(**a**)

(**b**)

Figure 13. *Cont.*

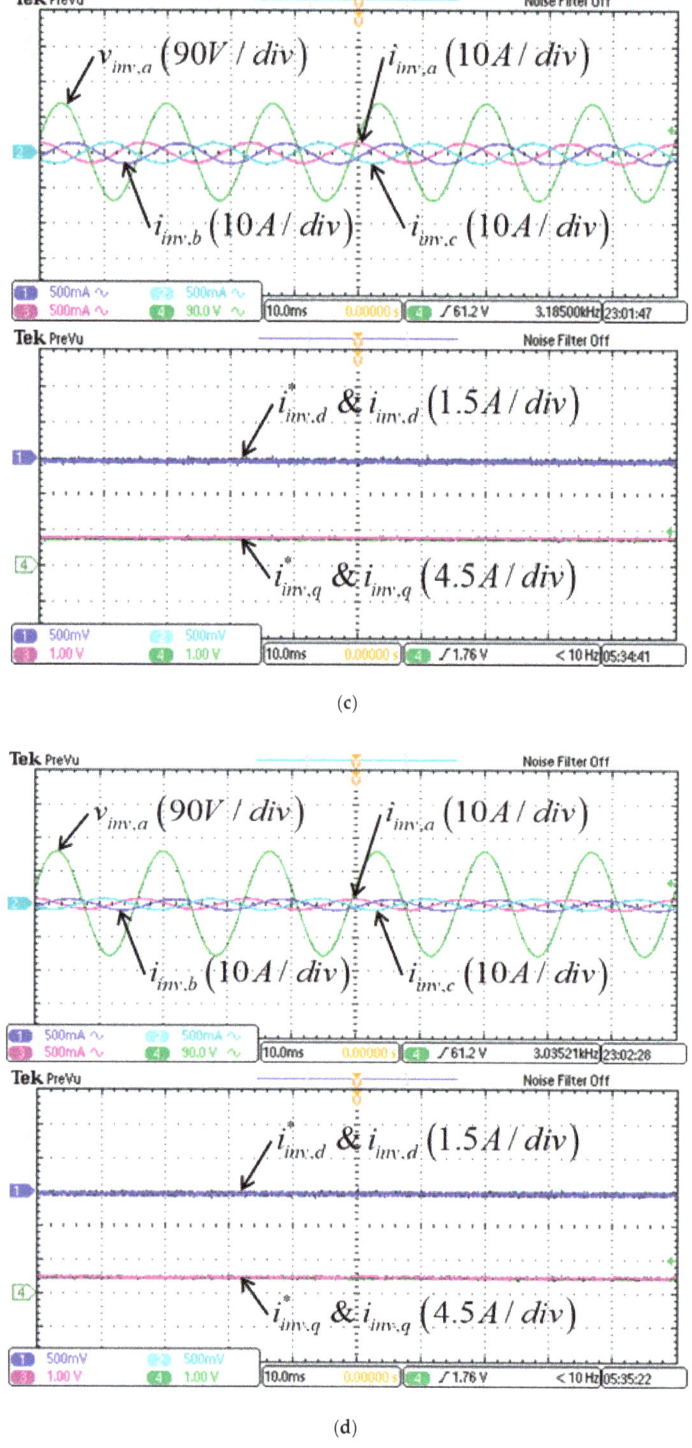

(c)

(d)

Figure 13. *Cont.*

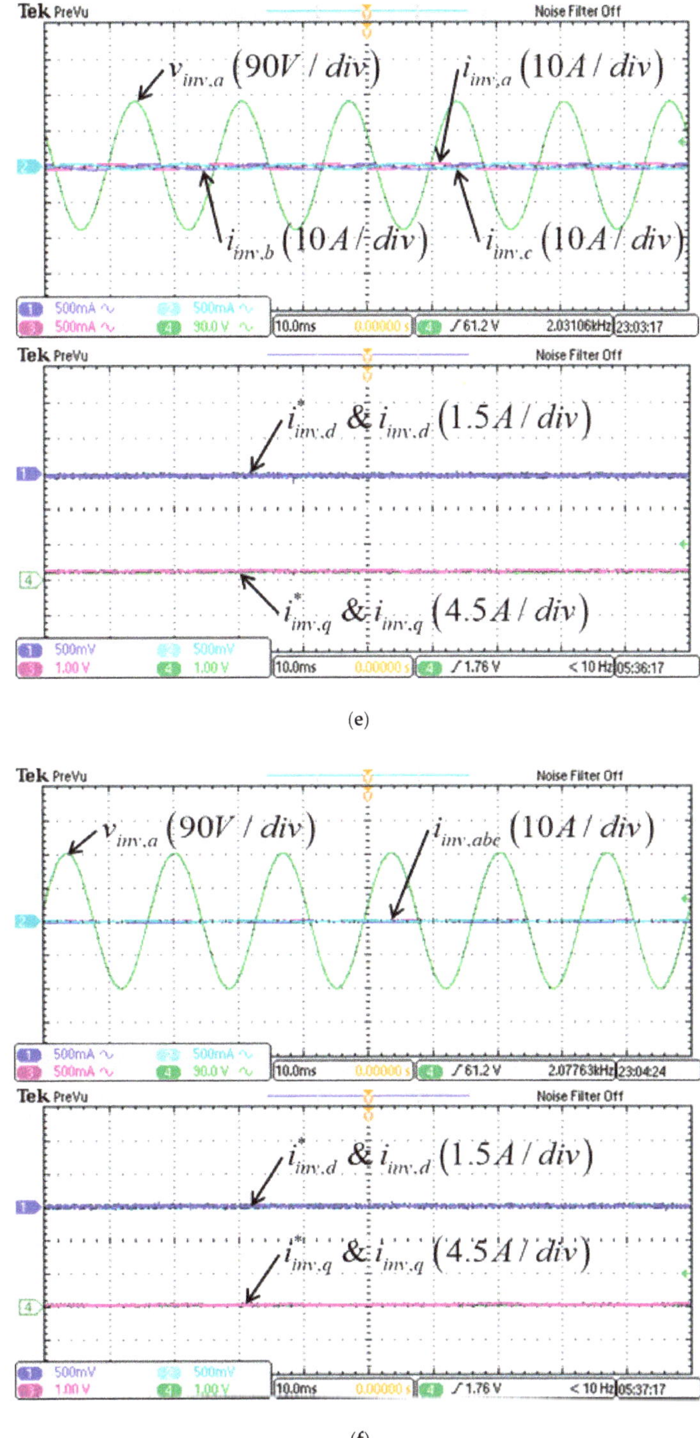

Figure 13. *Cont.*

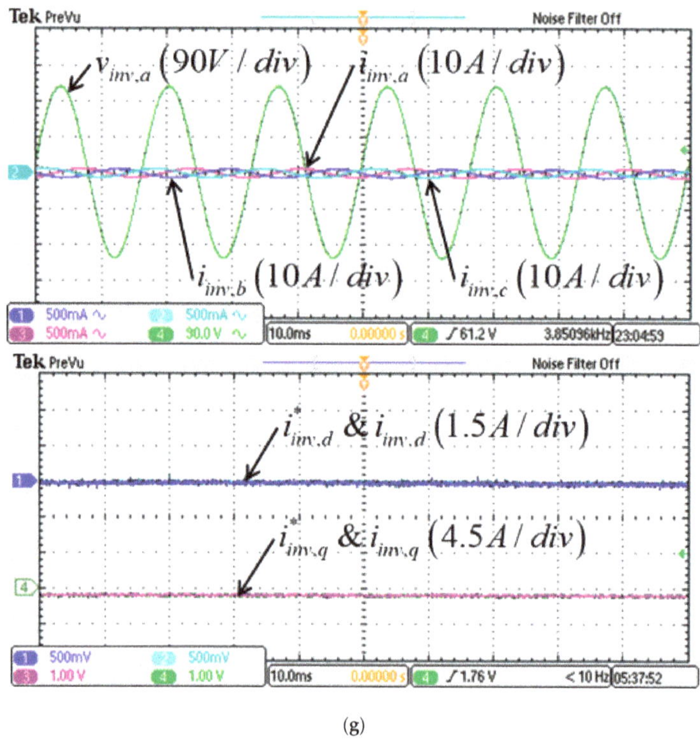

(g)

Figure 13. Detailed view of phase A voltage and three-phase currents/inverter dq-axis currents and their commands in Figure 12: (**a**) stage (1); (**b**) stage (2); (**c**) stage (3); (**d**) stage (4); (**e**) stage (5); (**f**) stage (6); (**g**) stage (7).

5. Discussion

With the growth of wind power generation over the past decade, various LVRT technologies have been proposed to ensure the safe and reliable operation of WTGs and power systems. Some advanced LVRT control algorithms reviewed in this paper and reported in many other similar papers in the literature can be categorized into two groups, i.e., the protection strategies- and the control strategies-based methodologies. However, most of the reported control strategies were implemented by the WTG built-in converters with complex controllers. It should be noted that with a limited hardware capacity, during the period of fault, the real power output capability of the built-in converters of WTGs must be dynamically limited to meet real-time reactive power requirements, and comply with strict grid codes. Based on the LVRT grid code, when the voltage dipped over 0.5 pu, the total current capacity of inverter system of the WTGs should be used to output the reactive power. This can strongly limit the control flexibility of a real power restoration scheme in the post-fault period. In this paper, we have demonstrated a distributed LVRT compensator (LVRTC) control scheme that can, to some degree, solve the above problem, and increase the system control flexibility. In addition, with the proposed distributed LVRTC control scheme, the WTG in a wind farm has greater freedom in performing its protection schemes, or regulating its power flow to ensure system stability.

6. Conclusions

This study has proposed a new control scheme regarding distributed LVRTC and real-time reactive power compensation functions for the operation of WTGs in a wind farm according to E.ON's specifications. The proposed compensation scheme using distributed LVRTCs can effectively enhance reliability, as well as reduce cost compared with centralized LVRT compensators. In addition, with the help of the proposed LVRTC scheme, WTGs can continuously output real power within their maximum current limit during the period of voltage drop, which is a great merit for the secure operation of modern power distribution systems embedded with WTGs. In this study, a complete design example of LVRTCs using silicon carbide (SiC), a wide band gap (WBG) semiconductor, based three-phase voltage source inverter has been demonstrated. The mathematical models of inverter systems have been derived using the dq-axis decoupling method to enable separate control of active and reactive powers. On the basis of the results obtained from the simulation and hardware tests, when a voltage fault occurs, the proposed control scheme can achieve a good regulation of LVRTCs' DC bus voltage, and the real-time compensation of appropriate reactive current according to the grid code to help restore the nominal grid voltage.

Author Contributions: The corresponding author, C.-T.M., conducted the research work, verified the results of the simulation cases and hardware tests, wrote the draft, and polished the final manuscript. Z.-H.S., a student who graduated from the department of EE, CEECS, National United University, Taiwan, assisted in paper searching and organized the test results. All authors have read and agreed to the published version of the manuscript.

Funding: This research was funded by MOST, Taiwan, with grant number MOST 110-2221-E-239-032.

Institutional Review Board Statement: Not applicable.

Informed Consent Statement: Not applicable.

Data Availability Statement: No new data were created or analyzed in this study. Data sharing is not applicable to this article.

Acknowledgments: The author would like to thank the Ministry of Science and Technology (MOST) of Taiwan for financially supporting the energy-related research regarding key technologies in microgrids, and the design of advanced power and energy systems.

Conflicts of Interest: The authors declare no conflict of interest.

References

1. Mishra, R.; Banerjee, U.; Sekhar, T.N.; Saha, T.K. Development and implementation of control of stand-alone PMSG-based distributed energy system with variation in input and output parameters. *IET Electr. Power Appl.* **2019**, *13*, 1497–1506. [CrossRef]
2. Shi, K.; Yin, X.; Jiang, L.; Liu, Y.; Hu, Y.; Wen, H.; Xin, Y. Perturbation Estimation Based Nonlinear Adaptive Power Decoupling Control for DFIG Wind Turbine. *IEEE Trans. Power Electron.* **2019**, *35*, 319–333. [CrossRef]
3. Zhang, F.; Yu, S.; Wang, H.; Wang, Y.; Wang, D. Overview of research and development status of brushless doubly-fed machine system. *Chin. J. Electr. Eng.* **2016**, *2*, 1–13.
4. Jamil, M.; Gupta, R.; Singh, M. A review of power converter topology used with PMSG based wind power generation. In Proceedings of the 2012 IEEE Fifth Power India Conference, Murthal, India, 19–22 December 2012.
5. Jayalakshmi, N.S.; Gaonkar, D.N.; Kumar, K.S.K. Dynamic modeling and performance analysis of grid connected PMSG based variable speed wind turbines with simple power conditioning system. In Proceedings of the 2012 IEEE International Conference on Power Electronics, Drives and Energy Systems (PEDES), Bengaluru, India, 16–19 December 2012.
6. Qais, M.H.; Hasanien, H.M.; Alghuwainem, S. Low voltage ride-through capability enhancement of grid-connected permanent magnet synchronous generator driven directly by variable speed wind turbine: A review. *J. Eng.* **2017**, *2017*, 1750–1754. [CrossRef]
7. Qais, M.H.; Hasanien, H.M.; Alghuwainem, S. A Grey Wolf Optimizer for Optimum Parameters of Multiple PI Controllers of a Grid-Connected PMSG Driven by Variable Speed Wind Turbine. *IEEE Access* **2018**, *6*, 44120–44128. [CrossRef]
8. Huang, C.; Xiao, X.Y.; Zheng, Z.; Wang, Y. Cooperative Control of SFCL and SMES for Protecting PMSG-Based WTGs under Grid Faults. *IEEE Trans. Appl. Supercond.* **2019**, *29*, 1–6.
9. Xing, P.; Fu, L.; Wang, G.; Wang, Y.; Zhang, Y. A compositive control method of low-voltage ride through for PMSG-based wind turbine generator system. *IET Gener. Transm. Distrib.* **2018**, *12*, 117–125. [CrossRef]
10. Yan, L.; Chen, X.; Zhou, X.; Sun, H.; Jiang, L. Perturbation compensation-based non-linear adaptive control of ESS-DVR for the LVRT capability improvement of wind farms. *IET Renew. Power Gener.* **2018**, *12*, 1500–1507. [CrossRef]
11. Jahanpour-Dehkordi, M.; Vaez-Zadeh, S.; Mohammadi, J. Development of a Combined Control System to Improve Performance of a PMSG Based Wind Energy Conversion System under Normal and Grid Fault Conditions. *IEEE Trans. Energy Convers.* **2019**, *34*, 1287–1295. [CrossRef]

12. Arani, M.F.M.; Mohamed, Y.A.R.I. Assessment and Enhancement of a Full-Scale PMSG-Based Wind Power Generator Performance under Faults. *IEEE Trans. Energy Convers.* **2016**, *31*, 728–739. [CrossRef]
13. Nasiri, M.; Mohammadi, R. Peak Current Limitation for Grid Side Inverter by Limited Active Power in PMSG-Based Wind Turbines during Different Grid Faults. *IEEE Trans. Sustain. Energy* **2016**, *8*, 3–12.
14. Yang, L.; Xu, Z.; Østergaard, J.; Dong, Z.Y.; Wong, K.P. Advanced control strategy of DFIG wind turbines for power system fault ride through. *IEEE Trans. Power Syst.* **2011**, *27*, 713–722. [CrossRef]
15. Xie, D.; Xu, Z.; Yang, L.; Østergaard, J.; Xue, Y.; Wong, K.P. A comprehensive LVRT control strategy for DFIG wind turbines with enhanced reactive power support. *IEEE Trans. Power Syst.* **2013**, *28*, 3302–3310. [CrossRef]
16. Alepuz, S.; Calle, A.; Busquets-Monge, S.; Kouro, S.; Wu, B. Use of stored energy in PMSG rotor inertia for low-voltage ride-through in back-to-back NPC converter-based wind power systems. *IEEE Trans. Ind. Electron.* **2012**, *60*, 1787–1796. [CrossRef]
17. Liu, R.; Yao, J.; Wang, X.; Sun, P.; Pei, J.; Hu, J. Dynamic Stability Analysis and Improved LVRT Schemes of DFIG-Based Wind Turbines during a Symmetrical Fault in a Weak Grid. *IEEE Trans. Power Electron.* **2019**, *35*, 303–318. [CrossRef]
18. Jin, Y.; Wu, D.; Ju, P.; Rehtanz, C.; Wu, F.; Pan, X. Modeling of Wind Speeds Inside a Wind Farm With Application to Wind Farm Aggregate Modeling Considering LVRT Characteristic. *IEEE Trans. Energy Convers.* **2019**, *35*, 508–519. [CrossRef]
19. Dey, P.; Datta, M.; Fernando, N. A coordinated control of grid connected PMSG based wind energy conversion system under grid faults. In Proceedings of the 2017 IEEE 3rd International Future Energy Electronics Conference and ECCE Asia (IFEEC 2017—ECCE Asia), Kaohsiung, Taiwan, 3–7 June 2017.
20. Dey, P.; Datta, M.; Fernando, N.; Senjyu, T. Fault-ride-through performance improvement of a PMSG based wind energy systems via coordinated control of STATCOM. In Proceedings of the 2018 IEEE International Conference on Industrial Technology (ICIT), Lyon, France, 20–22 February 2018.
21. Geng, H.; Liu, L.; Li, R. Synchronization and Reactive Current Support of PMSG-Based Wind Farm During Severe Grid Fault. *IEEE Trans. Sustain. Energy* **2018**, *9*, 1596–1604. [CrossRef]
22. Yao, J.; Guo, L.; Zhou, T.; Xu, D.; Liu, R. Capacity Configuration and Coordinated Operation of a Hybrid Wind Farm with FSIG-Based and PMSG-Based Wind Farms during Grid Faults. *IEEE Trans. Energy Convers.* **2017**, *32*, 1188–1199. [CrossRef]
23. Arunkumar, P.K.; Kannan, S.M.; Selvalakshmi, I. Low Voltage Ride Through capability improvement in a grid connected Wind Energy Conversion System using STATCOM. In Proceedings of the 2016 International Conference on Energy Efficient Technologies for Sustainability (ICEETS), Nagercoil, India, 7–8 April 2016.
24. Zheng, X.; Xu, G. Study on LVRT of DFIG under the asymmetric grid voltage based on fuzzy PID D-STATCOM. In Proceedings of the 2016 IEEE Advanced Information Management, Communicates, Electronic and Automation Control Conference (IMCEC), Xi'an, China, 3–5 October 2016.
25. Mosaad, M.I. Model reference adaptive control of STATCOM for grid integration of wind energy systems. *IET Electr. Power Appl.* **2018**, *12*, 605–613. [CrossRef]
26. Karoui, R.; Zoghlami, M.; Bacha, F. Impact of the STATCOM on the terminal voltage of a wind farm of Bizerte in Tunisia. In Proceedings of the 2016 7th International Renewable Energy Congress (IREC), Hammamet, Tunisia, 22–24 March 2016.
27. Tanaka, T.; Wang, H.; Ma, K.; Blaabjerg, F. Low voltage ride through peformance of a STATCOM based on modular multilevel cascade converters for offshore wind application. In Proceedings of the 2017 IEEE Energy Conversion Congress and Exposition (ECCE), Cincinnati, OH, USA, 1–5 October 2017.
28. Tanaka, T.; Wang, H.; Ma, K.; Blaabjerg, F. Reactive power compensation capability of a STATCOM based on two types of Modular Multilevel Cascade Converters for offshore wind application. In Proceedings of the 2017 IEEE 3rd International Future Energy Electronics Conference and ECCE Asia (IFEEC 2017—ECCE Asia), Kaohsiung, Taiwan, 3–7 June 2017.
29. Luo, X.; Wang, J.; Wojcik, J.D.; Wang, J.; Li, D.; Draganescu, M.; Li, Y.; Miao, S. Review of Voltage and Frequency Grid Code Specifications for Electrical Energy Storage Applications. *Energies* **2018**, *11*, 1070. [CrossRef]
30. Peng, Q.; Pan, H.; Liu, Y.; Lidan, X. Dual-loop Control Strategy for Grid-connected Inverter with LCL Filter. *Energy Power Eng.* **2013**, *5*, 97–101. [CrossRef]

 micromachines

Article

A Hybrid Active Neutral Point Clamped Inverter Utilizing Si and Ga_2O_3 Semiconductors: Modelling and Performance Analysis

Sheikh Tanzim Meraj [1], Nor Zaihar Yahaya [1], Molla Shahadat Hossain Lipu [2,3,*], Jahedul Islam [4], Law Kah Haw [5], Kamrul Hasan [6], Md. Sazal Miah [7], Shaheer Ansari [2] and Aini Hussain [2]

1 Department of Electrical and Electronic Engineering, Universiti Teknologi PETRONAS, Seri Iskandar 32610, Perak, Malaysia; sheikh_19001724@utp.edu.my (S.T.M.); norzaihar_yahaya@utp.edu.my (N.Z.Y.)
2 Department of Electrical, Electronic and Systems Engineering, Universiti Kebangsaan Malaysia, Bangi 43600, Selangor, Malaysia; p100855@siswa.ukm.edu.my (S.A.); draini@ukm.edu.my (A.H.)
3 Centre for Automotive Research (CAR), Universiti Kebangsaan Malaysia, Bangi 43600, Selangor, Malaysia
4 Department of Fundamental and Applied Sciences, Universiti Teknologi PETRONAS, Seri Iskandar 32610, Perak, Malaysia; jahedul_17010697@utp.edu.my
5 Faculty of Engineering, Universiti Teknologi Brunei, Bandar Seri Begawan 1410, Brunei; kahhaw.law@utb.edu.bn
6 School of Electrical Engineering, College of Engineering Studies, Universiti Teknologi MARA, Shah Alam 40450, Selangor, Malaysia; 2019984679@isiswa.uitm.edu.my
7 School of Engineering and Technology, Asian Institute of Technology, Pathumthani 12120, Thailand; st121577@ait.asia
* Correspondence: lipu@ukm.edu.my

Citation: Meraj, S.T.; Yahaya, N.Z.; Hossain Lipu, M.S.; Islam, J.; Haw, L.K.; Hasan, K.; Miah, M.S.; Ansari, S.; Hussain, A. A Hybrid Active Neutral Point Clamped Inverter Utilizing Si and Ga_2O_3 Semiconductors: Modelling and Performance Analysis. *Micromachines* **2021**, *12*, 1466. https://doi.org/10.3390/mi12121466

Academic Editor: Francisco J. Perez-Pinal

Received: 8 November 2021
Accepted: 25 November 2021
Published: 27 November 2021

Publisher's Note: MDPI stays neutral with regard to jurisdictional claims in published maps and institutional affiliations.

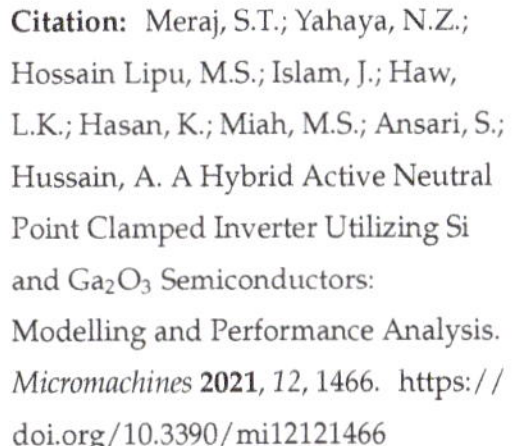

Abstract: In this paper, the performance of an active neutral point clamped (ANPC) inverter is evaluated, which is developed utilizing both silicon (Si) and gallium trioxide (Ga_2O_3) devices. The hybridization of semiconductor devices is performed since the production volume and fabrication of ultra-wide bandgap (UWBG) semiconductors are still in the early-stage, and they are highly expensive. In the proposed ANPC topology, the Si devices are operated at a low switching frequency, while the Ga_2O_3 switches are operated at a higher switching frequency. The proposed ANPC mitigates the fault current in the switching devices which are prevalent in conventional ANPCs. The proposed ANPC is developed by applying a specified modulation technique and an intelligent switching arrangement, which has further improved its performance by optimizing the loss distribution among the Si/Ga_2O_3 devices and thus effectively increases the overall efficiency of the inverter. It profoundly reduces the common mode current stress on the switches and thus generates a lower common-mode voltage on the output. It can also operate at a broad range of power factors. The paper extensively analyzed the switching performance of UWBG semiconductor (Ga_2O_3) devices using double pulse testing (DPT) and proper simulation results. The proposed inverter reduced the fault current to 52 A and achieved a maximum efficiency of 99.1%.

Keywords: power electronics; ultrawide bandgap; semiconductors; neutral point clamped; inverter; silicon; gallium trioxide; fabrication; hybridization

1. Introduction

Silicon-based devices have primarily been used and are still dominant in developing power inverters [1,2]. However, ultra-wide bandgap (UWBG) semiconductors have gained a significant amount of attention in recent years [3]. As a result, Ga_2O_3 being a strong candidate for UWBG devices have the potential to be profoundly applied in the various applications in the field of power electronics ranging from Photovoltaic (PV) inverters and UPS systems to inverters for traction and space applications, among others [4–6]. In these power inverters, UWBG semiconductors can contribute to high efficiency, inverter size reduction, and high-temperature environment operation, which are unlikely to be achieved otherwise [7]. These features of Ga_2O_3 devices are due to the specific properties of the UWBG material. Gallium trioxide (Ga_2O_3) devices are capable of achieving these because unlike their conventional counterpart, the blocking voltage that is rated for these devices is nearly a hundred times higher for the same width of the drift region [8]. In addition, the high thermal conductivity, along with the fast-switching speed are two main factors that are offered by Ga_2O_3 devices to gain this advantage [3]. Though the UWBG devices can be applied in medium power applications, the ongoing research has suggested that these devices have great potential to be applied in high power applications with modular multilevel inverters (MLIs) [9].

For medium-voltage-range Photovoltaics (PV), several DC-link voltages are proposed in recent years [10,11] considering different interests. However, when efficiency and reliability are the main concern, the 1.5kV DC-link-based PV generation system has gained significant attention along with systems [12,13]. In addition to that, for higher efficiency in PV systems, transformer-less configurations have shown better performance compared to other transformer-based configurations [6]. Though many inverter topologies had been proposed previously considering high voltage applications, a three-level neutral point clamped (NPC) inverter is one of the most optimal inverter choices for high voltage applications [14]. Since the clamped common-mode voltage (CMV) is enabled in this type of topology, it minimizes leakage-current-related issues [15]. That is why, for a transformer-less system, it is a better choice than other systems which are incorporated by leakage current. Despite this, due to the unequal loss distribution among the switches, the NPC inverter has issues related to neutral-point voltage imbalance, as well as a shoot-through fault in the switching devices [16].

Various types of control strategies along with modified inverter topologies have recently been proposed to overcome the inconveniences in the NPC inverter. Since NPC inverters are prone to the shoot-through problem, a split inductor configuration can be used to solve this issue [17]. It should also be noted that in addition to successfully protecting the shoot-through fault, reduced leakage current and the eradication of CMV transitions that are high frequency in nature, can be achieved through this inverter. However, even though this configuration removes most of the inconveniences, it operates in a unity power factor region. Therefore, this configuration is hardly suitable for high voltage applications that are designed specifically for supplying reactive power to the grid. In addition to that, the previously mentioned non-uniform loss distribution problem in the switches of the NPC inverters still exists in these topologies.

In [18], the non-uniform current distribution was addressed, and it proposed an active neutral point clamped (ANPC) inverter. As per the switching states of ANPC inverters, additional redundant zero states can be gained in the ANPC inverter topology. Therefore, unequal switching loss distribution can be mitigated if different zero states can be appropriately exploited in the switching states of the ANPC inverter. Considering these additional states, some notable PWM-based control techniques are employed previously in the ANPC inverter topology [19,20]. The use of current and voltage sensors for the selection of the redundant zero states that are available in the ANPC topology focused on power factors. So, how the states of the inverters will be chosen is largely related to the feedback signals of those current and voltage sensors. This solution is optimized to achieve high efficiency in ANPC, which provides the states for the hybrid Si/Ga_2O_3-devices-based ANPC topology. The high efficiency and the low cost relative to all Ga_2O_3 inverters can be obtained according to researchers [21]. Recent literature has also shown promising results using the aforementioned approach where the switching devices of the ANPC were mostly built using wide bandgap (WBG) materials or silicon carbide (SiC) [22]. However, one fact about their research is that they only considered low voltage applications, and the entire ANPC inverter was built using devices from the same bandgap materials. Furthermore, one fact about their research is that they considered the converters suitable for only low-power applications having low voltages. Because in the case of MV applications, unlike silicon devices, the body diodes of Ga_2O_3 MOSFETs are the cause of further switching losses along with overshoots that are significant in switching transient, the design criteria would be different [23]. In addition to that, as the dead-band time is declined in high-frequency devices, the severity of the shoot-through fault rises remarkably. It should also be noted that as high-frequency switching devices are employed at the output side, an increase has been seen in the voltage amplitude in the electromagnetic interference (EMI) frequency range, which ultimately contributes to the increased size and complexity in EMI filters [24].

Considering the issues stated above, this study proposes a hybrid ANPC inverter that utilizes both conventional Si and Ga_2O_3 devices. As a result of this hybridization, the switching losses of the inverter are reduced significantly. The hybridization also made the implementation of a split-output structure achievable. Thus, the proposed circuit can also handle the switching transient overshoots. In this structure, since the UWBG switch is decoupled externally by the parallel diode, both overshoot issues in the switching transient, as well as switching losses, are declined significantly. These reduced overshoots ultimately also lead to decreased voltage and current stresses on the UWBG devices. As this converter topology is capable of supplying reactive power to loads with a wide range of power factors, it can be used for grid-tied PV systems. The key contributions of the paper can be listed as follows:

- Incorporating UWBG semiconductors to utilize their various advantages such as reduced size, minimized switching transient overshoots, reduced current and voltage stress, high-frequency switching, and efficiency;
- Hybridization with conventional Si switches to prevent high leakage current and high-frequency switching losses;
- The split-output structure is adopted for the ANPC inverter to prevent shoot-through current fault, reduce electromagnetic interference (EMI) on the output, and enable it to operate under different ranges of power factors;
- Validating the performance enhancement by comparing with conventional ANPC in terms of power losses, efficiency, fault current, and EMI.

The rest of the paper is arranged as follows. The modeling of the proposed inverter topology is outlined in Section 2. Following this section, a characteristic and comparative analysis of the proposed inverter and conventional ANPC is presented in Section 3, including an analysis on fault currents, core losses, switching losses, efficiencies, EMI, and power factors. Section 4 discusses the summary and conclusion of the manuscript.

2. Modelling of Hybrid ANPC Inverter with Ga_2O_3 and Si Switches

2.1. Modelling of UWBG (Ga_2O_3) Semiconductors

In this article, the design of UWBG semiconductors is described briefly since the modeling and fabrication of the UWBG semiconductor is not the main objective of this study. The UWBG switches are modeled considering the drain current and source implementation [25], while the channel is isolated using the doping structure as shown in Figure 1. The Ga_2O_3 parameters that are used in this study to build the proposed inverter are demonstrated in Table 1. These parameters are only used in technology computer-aided design (TCAD) to evaluate the conduction behavior of the Ga_2O_3 devices. The I-V characteristics of these switches are shown in Figure 2.

Firstly, a Ga_2O_3 n-type epitaxial layer having 100 nm thickness is developed over a β-Ga_2O_3 (single crystal), which is semi-insulating in nature. Secondly, A dopant with a concentration of 2×10^{17} cm^{-3} is applied to dope the epitaxial layer. Tin (Si/Sn) implantation is used to form the 50 nm deep drain regions and the dopant concentration. Finally, a metal gate of 2 μm length and a work function of 5.93 eV is implanted on the top of a dielectric film gate with 20 nm length. The drain and gate are separated by a 4 μm gap [26].

To evaluate the performance of the Ga_2O_3 devices, accurate switching behavior is very crucial. However, since the switching behavior of the UWBG devices cannot be evaluated using TCAD, SPICE models of the Ga_2O_3 are required for further analysis [27]. In this regard, the level 1 Schichman–Hodges model parameters as shown in Table 2 are extracted from TCAD and were used to develop the SPICE model. The model parameters along with the switching, conduction, drain-source voltage, and drain current are implemented in LTSpice software to build a simulation model of the Ga_2O_3 switching device. The parameters that are used to build the LTSpice simulation model are shown in Table 2.

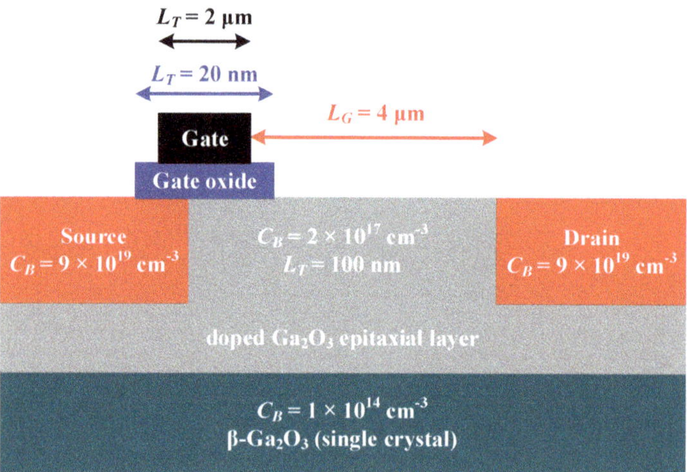

Figure 1. Modelling UWBG semiconductor switches.

Table 1. Parameters used in TCAD for analyzing conduction behavior of Ga_2O_3 switches.

Parameters	Values
Bandgap energy	4.8 eV
Effective density of states at 300 K	4.45×10^{18} cm^{-3}
Electron affinity	4 eV
Electron mobility	118 cm^2/Vs

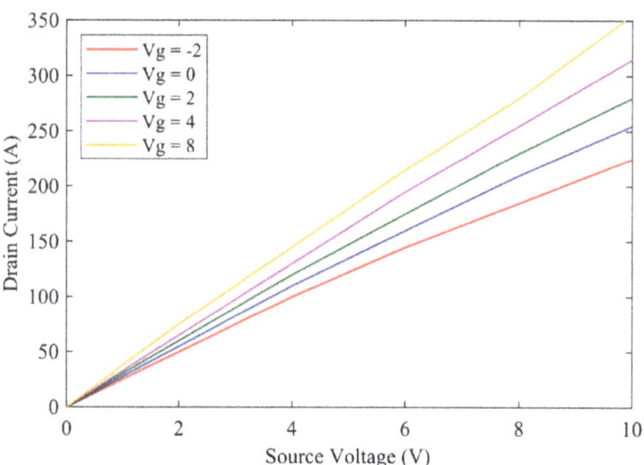

Figure 2. I-V characteristics of Ga_2O_3 semiconductor switches.

Table 2. Parameters used in SPICE for analyzing switching behavior of Ga_2O_3 switches.

Parameters	Values
Channel length	2 μm
Channel width	4.7×10^6 μm
Oxide thickness	20 nm
Electron mobility	118 cm^2/Vs
Substrate doping	2×10^{17} cm^{-3}
Zero-bias threshold voltage	−2.25 V
Transconductance	2.79×10^{-6} A/V^2
Gate-drain capacitance	4.3×10^{-11} F/m

2.2. Modelling of Hybrid ANPC Inverter

The schematic diagram of the proposed topology is depicted in Figure 3. The four switches, namely, S_1, S_4, S_5, and S_6, are constructed by using Si-based IGBTs, which are rated as 1.2 kV. On the other hand, the S_2 and S_3 switches are made by Ga_2O_3-based MOSFETs of 800 V rating. The utilization of both Si and Ga_2O_3 devices has ensured that the inductors can be split into L_1 and L_2 through these devices. In addition, it should be noted that the diodes D_2 and D_3 are both Ga_2O_3-based Schottky diodes [28]. As illustrated in Figure 3, the Ga_2O_3-based MOSFETs, i.e., S_2 and S_3 switches, are decoupled from D_2 and D_3, and this leads to the division of the inductors. Some capacitors are series-connected in the DC-link to make up neutral point '*n*'. There is a common portion of the two inductors between point '*a*' and the terminal '*n*', and the output is taken from this portion. As it is listed in Table 1, this inverter has six possible states. The states denoted by *P* and *N* represent positive and negative states, respectively, and null states are referred to as O_1 to O_4. S_2 and S_3 gallium trioxide (Ga_2O_3) switches are operated at a higher frequency, whereas Si IGBTs are operated in lower frequencies because it is required to maximize the output. To exploit this, only two null states, O_3 and O_2, as shown in Table 3, are utilized. More specifically, in case of the positive half cycle, the states *P* as well as O_3 are used, and the states *N* and O_2 are utilized for the operation of the negative half cycle. The UWBG is operated at a higher frequency of 100 kHz while the other four Si-based switches are operated at a lower fundamental frequency of 50 Hz. The gate pulses for switches are created using the level-shifted pulse width modulation (LSPWM) [29], which are depicted in Figure 4. The S_1 and S_6 switches will remain ON, while switches S_4 and S_5 will be turned OFF in case of positive cycle operation. On the contrary, the S_4 and S_5 switches will be ON and start conducting, while the S_1 and S_6 switches will be turned off for the negative half cycle.

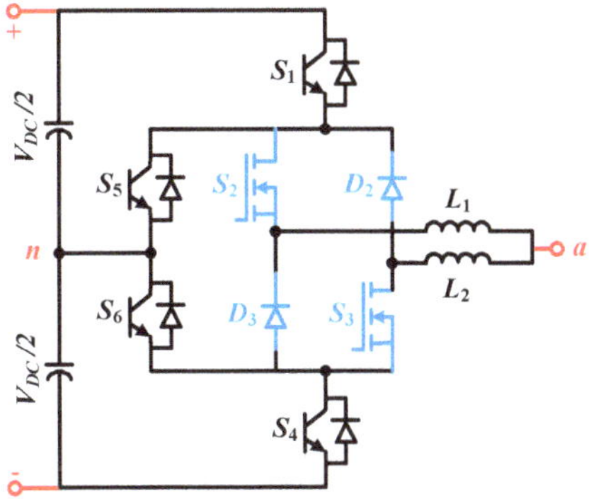

Figure 3. Schematic diagram of the hybrid ANPC inverter comprising UWBG switches.

Table 3. Switching states for the proposed HANPC inverter.

States	Switches					
	S_1	S_2	S_3	S_4	S_5	S_6
P	1	1	0	0	0	1
O_1	0	1	0	0	1	0
O_2	0	1	0	1	1	0
O_3	1	0	1	0	0	1
O_4	0	0	1	0	0	1
N	0	0	1	1	1	0

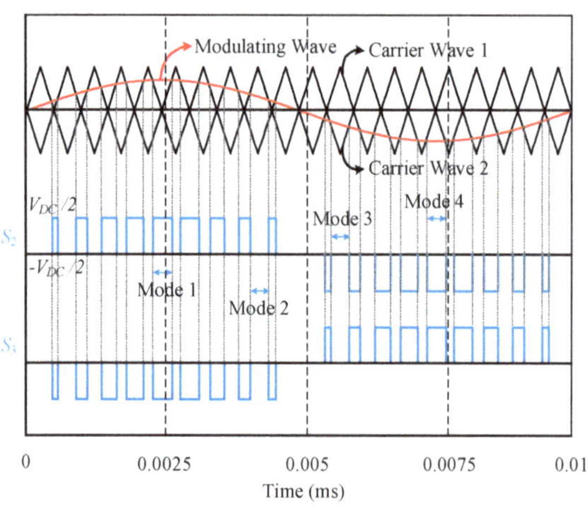

Figure 4. Switching pulses of the Ga_2O_3 devices of the hybrid ANPC using LSPWM.

As illustrated in Figure 4, the LSPWM is employed for the output voltage generation. In addition, there are three voltage levels, namely, 0.5 *Vdc*, 0, and 0.5 *Vdc*. It can be observed from Figure 4 that the proposed inverter has four modes of operation. Mode 1 and mode 2 are for the first half cycle whereas mode 3 and mode 4 are for the negative half cycle. As both cycles have a symmetrical operation, only mode 1 and mode 2 are discussed in this paper as depicted in Figure 5.

2.3. Modes of Operation

As the operation is dependent on the directions of the load current, each mode has two cases. The detailed circuit operation for mode 1 and mode 2 is shown in Figure 4.

Mode 1: In this mode, the output will be a positive voltage. A two-output load current is possible in this case, as shown in Figure 5a,b for for $i_L > 0$ and $i_L < 0$, respectively. During this mode, the gate pulse is received only by S_1, S_2, and S_6 switches while other switches remain OFF. In the case of $i_L > 0$, the current will flow through the split inductor L_1 because of the ON state of the switches S_1 and S_2. Similarly, when the load current direction is reversed, i.e., $i_L < 0$, the current flows through another split portion of the inductor in L_2. The inverter output voltage will be 0.5 *Vdc* in this mode irrespective of the load current direction, and it is depicted in Figure 4. Whether the load current is positive or negative, the current through the two split inductors, i.e., L_1 or L_2, will always be unidirectional. Therefore, unlike the split-NPC inverter that only can work for unity power factor because of one load direction current, the proposed hybrid ANPC inverter due to its two different load current direction can work on a wide range of power factors.

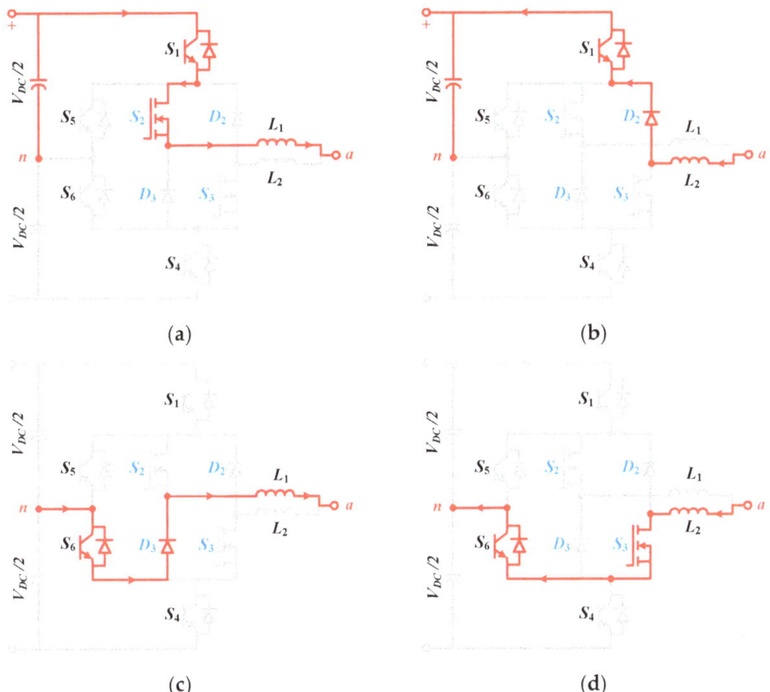

Figure 5. Switching paths of the hybrid ANPC inverter during positive half cycle: (**a**) mode 1 with $i_L > 0$, (**b**) mode 1 with $i_L < 0$, (**c**) mode 2 with $i_L > 0$, and (**d**) mode 2 with $i_L < 0$.

Mode 2: The operation of this mode is different from mode 1 because, here, the application of zero state, particularly O_3, is performed. In other words, the state of the S_1, S_4, S_5, and S_6 switches will be the same as mode 1, however, the state of S_2 and S_3 will be changed so that zero output voltage can be obtained. It is evident that at the neutral point, the output voltage will be clamped. The identical current flow path can be seen in Figure 5c,d. In this case, when the load current is positive, the current will flow through switch S_6 and the split portion of the inductor L_1. Similarly, for a negative load current, the current path will be through switch S_3 and other split portion L_2 of the inductor. The analytical study for the inverter will be presented in the next section.

3. Performance Analysis, Results, and Discussions of Hybrid ANPC Inverter

By analyzing the switching states given in Table 3, the value of the output voltage can be derived. Since the output terminal has split inductors, the voltage depends on the variation in the inductor current. This ultimately means that if the rate of current change is large, the output voltage will see a decline because of the losses associated with the inductors. Therefore, the output voltages of the proposed inverter can be derived by using (1) and (2) for the positive half cycle and negative half cycle, respectively:

$$V_{an} = 0.5 \times S_2 V_{DC} - L_1 \frac{di}{dt} \tag{1}$$

$$V_{an} = L_1 \frac{di}{dt} - 0.5 \times S_3 V_{DC} \tag{2}$$

It is noticeable that as the current passes through S_2 and S_3, the current stress (di/dt) is declined considerably because of using Ga_2O_3 based switches. If (1) is utilized, then the rate of change of current through S_2 for state transition from O_3 state to P state can be calculated by:

$$di = 0.5 \times S_2 V_{DC} \times \frac{dt}{L_1} \quad (3)$$

Here, dt is denoted for the time interval for the S_2 switch to transit from O_3 state to P state. Typically, the turn ON (t_{on}) time for each switch, including both Si and Ga_2O_3 switches will be comprised in this period. The nominal value of dt is acquired from the manufacturer's datasheets for Si-based switches, whereas for Ga_2O_3, the information is obtained from [30]. The summary is demonstrated in Table 4. It is clear from expression (3) that the di/dt stress is inversely proportional to the value of the first split inductor (L_1). This also pointed out the fact that as Ga_2O_3 switches have little t_{on} time (approximately 28.6 ns), the split inductance (L_1) value would be proportionally small to constrain the current stress of the ANPC inverter. Therefore, the voltage drop across L_1 would also be comparatively smaller than the DC-link voltage under steady-state operation. In addition to the reduced di/dt stress and voltage drop, under steady-state operating conditions, the inverter will experience reduced power loss across the inductor. Furthermore, as illustrated in Figure 1, the split inductors (L_1 and L_2) are contributing to decoupling Ga_2O_3 switches S_2 from D_2 as well as S_3 from D_3. The overshoots are significantly damped out because of this decoupling.

Table 4. Switching parameters of Si and Ga_2O_3 switches.

Model	Switching Parameters			
	Rated Voltage (V_r)	Rated Current (I_r)	Turn on Time (t_{on})	Turn off Time (t_{off})
IGW15T120FKSA1 (Si IGBT)	1200 V	15 A	50 ns	502 ns
Ga_2O_3 switch	800 V	20 A	28.6 ns	94 ns
Ga_2O_3 Schottky diode	1700 V	25 A	-	-

The common-mode voltage (CMV) of the hybrid ANPC inverter with 100 V DC-link can be calculated as follows:

$$V_{an} = 0.5 \times 100 - \frac{1 \times 10^{-6} di}{28.6 \times 10^{-9}} = 50 - 34.96di. \quad (4)$$

The CMV of the conventional ANPC inverter with 100 V DC-link can be calculated as follows:

$$V_{an} = 100 - \frac{1 \times 10^{-6} di}{50 \times 10^{-9}} = 100 - 20di. \quad (5)$$

It can be observed that for a certain value of di, the CMV of hybrid ANPC is almost 64.96% less than the CMV of conventional ANPC.

3.1. Analysis of Shoot through Fault Protection

In the proposed inverter, the complimentary operation of S_2 and S_3 at high switching frequency may result in the false turn-on of the switches [30]. Since Miller capacitance is present in all switches, the stored charge in it can cause the false turn ON of S_3. If both switches are in the ON state at the same time, the positive DC link voltage may become shorted in a positive half cycle of operation. The same thing is true for negative voltage during the negative half cycle. MOSFETs, in contrast to bipolar devices such as IGBTs, cannot endure overcurrent. Although shoot-through fault can happen in any switching device, since UWBG devices such as Ga_2O_3 switches are operating in this inverter at a very high frequency, they are more prone to this fault [31]. The issue is overcome by restricting the rate of the rising fault current using the split inductors. Hence, the proposed inverter configuration offers zero dead-band between S_2 and S_3.

To observe the impact, the shoot-through fault is allowed to happen on purpose when transitioning from the zero state O_3 to the state P. The fault current (I_f) is allowed to pass through S_2 and can be determined by:

$$I_f(t) = \frac{0.5 \times V_{DC}}{R_{eq} + R_1 + R_2}\left(1 - e^{\frac{-t(R_{eq}+R_1+R_2)}{L_{eq}+L_1+L_2}}\right) \tag{6}$$

Here, t is the time interval when shooting through the fault is allowed to happen, the resistances of L_1 and L_2 are denoted by R_1 and R_2, respectively, and, R_{eq} and L_{eq} are the equivalent resistance and inductance of the printed circuit board (PCB) path. R_{eq} and L_{eq} are required to calculate the maximum allowable time of shoot-through fault for a selected PCB.

The values of R_{eq} and L_{eq} are calculated to be 0.245 Ω and 187 nH, respectively, from the information given for PCB in [32,33]. Thus, the maximum allowable time is 21.06 ns for the selected design which is, in fact, lower than the turn OFF time of the Ga_2O_3 devices. In addition, the overcurrent limit for the design is 80 A. Therefore, before the switch S_2 is turned off (with t_{off} = 94 ns), the switch S_3 will be turned on falsely and can cause device failure. This issue is resolved by allowing a shoot-through time which is almost twice the turn OFF time of the Ga_2O_3 devices by using 1 uH split inductors. The numerical calculations can be realized by:

For conventional ANPC with split inductors,

$$I_f(t) = \frac{600}{0.245\ \Omega}\left(1 - e^{\frac{-1004\ \text{ns}\ \times 0.245\ \Omega}{2.187\ \mu\text{H}}}\right) = 260.52\ \text{A} \tag{7}$$

For the proposed hybrid ANPC with split inductors,

$$I_f(t) = \frac{600}{0.245\ \Omega}\left(1 - e^{\frac{-188\ \text{ns}\ \times 0.245\ \Omega}{2.187\ \mu\text{H}}}\right) = 51.04\ \text{A} \tag{8}$$

A simulation is conducted to determine the shoot through fault current of the proposed inverter by taking into consideration all the parasitic elements of the presented inverter circuit. Accordingly, the shoot-through fault's current paths are illustrated for both the positive and negative half cycle in Figure 6a,b, respectively. The simulation results are shown in Figure 7, and it can be observed that they are almost similar to the calculated values. It can be observed that in the case of the proposed inverter, the fault current is within the limit. This validates the predominance of the UWBG device as well as the hybridization that has been utilized in this article. It is worth noting that the fault current can be reduced for the conventional ANPC by increasing the value of the split inductors. However, it will incur more inductor core losses into the system and will eventually reduce the inverter's efficiency, making it radically unsuitable for industrial applications.

3.2. Analysis of Core Losses

The core losses of the proposed inverter are calculated in this section by considering the split inductors. The parameters which are considered for the proposed inverter's inductor design are listed in Table 5. The permissible losses in the copper winding are computed for the chosen core, with the required product area, which is the product of the window area (W_a) and the core area (A_c):

$$W_a \times A_c = \frac{L I_{max} I_{rms}}{K_t B_{max} j_{max}} \tag{9}$$

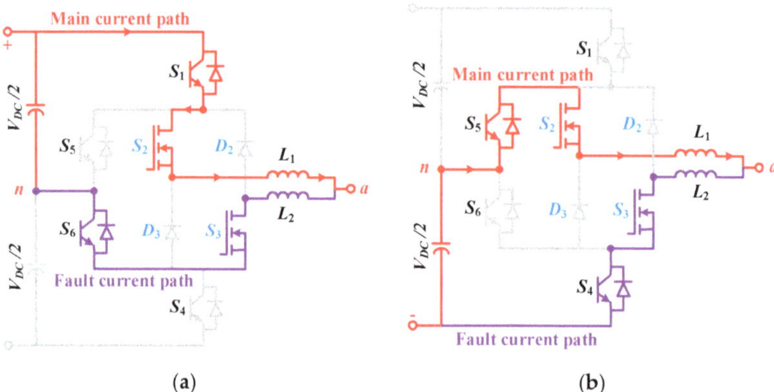

Figure 6. Shoot through fault current paths of the hybrid ANPC inverter during (**a**) state O_3 to state P for positive half cycle, (**b**) state O_2 to state N for the negative half cycle.

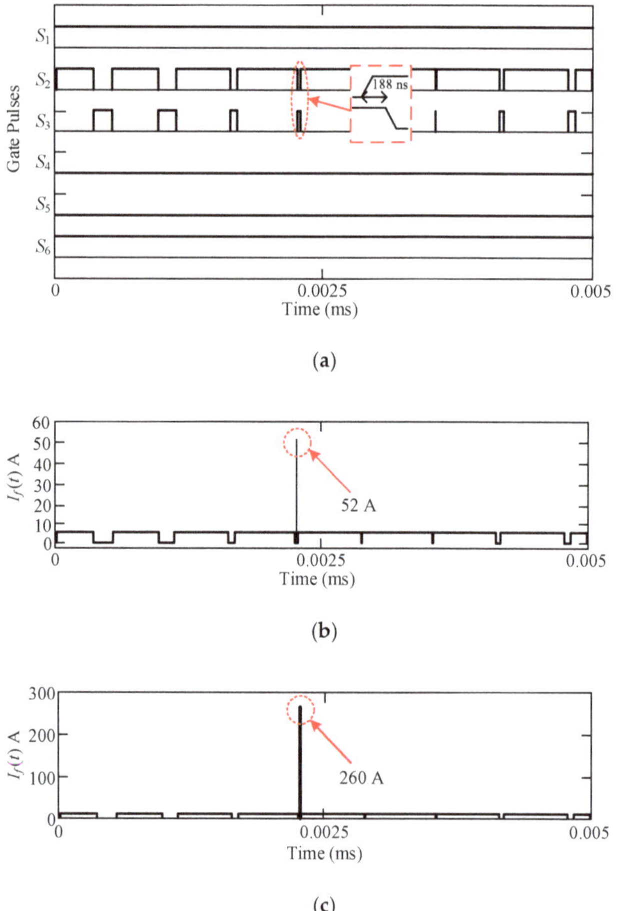

Figure 7. Shoot through fault current analysis: (**a**) gate pulses of S_2 and S_3 overlapping and causing shoot through fault, (**b**) hybrid ANPC, (**c**) conventional ANPC.

Table 5. Parameters for designing the split inductors.

Parameters	Nomenclature	Values
Inductor	$L_1 = L_2$	1 µH
Maximum current	I_{max}	42 A
RMS current	I_{rms}	35 A
Topological constant	K_t	0.3
Maximum flux density	B_{max}	160 mT
Maximum current density	j_{maz}	5 A/mm^2
Product area	$W_a \times A_c$	0.002 cm^4

Here, L is one of the split inductors, I_{max} is the maximum current flowing through the inductor, I_{rms} is the rated RMS current, K_t is the topological constant, B_{max} is the maximum flux density, and j_{max} is the maximum current density of the inductor. Although for complete accuracy the optimum loss for copper should be measured, the maximum permissible copper loss is calculated in this section because of the minimal difference between the accurate and approximate values, as well as for simplicity. Thus, the maximum allowable copper loss is used to measure the efficiency. The product area value obtained from (9) is used to determine the thermal resistance R_{th} by utilizing the data from [33], assuming that the core temperature is increasing by 50 °C:

$$R_{th} = 17.45(W_a \times A_c)^{-0.509} + 0.416\ °C/W \tag{10}$$

After the thermal resistance is calculated, this can lead to the measurement of maximum possible core loss (P_{Cu}) for a particular temperature rise ΔT, and it can be determined by the following equation:

$$P_{Cu} = \frac{\Delta T}{R_{th}} \tag{11}$$

The measurement of the copper winding loss can be performed for the split inductors by utilizing (9) to (11). Because of the minimal values of the product area, a large core size is selected for the practical design. The core losses for the selected material from Magnetics [34] are plotted using the values given in [33] in Figure 8 for the selected core volume.

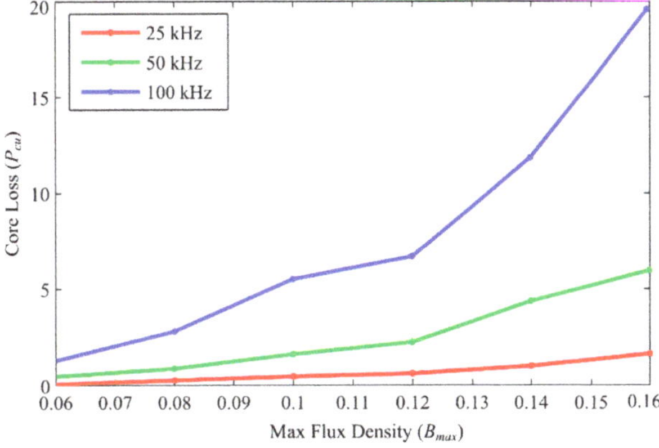

Figure 8. Core loss induced by the inductors under different switching frequencies.

3.3. Analysis of Switching Losses

Although it is already clear that the use of split inductors in the hybrid ANPC module is a major source of loss in steady-state operation, the inherent nature of the hybrid ANPC inverter is also responsible for the additional losses. The use of Ga_2O_3 switches S_2 and S_3 is a viable solution for this topology because these UWBG switches help to reduce the switching losses. Therefore, to quantify the improvement, it is essential to know how much loss is reduced after the addition of the UWBG switches.

For switching loss measurement, double pulse testing (DPT) [30] is conducted. The DPT circuit used for the switching measurement is illustrated in Figure 9. The parasitic inductors in the PCB path are denoted by L_{p1}, L_{p2}, and L_{p3}; the series inductor in the DC link is denoted by L_s; and the output inductor is denoted by L_o. Similarly, the drain to source capacitance of Ga_2O_3 switches and the anode–cathode capacitance of the Ga_2O_3 Schottky diode are indicated as C_{ds} and C_{ac}, respectively. The output inductance is measured following [30] while L_{p1}, L_{p2}, and L_{p3} are measured following [33]. All the calculated values are listed in Table 6.

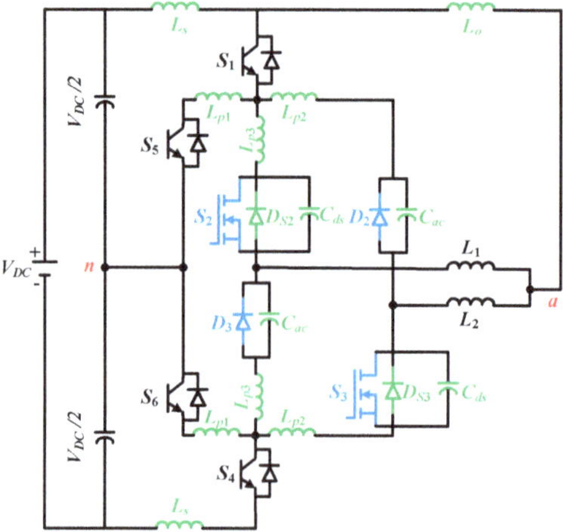

Figure 9. DPT circuit of the hybrid ANPC inverter with parasitic elements.

Table 6. Parameters for DPT testing.

Equipment	Nomenclature	Value
Inductors	L_s	36.15 nH
	L_{p1}	11.6 nH
	L_{p2}	19.16 nH
	L_{p3}	11.6 nH
	L_o	1200 uH
Capacitors	C_{ds}	171 pF
	C_{ac}	80 pF

Table 6 after putting these values in LT Spice, the simulation is conducted and switching transients are calculated.

The DPT test is performed repeatedly for different load currents and switching voltages to emulate practical scenarios. The data obtained from DPT are used to measure the energies required for the turning ON and turning OFF of the switches by using simulation, and they are referred to as E_{on} and E_{off}, respectively. Figure 10 illustrates the measured switching energies for both the conventional and the proposed inverter topologies. Though the energy consumption in the ideal switch should be zero, the semiconductor switches are hardly ideal, and thus, from these curves, it can be observed how switching energies rise when the load current increases. In addition, it is evident from these curves that the use of Ga_2O_3 switches has greatly contributed to reducing both the turn-on and turn-off switching energies. The simulated waveform shown in Figure 11 represents the minimization of switching losses with the utilization of Ga_2O_3 switches. It can be observed from Figure 11a that when S_2 is turned on, the switching current has increased as soon as the gate pulse is applied. In other words, since conventional Si switches have a slow turn-on time, an overshoot current of 43 A is caused by C_{ac} of D_3. On the contrary, the Ga_2O_3 switches have a very fast turn-on time, which is why the overshoot current in this case significantly declined as shown in Figure 11c. This phenomenon also implies that due to the decreased overshoot, a faster decrease in switching voltage across the switch S_2 in the case of the proposed inverter leads to decreased loss. In the case of turn-off, an almost similar event occurs in both case 1 and case 2, which are illustrated in Figure 11b,d, respectively. In this case, it can be observed that an overvoltage spike of almost 630 V is experienced by the conventional inverter compared to the 560 V spile of the hybrid.

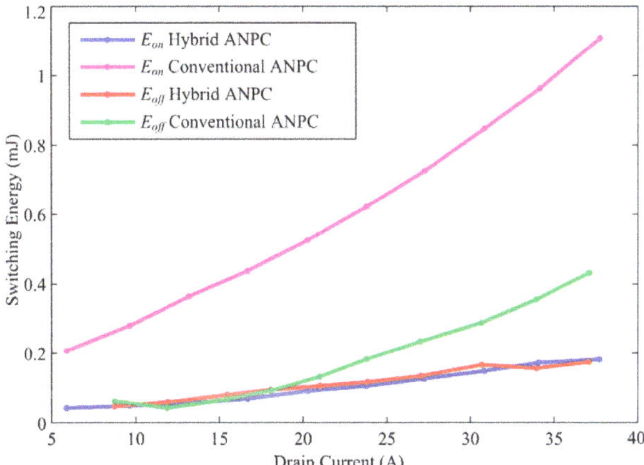

Figure 10. Characteristics curves highlighting the energies required for switches to turn ON and turn OFF with respect to the switching current for conventional ANPC and hybrid ANPC inverters.

ANPC inverter. This has resulted in higher turn OFF losses incurred by the conventional inverter. Although the margin of differences between the conventional inverter and the hybrid ANPC for turn OFF losses is very close, the overall switching losses of hybrid ANPCs are significantly lower because the turn ON losses are more dominant.

3.4. *Analysis of Efficiency*

The efficiencies of switching losses, conduction losses, and split-inductors losses are considered. The switching energies obtained from the DPT test are used for switching loss calculation. In case of switching loss, turn ON loss P_{on} and turn OFF loss P_{off} are determined by:

$$P_{on} = f_s \times E_{on} \tag{12}$$

$$P_{off} = f_s \times E_{off} \tag{13}$$

where the switching energies E_{on} and E_{off} can be determined by:

$$E_{on} = I_s \times x_{on} \tag{14}$$

$$E_{off} = I_s \times x_{off} \tag{15}$$

The equations for x_{on} and x_{off} can be mathematically expressed by:

$$x_{on} = x_{1on} \times I_s{}^2 + x_{2on} \times I_s + x_{3on} \tag{16}$$

$$x_{off} = x_{1off} \times I_s{}^2 + x_{2off} \times I_s + x_{3off} \tag{17}$$

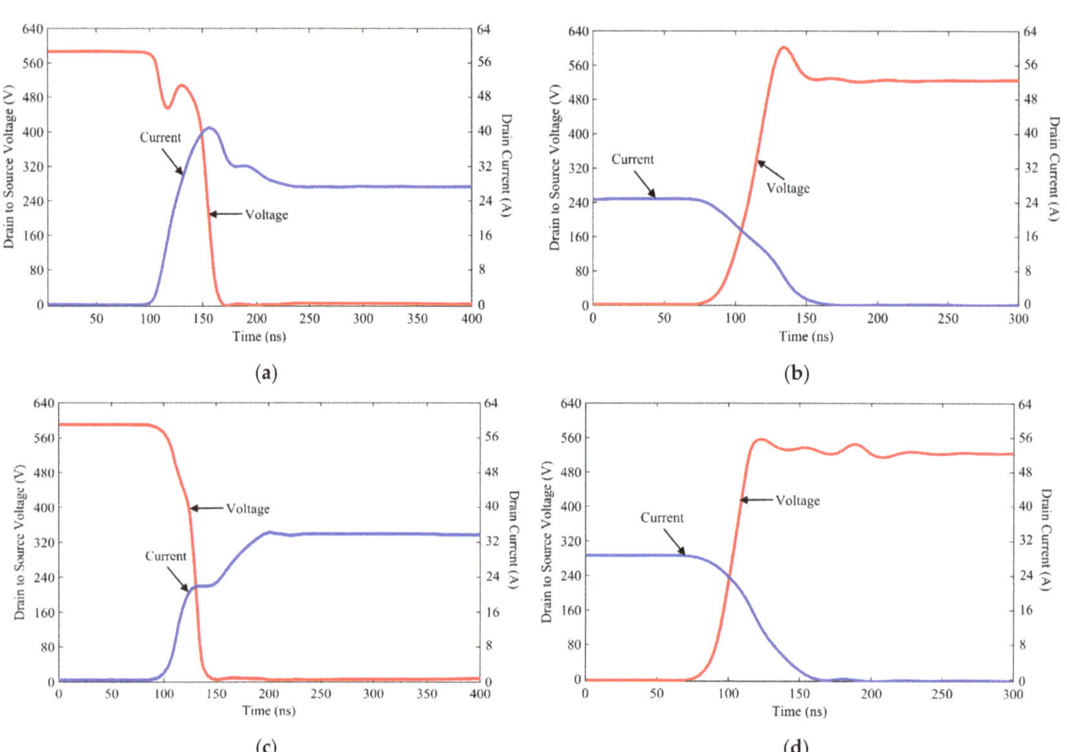

Figure 11. Switching curves of S_2 obtained from LTSpice: (**a**) switch turn ON for conventional ANPC, (**b**) switch turn OFF for conventional ANPC, (**c**) switch turn ON for hybrid ANPC, (**d**) switch turn OFF for hybrid ANPC.

Here, the constants x_{1on}, x_{2on}, x_{3on}, …… are representative of the constants that are used for curve fitting shown in Figure 10. Additionally, the conduction losses are calculated using the manufacturer's datasheet curves for different load currents in the case of conventional Si switches, whereas, for Ga_2O_3 switches, it has been obtained from the information provided in [35]. The expressions obtained from these curves are:

$$P_c = x_4 \times I_s{}^2 + x_5 \times I_s \tag{18}$$

where x_4 and x_5 are the constants for the curve fitting of Figure 10. Furthermore, the core losses from the split inductors are determined using the curves shown in Figure 8 and the information provided in [34].

In this paper, the losses of both conventional ANPCs as well as the proposed hybrid ANPC inverter are calculated considering different loads. In addition, three switching frequencies are considered to compare the loss behavior of the configurations, as shown in Figure 12. It can be validated from Figure 12 that because of using UWBG switches and due to reduced switching losses, the proposed inverter's efficiency in all cases is much higher compared to the conventional Si-based ANPC inverter.

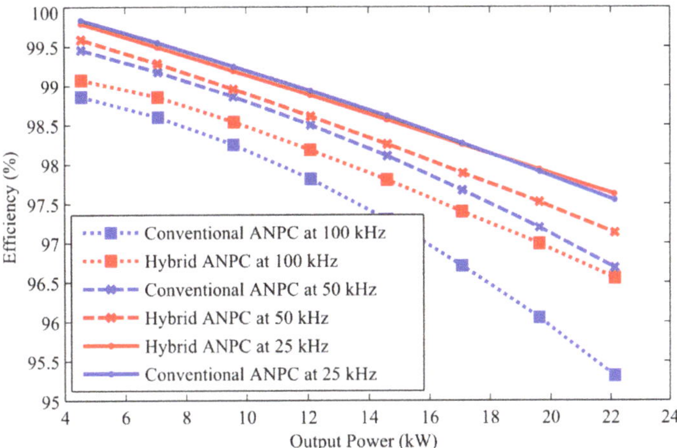

Figure 12. Efficiency comparison between conventional ANPC and hybrid ANPC under various switching frequencies.

3.5. Analysis of High-Frequency Transient in Output Voltage

Along with the advantages of the conventional ANPC inverter, the proposed inverter can reduce high-frequency switching noise in the output voltage. This high-frequency noise primarily contributes to electromagnetic interference (EMI) issues and also has some impacts on the operation of the gate driver [24]. In addition, the incorporation of the two split inductors, i.e., L_1 and L_2, in the proposed inverter topology makes it possible to decrease the high-frequency transients considerably because of the filter of the transients by the inductances. Thus, the size of the electromagnetic compatibility (EMC) filter becomes significantly smaller. This statement can be validated by using (1) and (2). If any sudden change has occurred in the output voltage of the presented inverter, that impact will be damped by the inductance's inherent capability to oppose any sudden change in current. The blocking voltage is tuned according to the values of the split inductor. For the proposed design, as the inductance value was 1 uH for the split inductor, the output voltages' harmonic spectra can be illustrated for both conventional ANPCs and the proposed hybrid ANPC inverter through LT Spice simulation, as is illustrated in Figure 13. It can be seen that the final range of the high-frequency transient will be 5 to 15 MHz. This is due to the damped high-frequency voltage in this frequency range by the split-inductors. Thus, the added split inductors for the shoot-through protection also help to reduce the EMI filter size.

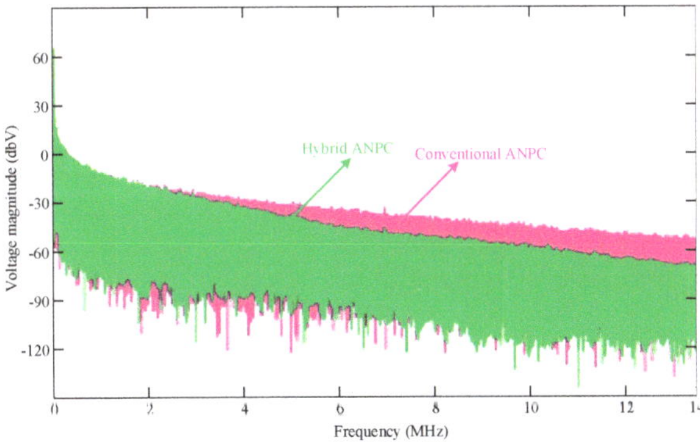

Figure 13. High-frequency voltage spectrum of the conventional ANPC and hybrid ANPC.

The cross-sectional area (A) of an EMI Filter for the hybrid ANPC with 100 kHz switching frequency can be determined by:

$$A = \frac{2\pi rL}{\mu_0 \mu_r N^2} = \frac{2\pi \times 0.1 \times 0.5 \times 10^{-6}}{1.2566 \times 10^{-6} \times 6000 \times 400} = 1.04 \times 10^{-7} \text{m}^2 \quad (19)$$

Here, r, μ_0, μ_r, and N represent the toroid radius to centerline, the magnetic constant, the relative permeability of Mn–Zn ferrite, and the number of turns, respectively. Similarly, the cross-sectional area of the EMI filter for a conventional ANPC can be calculated as follows:

$$A = \frac{2\pi rL}{\mu_0 \mu_r N^2} = \frac{2\pi \times 0.1 \times 20 \times 50 \times 10^{-9}}{1.2566 \times 10^{-6} \times 6000 \times 400} = 2.08 \times 10^{-7} \text{m}^2 \quad (20)$$

Thus, it can be observed that the size of the EMI filter for the proposed ANPC inverters becomes halved compared to the conventional ANPC inverter due to the usage of split inductors. Furthermore, the relative permeability versus the switching frequency curve for Mn–Zn ferrite is shown in Figure 14. It is noticeable that with higher switching frequency, the relative permeability tends to decrease logarithmically. Therefore, the cross-sectional area of the EMI filter will increase with a higher switching frequency.

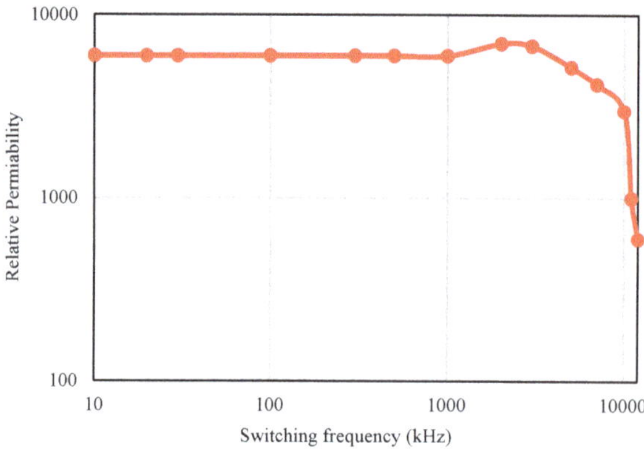

Figure 14. Relative permeability of Mn Zn ferrite under different switching frequencies.

3.6. *Analysis of Operation at Various Range of Power Factors*

The MATLAB/Simulink version of the proposed hybrid ANPC inverter is developed in this section to validate that it can operate in various ranges of power factors. LTSpice simulation is not required in this case since this feature is embraced by the proposed inverter due to implementing the split-inductors-based design, and this feature is not associated with using UWBG switches. Thus, for operational simplicity, MATLAB Simulink along with ideal MOSFETs and IGBTs are used to develop the proposed inverter. The output voltage and current waveforms are obtained for the proposed topology using a 200 V DC link. Thus, a voltage of 100 V will come across each DC-link capacitor. The simulation tests are repeated with the loads with non-unity power factor. To show the applicability of the proposed converter compared to the existing topologies. The results show the non-distorted waveforms for voltage and currents. The results for output voltage V_{an} and load current I_{an} are shown in Figure 15. Furthermore, the voltage across one DC-link capacitor is also shown, which indicates the nature of the common-mode voltage (CMV). It can be observed that the CMV is always constant at 100 V and it does not contain any ripples of high frequency. Thus, the leakage-current-related issues can also be solved using this topology.

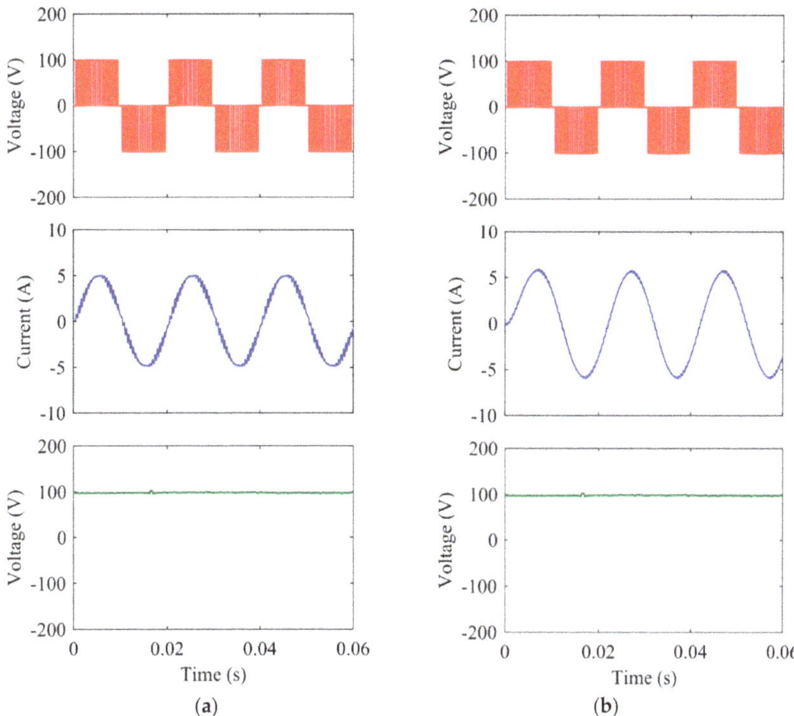

Figure 15. Simulation results for the hybrid ANPC inverter for the output voltage (V_{an}), current (I_{an}) and common-mode voltage (CMV) with (**a**) unity power factor, (**b**) non-unity power factor.

4. Conclusions

To sum up, this paper presents a three-level hybrid ANPC topology that includes Ga_2O_3-based MOSFET as well as Si-based IGBTs. This inverter has split inductors at the output, which are not only capable of protecting against the shoot-through fault but can also contribute to the reduced EMI in the output voltage. To maximize the efficiency of our converter, as well as to maximize the benefit of the Ga_2O_3 switches, both the modulation technique as well as four modes of operation are discussed in this paper. The efficiency of both the conventional ANPC and the proposed hybrid ANPC inverter is measured and compared through LT Spice and MATLAB simulations. It was observed that under various switching frequencies and output power, the minimum efficiency was 96.8%, whereas a 99.1% maximum efficiency was obtained by the proposed inverter. The employability of the proposed module is analyzed by taking into consideration the reduced overshoots in switching waveforms, higher efficiency, lower current, voltage stress, minimized shoot-through current, and EMI. Eliminating the dominating switching losses, especially turn-on losses, as well as the addition of UWBG switches, contributes to an increase in efficiency. In addition, to validate the inverter's capability to supply reactive power, the module was operated under both various load conditions by changing the power factors. The simulation result acquired from the proposed module coincides with the theoretical results. The following is a list of the manuscript's concluding statements:

- The proposed inverter incorporated UWBG-based Ga_2O_3 switches, which contributed to its enhanced efficiency and reduced switching losses.
- The Ga_2O_3 switches of the inverter make it a suitable candidate for high voltage, high temperature, and high switching operation.

- A maximum efficiency of 99.1% is obtained, making this inverter suitable for applications in grid-tied PV structures.
- The minimized EMI and fault current, because of the split-inductors-based design, allowed this inverter to be utilized in sophisticated industrial applications.

This study applies UWBG switches for ANPC inverters considering the technical pros and cons. Since the fabrication and production of UWBG semiconductors are still in their early phase industrially, experimental verification of the proposed inverter will be considered in the future. In the future, UWBG devices have great potential in the field of power electronics because of their superior characteristics over wide bandgap (WBG) and conventional semiconductors. Thus, researchers can utilize this opportunity to incorporate UWBG devices in other inverters/converter topologies and power electronic applications.

Author Contributions: Conceptualization, S.T.M. and J.I.; methodology, S.T.M. and N.Z.Y.; formal analysis, S.T.M. and M.S.H.L.; investigation, S.T.M., N.Z.Y. and M.S.H.L.; resources, N.Z.Y., J.I. and L.K.H.; data curation, S.T.M., L.K.H. and K.H.; writing—original draft preparation, S.T.M.; writing—review and editing, M.S.H.L., K.H. and M.S.M.; S.A. and A.H.; supervision, N.Z.Y. and M.S.H.L.; project administration, M.S.H.L.; funding acquisition, M.S.H.L. All authors have read and agreed to the published version of the manuscript.

Funding: This work was supported by Universiti Kebangsaan Malaysia under Grant Code GP-2021-K023221.

Conflicts of Interest: The authors declare no conflict of interest.

References

1. She, X.; Huang, A.Q.; Lucia, O.; Ozpineci, B. Review of Silicon Carbide Power Devices and Their Applications. *IEEE Trans. Ind. Electron.* **2017**, *64*, 8193–8205. [CrossRef]
2. Wang, F.; Zhang, Z. Overview of Silicon Carbide Technology: Device, Converter, System, and Application. *CPSS Trans. Power Electron. Appl.* **2016**, *1*, 13–32. [CrossRef]
3. Tsao, J.Y.; Chowdhury, S.; Hollis, M.A.; Jena, D.; Johnson, N.M.; Jones, K.A.; Kaplar, R.J.; Rajan, S.; Van de Walle, C.G.; Bellotti, E.; et al. Ultrawide-Bandgap Semiconductors: Research Opportunities and Challenges. *Adv. Electron. Mater.* **2018**, *4*, 1600501. [CrossRef]
4. Wang, Z.; Li, G.; Tseng, M.-L.; Wong, W.-P.; Liu, B. Distributed Systematic Grid-Connected Inverter Using IGBT Junction Temperature Predictive Control Method: An Optimization Approach. *Symmetry* **2020**, *12*, 825. [CrossRef]
5. Meraj, S.T.; Hasan, M.K.; Islam, J.; Baker El-Ebiary, Y.A.; Nebhen, J.; Hossain, M.M.; Alam, M.K.; Vo, N. A Diamond Shaped Multilevel Inverter with Dual Mode of Operation. *IEEE Access* **2021**, *9*, 59873–59887. [CrossRef]
6. Shayestegan, M.; Shakeri, M.; Abunima, H.; Reza, S.M.S.; Akhtaruzzaman, M.; Bais, B.; Mat, S.; Sopian, K.; Amin, N. An overview on prospects of new generation single-phase transformerless inverters for grid-connected photovoltaic (PV) systems. *Renew. Sustain. Energy Rev.* **2018**, *82*, 515–530. [CrossRef]
7. Xue, H.W.; He, Q.M.; Jian, G.Z.; Long, S.B.; Pang, T.; Liu, M. An Overview of the Ultrawide Bandgap Ga2O3 Semiconductor-Based Schottky Barrier Diode for Power Electronics Application. *Nanoscale Res. Lett.* **2018**, *13*, 290. [CrossRef] [PubMed]
8. Liao, M.; Shen, B.; Wang, Z. Progress in semiconductor β-Ga_2O_3. In *Ultra-Wide Bandgap Semiconductor Materials*; Elsevier: Amsterdam, The Netherlands, 2019; pp. 263–345. ISBN 9780128172568.
9. He, J. Comparison between The ultra-wide band gap semiconductor AlGaN and GaN. In *IOP Conference Series: Materials Science and Engineering*; IOP Publishing: Kuala Lumpur, Malaysia, 2020; pp. 1–5.
10. Meraj, S.T.; Yahaya, N.Z.; Hasan, K.; Lipu, M.S.H.; Masaoud, A.; Ali, S.H.M.; Hussain, A.; Othman, M.M.; Mumtaz, F. Three-Phase Six-Level Multilevel Voltage Source Inverter: Modeling and Experimental Validation. *Micromachines* **2021**, *12*, 1133. [CrossRef]
11. Hamidi, M.N.; Ishak, D.; Zainuri, M.A.A.M.; Ooi, C.A. Multilevel inverter with improved basic unit structure for symmetric and asymmetric source configuration. *IET Power Electron.* **2020**, *13*, 1445–1455. [CrossRef]
12. Serban, E.; Ordonez, M.; Pondiche, C. DC-Bus Voltage Range Extension in 1500 v Photovoltaic Inverters. *IEEE J. Emerg. Sel. Top. Power Electron.* **2015**, *3*, 901–917. [CrossRef]
13. Gkoutioudi, E.; Bakas, P.; Marinopoulos, A. Comparison of PV systems with maximum DC voltage 1000V and 1500V. In Proceedings of the 2013 IEEE 39th Photovoltaic Specialists Conference (PVSC), Tampa, FL, USA, 16–21 June 2013; pp. 2873–2878.
14. Zhang, L.; Sun, K.; Feng, L.; Wu, H.; Xing, Y. A family of neutral point clamped full-bridge topologies for transformerless photovoltaic grid-tied inverters. *IEEE Trans. Power Electron.* **2013**, *28*, 730–739. [CrossRef]
15. Guo, X.; Wang, N.; Zhang, J.; Wang, B.; Nguyen, M.K. A Novel Transformerless Current Source Inverter for Leakage Current Reduction. *IEEE Access* **2019**, *7*, 50681–50690. [CrossRef]

16. López, I.; Ceballos, S.; Pou, J.; Zaragoza, J.; Andreu, J.; Ibarra, E.; Konstantinou, G. Generalized PWM-Based Method for Multiphase Neutral-Point-Clamped Converters with Capacitor Voltage Balance Capability. *IEEE Trans. Power Electron.* **2017**, *32*, 4878–4890. [CrossRef]
17. Ramasamy, P.; Krishnasamy, V.; Sathik, M.A.J.; Ali, Z.M.; Aleem, S.H.E.A. Three-Dimensional Space Vector Modulation Strategy for Capacitor Balancing in Split Inductor Neutral-Point Clamped Multilevel Inverters. *J. Circuits Syst. Comput.* **2018**, *27*, 1850232–1850248. [CrossRef]
18. Lee, S.S.; Lee, K.B. Dual-T-Type Seven-Level Boost Active-Neutral-Point-Clamped Inverter. *IEEE Trans. Power Electron.* **2019**, *34*, 6031–6035. [CrossRef]
19. Jiang, W.; Huang, X.; Wang, J.; Wang, J.; Li, J. A carrier-based PWM strategy providing neutral-point voltage oscillation elimination for multi-phase neutral point clamped 3-level inverter. *IEEE Access* **2019**, *7*, 124066–124076. [CrossRef]
20. Li, Y.; Tian, H.; Li, Y.W. Generalized Phase-Shift PWM for Active-Neutral-Point-Clamped Multilevel Converter. *IEEE Trans. Ind. Electron.* **2020**, *67*, 9048–9058. [CrossRef]
21. Feng, Z.; Cai, Y.; Li, Z.; Hu, Z.; Zhang, Y.; Lu, X.; Kang, X.; Ning, J.; Zhang, C.; Feng, Q.; et al. Design and fabrication of field-plated normally off β-Ga2O3 MOSFET with laminated-ferroelectric charge storage gate for high power application. *Appl. Phys. Lett.* **2020**, *116*, 243503–243516. [CrossRef]
22. He, J.; Zhang, D.; Pan, D. An Improved PWM Strategy for "SiC+Si" Three-Level Active Neutral Point Clamped Converter in High-Power High-Frequency Applications. In Proceedings of the 2018 IEEE Energy Conversion Congress and Exposition (ECCE), Portland, OR, USA, 23–27 September 2018; pp. 5235–5241. [CrossRef]
23. Jahdi, S.; Alatise, O.; Bonyadi, R.; Alexakis, P.; Fisher, C.A.; Gonzalez, J.A.O.; Ran, L.; Mawby, P. An analysis of the switching performance and robustness of power MOSFETs body diodes: A technology evaluation. *IEEE Trans. Power Electron.* **2015**, *30*, 2383–2394. [CrossRef]
24. Fang, Z.; Jian, D.; Zhang, Y. Study of the Characteristics and Suppression of EMI of Inverter with SiC and Si Devices. *Chinese J. Electr. Eng.* **2018**, *4*, 37–46. [CrossRef]
25. Ping, L.K.; Mohamed, M.A.; Mondal, A.K.; Taib, M.F.M.; Samat, M.H.; Berhanuddin, D.D.; Susthitha Menon, P.; Bahru, R. First-principles studies for electronic structure and optical properties of strontium doped β-Ga_2O_3. *Micromachines* **2021**, *12*, 604. [CrossRef]
26. Barthel, A.; Roberts, J.; Napari, M.; Frentrup, M.; Huq, T.; Kovács, A.; Oliver, R.; Chalker, P.; Sajavaara, T.; Massabuau, F. Ti Alloyed α-Ga_2O_3: Route towards Wide Band Gap Engineering. *Micromachines* **2020**, *11*, 1128. [CrossRef] [PubMed]
27. Hu, Z.; Nomoto, K.; Li, W.; Zhang, Z.; Tanen, N.; Thieu, Q.T.; Sasaki, K.; Kuramata, A.; Nakamura, T.; Jena, D.; et al. Breakdown mechanism in 1 kA/cm2 and 960 V E-mode β-Ga_2O_3 vertical transistors. *Appl. Phys. Lett.* **2018**, *113*, 122103–122116. [CrossRef]
28. Hu, Z.; Zhou, H.; Dang, K.; Cai, Y.; Feng, Z.; Gao, Y.; Feng, Q.; Zhang, J.; Hao, Y. Lateral β-Ga2O3 Schottky Barrier Diode on Sapphire Substrate with Reverse Blocking Voltage of 1.7 kV. *IEEE J. Electron Devices Soc.* **2018**, *6*, 815–820. [CrossRef]
29. Meraj, S.T.; Kah Haw, L.; Masaoud, A. Simplified Sinusoidal Pulse Width Modulation of Cross-switched Multilevel Inverter. In Proceedings of the 2019 IEEE 15th International Colloquium on Signal Processing & Its Applications (CSPA), Batu Ferringhi, Malaysia, 8–9 March 2019; pp. 1–6.
30. Dong, H.; Long, S.; Sun, H.; Zhao, X.; He, Q.; Qin, Y.; Jian, G.; Zhou, X.; Yu, Y.; Guo, W.; et al. Fast Switching β-Ga2O3 Power MOSFET With a Trench-Gate Structure. *IEEE Electron Device Lett.* **2019**, *40*, 1385–1388. [CrossRef]
31. Sam, D.; Power Management Chapter 11: Wide Bandgap Semiconductors | Power Electronics. Power Electronics (Endeavor Business Media). 31 May 2018. Available online: https://www.powerelectronics.com/technologies/power-management/article/21864166/power-management-chapter-11-wide-bandgap-semiconductors (accessed on 8 November 2021).
32. Terman, F.E. *Radio Engineers' Handbook*, 1st ed.; McGraw-Hill Book: London, UK, 1943.
33. Abraham, I.P.; Billings, K.; Taylor, M. Transformers and Magnetic Design. In *Switching Power Supply Design*, 3rd ed.; McGraw-Hill Education: London, UK, 2009; pp. 285–421. ISBN 9780071482721.
34. FERRITE CORES Toroids | Shapes | Pot Cores. Magnetics 2017. Available online: www.mag-inc.com (accessed on 8 November 2021).
35. Lee, I.; Kumar, A.; Zeng, K.; Singisetti, U.; Yao, X. Modeling and power loss evaluation of ultra wide band gap Ga_2O_3 device for high power applications. In Proceedings of the 2017 IEEE Energy Conversion Congress and Exposition (ECCE), Cincinnati, OH, USA, 1–5 October 2017; pp. 4377–4382.

Article

Three-Phase Six-Level Multilevel Voltage Source Inverter: Modeling and Experimental Validation

Sheikh Tanzim Meraj [1], Nor Zaihar Yahaya [1], Kamrul Hasan [2], Molla Shahadat Hossain Lipu [3,4,*], Ammar Masaoud [5], Sawal Hamid Md Ali [3], Aini Hussain [3], Muhammad Murtadha Othman [2] and Farhan Mumtaz [1]

1 Department of Electrical and Electronic Engineering, Universiti Teknologi PETRONAS, Seri Iskandar 32610, Perak, Malaysia; sheikh_19001724@utp.edu.my (S.T.M.); norzaihar_yahaya@utp.edu.my (N.Z.Y.); fahan_19001785@utp.edu.my (F.M.)

2 Solar Research Institute (SRI), Universiti Teknologi MARA, Shah Alam 40450, Selangor, Malaysia; 2019984679@isiswa.uitm.edu.my (K.H.); m_murtadha@uitm.edu.my (M.M.O.)

3 Department of Electrical, Electronic and Systems Engineering, Universiti Kebangsaan Malaysia, UKM Bangi 43600, Selangor, Malaysia; sawal@ukm.edu.my (S.H.M.A.); draini@ukm.edu.my (A.H.)

4 Centre for Automotive Research (CAR), Universiti Kebangsaan Malaysia, UKM Bangi 43600, Selangor, Malaysia

5 Faculty of Mechanical and Electrical Engineering, Al Baath University, Homs 22743, Syria; ammarshz123@gmail.com

* Correspondence: lipu@ukm.edu.my

Citation: Meraj, S.T.; Yahaya, N.Z.; Hasan, K.; Hossain Lipu, M.S.; Masaoud, A.; Ali, S.H.M.; Hussain, A.; Othman, M.M.; Mumtaz, F. Three-Phase Six-Level Multilevel Voltage Source Inverter: Modeling and Experimental Validation. *Micromachines* **2021**, *12*, 1133. https://doi.org/10.3390/mi12091133

Academic Editor: Francisco J. Perez-Pinal

Received: 4 September 2021
Accepted: 16 September 2021
Published: 21 September 2021

Abstract: This research proposes a three-phase six-level multilevel inverter depending on twelve-switch three-phase Bridge and multilevel DC-link. The proposed architecture increases the number of voltage levels with less power components than conventional inverters such as the flying capacitor, cascaded H-bridge, diode-clamped and other recently established multilevel inverter topologies. The multilevel DC-link circuit is constructed by connecting three distinct DC voltage supplies, such as single DC supply, half-bridge and full-bridge cells. The purpose of both full-bridge and half-bridge cells is to provide a variable DC voltage with a common voltage step to the three-phase bridge's mid-point. A vector modulation technique is also employed to achieve the desired output voltage waveforms. The proposed inverter can operate as a six-level or two-level inverter, depending on the magnitude of the modulation indexes. To guarantee the feasibility of the proposed configuration, the proposed inverter's prototype is developed, and the experimental results are provided. The proposed inverter showed good performance with high efficiency of 97.59% following the IEEE 1547 standard. The current harmonics of the proposed inverter was also minimized to only 5.8%.

Keywords: multilevel inverter; power electronics; staircase modulation; vector modulation

1. Introduction

In the sector of power engineering, multilevel inverters (MLIs) have been developed and their applications have been expanded in a rapid manner in recent years. The composition of MLIs basically includes a power components' array and DC voltage supplies/capacitors that produce a variable and controllable stepped voltage waveform. Over the years, various MLIs topologies have been proposed and used due to their importance and roles in power conversion systems, with some of the most well-known and conventional topologies including the flying capacitor (FC), the cascaded H-bridge (CHB) and the neutral point clamped (NPC) inverter. Compared to the conventional two-level inverter, the CHB, FC and NPC multilevel inverters exhibit several advantageous features such as low voltage stress, low total harmonic distortions (THD), low switching losses, high electro-magnetic compatibility and high-quality output voltage with high efficiency [1–4].

To date, MLIs have been extensively used to control variable frequency drivers [5], static VAR compensators [6,7], HVDC systems [8], un-interrupted power supplies (UPS)

and PV grid connected systems and renewable energy resources. By increasing the resolution of stepped voltage, the power quality of MLIs can be significantly improved. However, this process can significantly increase the number of power electric components count, installation area along with increased cost and control complications. Thus, the increased cost of the system and complexity of the inverter generally can limit the amount of voltage levels produced by any MLI structure.

To mitigate and balance out this particular issue, many variations of MLI topologies have been proposed in recent years to enhance their capability and utilize them in various industrial applications. Advanced NPC (ANPC) [9], the stacked MC [10], symmetrical and asymmetrical CHB are some notable examples [11–14]. Symmetrical MLIs utilize DC voltage sources of equal magnitudes and can have certain beneficial features such as, standardized control, lower voltage stresses on semiconductor devices, modular structure, etc. Hybrid MLIs are some alternative inverters that are built using the combination of two or more MLI topologies. These inverters are generally designed using DC supplies of unequal magnitudes. These MLIs can generate higher voltage levels utilizing less amount of power equipment and using a high voltage ratio of DC supplies [15,16]. Although asymmetry and hybridization can increase the number of voltage levels, these MLI topologies usually generate a high total standing voltage (TSV) on power electric components which in turn can increase the overall system cost, increase the voltage ratings of switches, decrease the switching redundancies and also limit the industrial applications. However, in [17], a novel family of MLI structures was designed using half-bridge inverter and multilevel DC-link. These variations in inverter topologies may produce even higher voltage levels utilizing abridged amount of power components.

The goal of the multilevel DC-link is to provide the H-bridge with successive positive voltage levels starting from zero voltage, whereas the purpose of the half-bridge is to provide the desired outputs comprising of negative and positive voltage. The aforementioned method has been utilized to propose different three-phase cascaded MLI topologies which are presented in [18–20]. To further decrease the overall amount of DC supplies and switches in these topologies, a new multilevel inverter configuration has been recently suggested in [21–24]. These MLIs were built by utilizing the combination of a twelve-switch three-phase Bridge and a multilevel DC-link. The multilevel DC-link is constructed by a fixed supply of cascaded H-bridge power modules and DC voltage. Comparing with the available MLI topologies, the projected configuration comes with more voltage levels and lesser power components.

In this study, an improved MLI is configured that can synthesize six-level output by using the combination of a conventional twelve-switch three-phase bridge and a modified multilevel DC-link. The multilevel DC-link is composed of only one DC voltage supply, H-bridge and full-bridge power cells. Both of the H-bridge and full-bridge modules are controlled, and the proposed inverter generates a stair-case waveform of eleven symmetrical line-to-line voltage levels. The proposed inverter needs to be operated under a certain condition where the modulation index is more than 0.98 to generate 11 voltage levels. On the other hand, the inverter can generate a stair-case waveform of nine asymmetrical lines to line voltage levels if the modulation index is equal to 0.98. Furthermore, if the modulation index decreases below 0.98, the MLI operates identically to a traditional two-level three-phase inverter and accordingly three symmetrical voltage levels are seen in its line-to-line output voltage waveforms. This research offers the following key contributions:

- The advantages of both multilevel DC-links and H-bridge circuits have been inherited by the suggested topology. The proposed MLI has been able to lower total voltage stress and generate greater voltage levels with fewer components because of this smart design.
- Since it does not contain switched capacitors, the proposed MLI does not require a complex control approach or additional circuits to deal with voltage balancing or excessive power losses.

- Depending on the application needs, the proposed topology can be modified to achieve higher voltage levels.
- The suggested MLI's modularity and smart switching arrangements have allowed it to perform in four different ways depending on the modulation indices. This has provided the MLI with more flexibility and reliability.

The paper is organized into 6 sections. The MLI topology and its operation are studied in Section 2. Section 2 further discusses the efficiency, extended version of the MLI and the implementation of vector modulation technique. Section 3 shows the detailed simulation results of the proposed MLI. The comparison of this MLI with classical and recent MLI topologies is discussed in Section 4. Section 5 contains the experimental results, and Section 6 concludes the paper.

2. Operating Principle of Three-Phase MLI

2.1. Proposed Topology

The circuit diagram of the proposed three-phase six-level MLI topology is depicted in Figure 1. Six unidirectional switches formed by (S1~S6) and (Da1~Dc2) diodes are connected with a classical three-phase six-switch (Q1~Q6) to build the main bridge. On the other hand, a single DC voltage supply, a half-bridge cell consisting of two switches (Ta1, Ta2) and a single DC supply along with a full-bridge power cell comprising four switches (Tb1~Tb4) and a DC supply are connected together to form a multilevel DC-link voltage. The DC voltages for half-bridge and full-bridge power cells are 3Vdc and Vdc, respectively, while the DC voltage for a single supply is 2Vdc. The power cell switches are turned on/off to synthesize four consecutive voltage levels: 4Vdc, 3Vdc, 2Vdc and Vdc at the middle point (o) relating to the inverter's ground. Table 1 illustrates the switching operation of the proposed inverter.

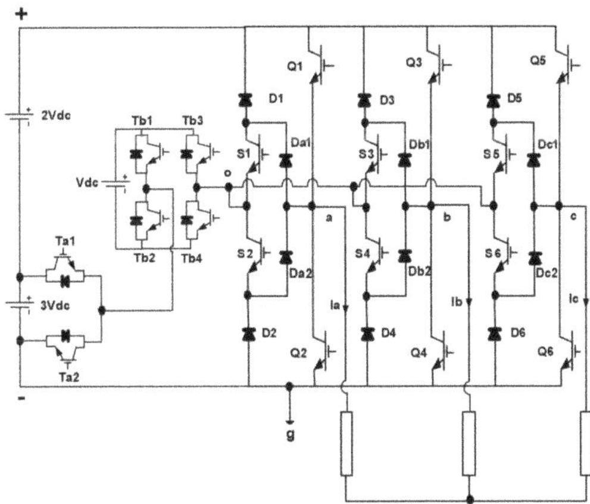

Figure 1. Proposed structure of a three-phase six-level inverter.

Table 1. Switching states for the proposed six-level inverter ($Ma \geq 0.98$).

Sa	Q1	S1	S2	Q2	Ta1	Ta2	Tb1	Tb2	Tb3	Tb4	Vag
5	1	0	0	0	0	0	0	0	0	0	+5*Vdc*
4	0	1	1	0	1	0	0	1	1	0	+4*Vdc*
3	0	1	1	0	1	0	0	1	0	1	+3*Vdc*
2	0	1	1	0	1	0	1	0	0	1	+2*Vdc*
1	0	1	1	0	0	1	0	1	1	0	+*Vdc*
0	0	0	0	1	0	0	0	0	0	0	0

It is worth mentioning that the switching states demonstrated in Table 1 are applicable for all three phases. For phase b and c, it should be Q3, Q4, S3, S4 and Q5, Q6, S5, S6, respectively instead of Q1, Q2, S1, S2. The switching configuration of the proposed MLI is given in the following equation:

$$\begin{bmatrix} Vag \\ Vbg \\ Vcg \end{bmatrix} = \frac{5Vdc}{N-1} \times \begin{bmatrix} Sa \\ Sb \\ Sc \end{bmatrix} \tag{1}$$

where *Sa*, *Sb* and *Sc* are switching positions. *Vag*, *Vbg* and *Vcg* are the MLI's line to ground voltages and $N = 6$ is the number of voltage levels. Since the topology employed for the proposed MLI is modeled to obtain the three-phase well-adjusted line to line output voltages producing the highest number of voltage levels, a suitable switching structure is needed for generating the MLI's gate pulses. Consecutively, these switching states generate the gate pulses and drive the studied MLI under all modulation indexes. Different modulation techniques have been suggested to control the multilevel inverters. These modulation methods are categorized depending on the operating switching frequency [25]. The low frequency modulation method known as selective harmonic elimination (SHE-PWM) is usually employed to predefine and pre-calculate the optimal switching angles. To obtain a solution for the angles, a set of equations is normally solved offline using numerical methods [26,27]. The proposed MLI can be easily controlled, and the desired outputs can be simply achieved with regard to the MLI's switching states *Sa*, *Sb* and *Sc*. Therefore, a new modulation method was recently suggested in [28] where each phase in the proposed inverter is controlled independently. For a given modulation index *Ma*, the inverter's switching states *Sa*, *Sb* and *Sc* are determined with respect to the MLI's line to ground reference voltages *Vag_ref*, *Vbg_ref* and *Vcg_ref*. The correlation among the MLI's reference line to ground voltages and the respective switching sequences is:

$$\begin{bmatrix} Sa \\ Sb \\ Sc \end{bmatrix} = \left(\frac{N-1}{5Vdc} \times \begin{bmatrix} Vag\_ref \\ Vbg\_ref \\ Vcg\_ref \end{bmatrix} \right) \tag{2}$$

$$\begin{bmatrix} Vag\_ref \\ Vbg\_ref \\ Vcg\_ref \end{bmatrix} = \frac{Ma \times 5Vdc}{2} \times \begin{bmatrix} \cos(\omega t) \\ \cos(\omega t - \frac{2\pi}{3}) \\ \cos(\omega t + \frac{2\pi}{3}) \end{bmatrix} + \frac{5Vdc}{2} \times \left[1 - \frac{Ma}{6}\cos(3\omega t) \right] \times \begin{bmatrix} 1 \\ 1 \\ 1 \end{bmatrix}. \tag{3}$$

According to (3), the 3rd harmonic element is added to MLI's line to ground reference voltage waveforms. Therefore, *Ma* can extend to 1.15 without instigating over modulation.

2.2. Extended Structure

The multistep DC-link circuit can be extended by keeping the proposed MLI's main bridge construction. This enables the proposed inverter to achieve a higher amount of voltage steps using a small quantity of power equipment. To extend the multistep DC-link circuit, the half-bridge power module and (n) multiple units of full-bridge power modules are connected following series connection to generate (N – 1) sequent voltage steps as depicted in Figure 2. For full-bridge power cells:

$$Vdc\_fb1 = 3^{(0)}Vdc \tag{4}$$

$$Vdc\_fbn = 3^{(n-1)}Vdc. \tag{5}$$

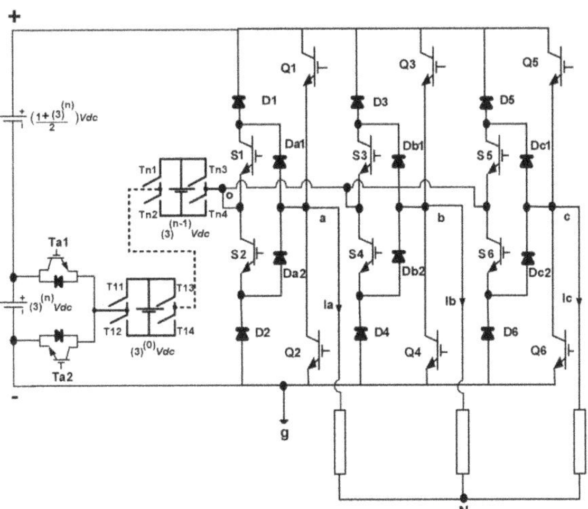

Figure 2. Proposed structure of three-phase *N*-level multilevel inverter.

For half-bridge power cells:

$$Vdc\_hb = 3^{(n)} Vdc. \tag{6}$$

For a single DC voltage supply:

$$Vdc\_s = Vdc + \sum_{i=1}^{i=n} 3^{(n-1)} Vdc = \left(\frac{1+3^{(n)}}{2}\right) Vdc. \tag{7}$$

Hence, the highest number of voltage steps that can be achieved are:

$$N = \frac{3}{2}\left(1+3^{(n)}\right). \tag{8}$$

Here, N = [6, 15, 42, 123, ...]. The total required numbers of IGBTs (M_sw) and DC supplies (M_DC) in terms of voltage steps are given as:

$$M\_sw = 4Log_3\left(\frac{2}{3}N - 1\right) + 14 \tag{9}$$

$$M\_DC = Log_3\left(\frac{2}{3}N - 1\right) + 2. \tag{10}$$

2.3. *Efficiency Calculation*

It is important to validate the two principal losses that generally occur during the operation of switching devices. These losses are classified as conduction losses and switching losses. Conduction losses ($P_{Conduction}$) are defined as the losses of power electronic devices during on-state. In this case, both the diode and the switches are taken into consideration. The instantaneous conduction losses of the IGBT and diodes can be determined using (11) and (12), respectively.

$$P_{Conduction\_IGBT}\,(t) = [V_{IGBT} + R_{IGBT}\, i(t)] \times i(t) \tag{11}$$

$$P_{Conduction\_Diode}\,(t) = [V_{Diode} + R_{Diode}\, i(t)] \times i(t) \tag{12}$$

Here, V_{IGBT} is the voltage of IGBTs while V_{Diode} is the voltage of the diodes when these devices are activated. Again, R_{IGBT} is the corresponding resistance of IGBTs while R_{Diode} represents the corresponding resistance of the diodes. The subsequent results are then summed to calculate the conduction losses of the proposed MLI.

Switching losses ($P_{Switching}$) are defined as the losses of power electronic modules when they are activated and deactivated. Switching loss is proportional to the switching frequency. The activation (W_{on}) and deactivation (W_{off}) energy loss of these modules are obtained by:

$$W_{on} = \int_0^{t_{on}} \left[\left(\frac{v_{switch}}{t_{on}} t \right) \left(-\frac{I}{t_{on}} (t - t_{on}) \right) \right] dt. \tag{13}$$

$$W_{off} = \int_0^{t_{off}} \left[\left(\frac{v_{switch}}{t_{off}} t \right) \left(-\frac{I}{t_{off}} (t - t_{off}) \right) \right] dt. \tag{14}$$

Here, t_{on} and t_{off} are the turn-on time and turn-off time of the IGBTs consecutively. I represents the current through the IGBT devices before/after they are turned off/on. v_{switch} demonstrates the forward voltage drop for the IGBTs. Thus,

$$P_{switch} = f \left[\sum_{i=1}^{N_{switch}} \left(\sum_{i=1}^{N_{on}} W_{on,i} + \sum_{i=1}^{N_{off}} W_{off,i} \right) \right] \tag{15}$$

where f, N_{on} and N_{off} are the fundamental frequency and the number of times each switch is turned on and off during a time period of t, respectively. Again, $W_{on,i}$ and $W_{off,i}$ are the energy loss of each switch turning on and off for ith time. Thus, the total loss (P_{Total}) can be calculated as follows:

$$P_{Total} = P_{conduction} + P_{switching}. \tag{16}$$

The following parameters are considered for calculating the power loss of the proposed module: $V_{IGBT} = 2.4$ V, $R_{IGBT} = 0.052\ \Omega$, $V_{Diode} = 2$ V and $R_{Diode} = 0.1\ \Omega$, $f = 50$ Hz, $t_{on} = t_{off} = 1$. The proposed inverter is running under the conditions: Total DC-link voltage = 450 V, $Ma = 1$, $f = 50$ Hz, and Pout = 3.6 kW. Again, a single-phase resistive inductive load (237 Ω-0.53H) is connected with the module as the output. The calculation is conducted for each switch and diode for one period and applying (11) to (16) the total losses (P_{Total}) of the inverter is evaluated to be 88.75 W operating for 1 s. Therefore, the proposed inverter efficiency is 97.59%. This has passed the IEEE 1547 standard for interconnected devices of power grids.

2.4. Switching Pulses Generation Using Nearest Vector Modulation

The goal of the modulation strategy is to produce the appropriate switching pulses by determining the MLI's switching states. The operating block diagram of the nearest vector modulation (NVM) technique is shown in Figure 3. The space vector illustration of the operation of the projected MLI in stationary $\alpha-\beta$ location frame is obtained as follows:

$$V\alpha = \frac{5Vdc}{3(N-1)} \times (2Sa - Sb - Sc) \tag{17}$$

$$V\beta = \frac{5Vdc}{\sqrt{3}(N-1)} \times (Sc - Sb) \tag{18}$$

$$V = V\alpha - jV\beta. \tag{19}$$

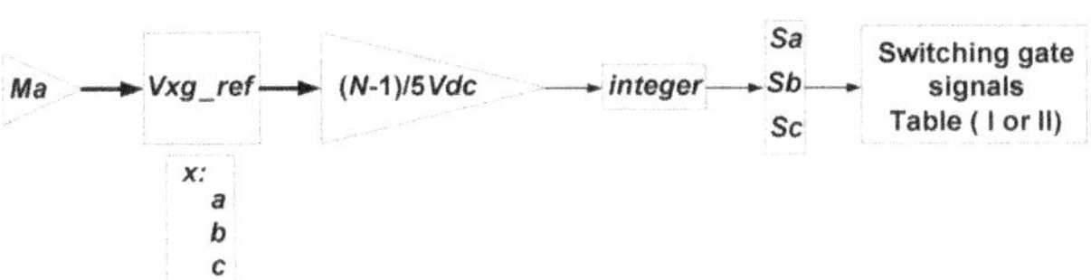

Figure 3. Operating principle of the NVM technique.

Four different space vector diagrams representing the function of the three-phase MLI with four valid switching sequences are shown in Figure 4.

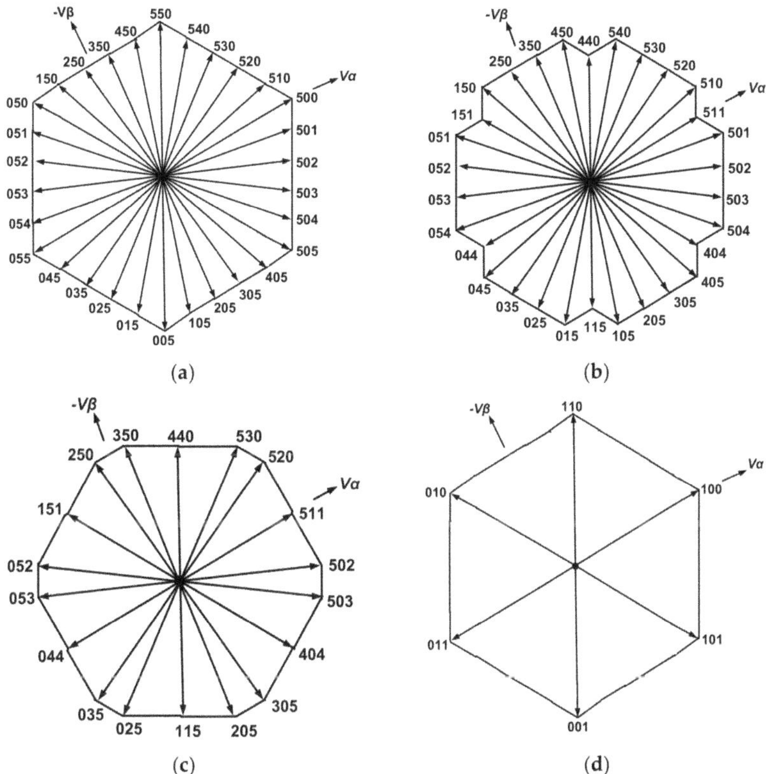

Figure 4. Space vector diagrams for the proposed MLI at: (**a**) $Ma = 1.3$, (**b**) $Ma = 1.15$, (**c**) $Ma = 0.98$ and (**d**) $Ma = 0.8$.

3. Simulation Results

According to these space vector diagrams shown in Figure 4, the switching states of the proposed MLI can be classified into four categories.

(1) In case of $Ma \geq 1.2$, the resultant switching state compromises of 30 various groupings of switching sequences (active vectors) are as follows: {055-054-053-052-051-050-150-250-350-450-550-540-530-520-510-500-501-502-503-504-505-405-305-205-105-005-015-025-035-045}.

For instance: at $Ma = 1.3$, Figure 5a shows the instantaneous values of *Sa*, *Sb* and *Sc* and switching pulses arrangements within a full cycle of operation. These switching states allow the proposed MLI to produce a three-phase stable staircase 11 level line to a line voltage with mutual variance of *Vdc*, which is as follows: 5*Vdc*, +4*Vdc*, +3*Vdc*, +2*Vdc*, +*Vdc*,

0, $-Vdc$, $-2Vdc$, $-3Vdc$, $-4Vdc$, $-5Vdc$) where Vab, Vbc and Vca are related to Sa, Sb and Sc by:

$$\begin{bmatrix} Vab \\ Vbc \\ Vca \end{bmatrix} = \frac{5Vdc}{N-1} \times \begin{bmatrix} 1 & -1 & 0 \\ 0 & 1 & -1 \\ -1 & 0 & 1 \end{bmatrix} \times \begin{bmatrix} Sa \\ Sb \\ Sc \end{bmatrix} \tag{20}$$

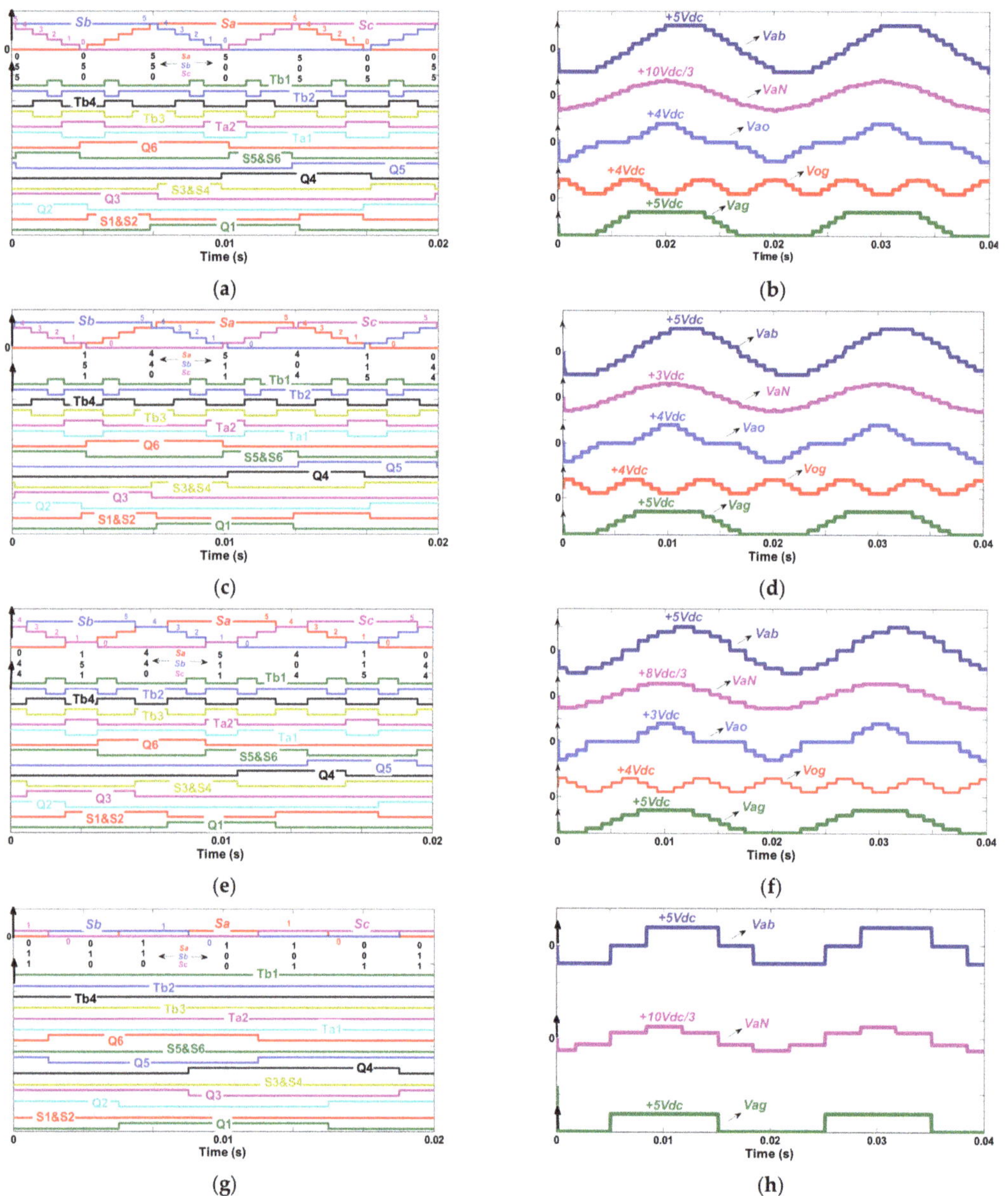

Figure 5. Simulation results: (**a**) switching states at Ma = 1.3, (**b**) simulated waveforms at Ma = 1.3 (**c**) switching states at Ma = 1.15, (**d**) simulated waveforms at Ma = 1.15, (**e**) switching states at Ma = 0.98, (**f**) simulated waveforms at Ma = 0.98, (**g**) switching states at Ma = 0.8 and (**h**) simulated waveforms Ma = 0.8.

The simulated outputs including the MLI's line to line (V_{ab}), line to neutral (V_{aN}), line to ground (V_{ag}), line to center (V_{ao}) and center to ground (V_{og}) voltage outputs are illustrated in Figure 5b.

(2) For 1.2 > Ma > 0.98, the maneuver of the proposed MLI is organized inside 30 other combinations of switching sequences (active vectors) as follows: {044 (155)-054-053-052-051-151 (040)-150-250-350-450-440-(551)-540-530-520-510-511 (400)-501-502-503-504-404 (515)-405-305-205-105-115 (004)-015-025-035-045}.

The switching sequences of the proposed MLI and its equivalent switching pulses at (for instance: Ma = 1.15) are shown in Figure 5c. Reducing the modulation index affects the peak value voltage. Thus, the amount of line to neutral (V_{aN}) voltage levels of the MLI, in this case, is decreased to 14 voltage levels, as shown in Figure 5d. These voltage levels are as follows: (+9Vdc/3, +8Vdc/3, +7Vdc/3, +6Vdc/3, +4Vdc/3, +3Vdc/3, +Vdc/3, −Vdc/3, −3Vdc/3, −4Vdc/3, −6Vdc/3, −7Vdc/3, −8Vdc/3 and −9Vdc/3). Although this switching category has decreased the peak voltage and the line to neutral voltage, the number of line to line voltage (V_{ab}) outputs is maintained at 5Vdc with 11 voltage levels (+5Vdc, +4Vdc, +3Vdc, +2Vdc, +Vdc, 0, −Vdc, −2Vdc, −3Vdc, −4Vdc, −5Vdc).

From Figure 5d, it can be further observed that utilizing the synthesized switching states, six voltage steps can be generated using multiple combinations of switching sequences. For example, the voltage levels Vab = −4Vdc, Vbc = 0 and Vca = +4Vdc can be produced applying two valid redundant switching states 044 or (155). Since the switching sequence design should minimize the number of switching instances, the redundant switching states: {(155)-(040)-(551)-(400)-(515)-(004)} are not utilized.

(3) For Ma = 0.98, 18 different combinations of switching states (active vectors) are generated to form the new valid switching sequence which is as follows: {044 (155)-053-052-151 (040)-250-350-440 (551)-530-520-511 (400)-502-503-404 (515)-305-205-115 (004)-025-035}.

Figure 5e depicts the switching sequences of the projected MLI and switching pulses under the condition of Ma = 0.98. It can be observed that the MLI's line to line voltage (V_{ab}) waveforms reach their highest values of +5Vdc compromising nine voltage levels (+5Vdc, +4Vdc, +3Vdc, +2Vdc, 0, −2Vdc, −3Vdc, −4Vdc, −5Vdc), while the inverter line to neutral voltages (V_{aN}) reach their maximum value of +8Vdc/3 with eight voltage steps (+8Vdc/3, +7Vdc/3, +4Vdc/3, +Vdc/3, −Vdc/3, −4Vdc/3, +7Vdc/3, −8Vdc/3). It is worth mentioning that the operation under the condition of Ma = 0.98 makes the common voltage deference in staircase waveform of nine line to line voltage levels fluctuate between Vdc and 2Vdc. As a result, the MLI functions like a three-phase five-level asymmetrical MLI.

The simulated staircase waveforms of voltages at Ma = 0.98 are shown in Figure 5f. Similar to the previous case, the six redundant switching states in this switching sequence: {(155)-(040)-(551)-(400)-(515)-(004)} are also not utilized.

(4) For Ma < 0.98, the MLI functions like a traditional three-phase two-level MLI. The novel switching sequences are demonstrated in Table 2 (For instance: at Ma = 0.8). The switching sequences tabulated in Table 2 are also applicable for other phases.

Table 2. Switching states for the proposed six-level inverter (Ma < 0.98).

Sa	Q1	S1	S2	Q2	Ta1	Ta2	Tb1	Tb2	Tb3	Tb4	Vag
1	1	0	0	0	0	0	0	0	0	0	+5Vdc
0	0	0	0	1	0	0	0	0	0	0	0

In this case, the MLI's switching pulses are generated utilizing the switching sequence shown in Figure 5g. It is identified that the final switching state comprises of six groupings of active vectors: {011-010-110-100-101-001}. Following this switching state leads the projected MLI to obtain the desired output voltage of Vab, VaN and Vag as depicted in Figure 5h.

The extended configuration is also verified by simulating a model for (n = 2). The simulation is done utilizing 10 V voltage supply and a three-phase resistive-inductive load of 237 Ω–0.53 H as output. The nearest vector modulation technique is implemented with a nominal frequency of 50 Hz. The simulation results of fifteen-level line to line output voltage (V_{ab}) built on the extended model and its equivalent switching pulses are illustrated in Figure 6a. In addition, the output voltages of *VaN*, *Vao*, *Vog* and *Vag* are also shown in Figure 6b.

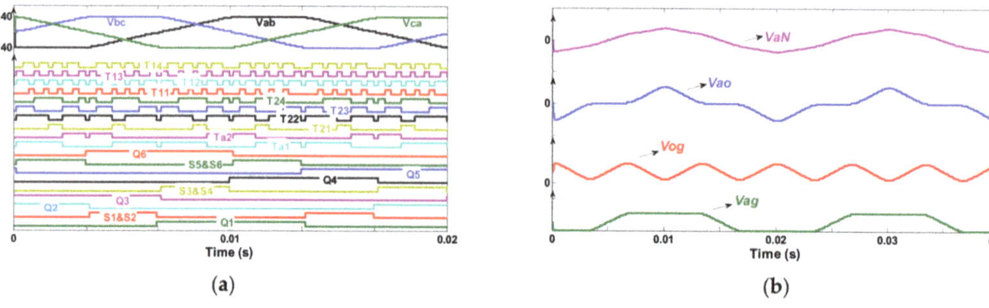

Figure 6. For (n = 2, N = 15, f = 50 Hz) (**a**) Switching gate signals and *Vab*, *Vbc* and *Vca*, (**b**) Simulated waveforms of *VaN*, *Vao*, *Vog* and *Vag*.

4. Comparative Analysis

The major advantage of the proposed MLI configuration is the ability to generate a high amount of voltage steps utilizing a small number of power electric components, as depicted in Table 3 and Figure 7. To efficiently achieve the required output voltage and current, the investigated MLI requires appropriate power devices. Thus, it is necessary to determine their rated power, voltage and current. Since the proposed inverter is built by a multilevel DC-link and three-phase Bridge, it uses switches with different voltage ratings, as illustrated in Table 4. Furthermore, it was assumed that all power devices of the inverter have the same current rating. Compared to other multilevel inverters listed in Table 5, the proposed inverter requires high voltage rating switches. As a result of this comparison, the total average cost per switch for the proposed inverter is higher than other NPC, FC, CHB and hybrid multilevel inverter switches, and approximately equal to that calculated in [29,30]. This additional cost can be mostly compensated for, since the proposed inverter requires a lower amount of DC supplies, IGBTs, diodes and gate drivers. In general, when matched by additional MLI structures, the proposed MLI's major benefits are: (1) a higher number of voltage levels applying power components in small numbers, (2) a low-cost heat sink, smaller installation area and low gate drivers being required, (3) inheriting certain benefits of traditional two-level inverters such as a lesser number of control signals, simple operating standard and low conduction loss. It also has similar advantages to multilevel inverters such as reduced switching losses, small harmonic distortions and improved performance. In contrast, the disadvantages are: (1) IGBT switches (Q1~Q6) must sustain the complete DC-link voltage once the MLI generates the maximum voltage $\pm 5Vdc$, whereas the bidirectional, half-bridge and full-bridge IGBTs required to sustain $\pm 4Vdc$, $\pm 3Vdc$ and $\pm Vdc$, respectively. As a result, different power ratings on different types of switches are required for building the proposed inverter. (2) Voltage levels (six) in maximum number can only be obtained when the modulation index is greater than 0.98, otherwise the behavior of the MLI becomes akin to a classical two-level inverter. Therefore, the projected MLI is highly appropriate for PV applications working on medium voltage conditions [31] and following fundamental switching frequency at the fixed modulation index Ma > 0.98.

Table 3. Comparative analysis between different multilevel inverter topologies in terms of voltage levels.

Components	NPC	FC	CHB	[9]	[15]	Proposed
Switches	$6(N-1)$	$6(N-1)$	$6(N-1)$	$2\log_2(N-1)+12$	$12[\log_3(N/2)-1]+18$	$4\log_3[2/3(N)-1]+14$
Diodes	$6(N-1)$	$6(N-1)$	$6(N-1)$	$2\log_2(N-1)+12$	$12[\log_3(N/2)-1]+18$	$4\log_3[2/3(N)-1]+14$
Clamping diodes	$6(N-2)$	0	0	0	0	0
Gate drivers	$6(N-1)$	$6(N-1)$	$6(N-1)$	$2\log_2(N-1)+9$	$12[\log_3(N/2)-1]+18$	$4\log_3[2/3(N)-1]+11$
DC supplies	$N-1$	$N-1$	$3(N-1)/2$	$1+\log_2(N-1)$	$3[\log_3(N/2)-1]+4$	$\log_3[2/3(N)-1]+2$
Capacitors	0	$3(N-2)$	0	0	0	0
Control signals	$6(N-1)$	$6(N-1)$	$6(N-1)$	$2\log_2(N-1)+9$	$12[\log_3(N/2)-1]+18$	$4\log_3[2/3(N)-1]+11$

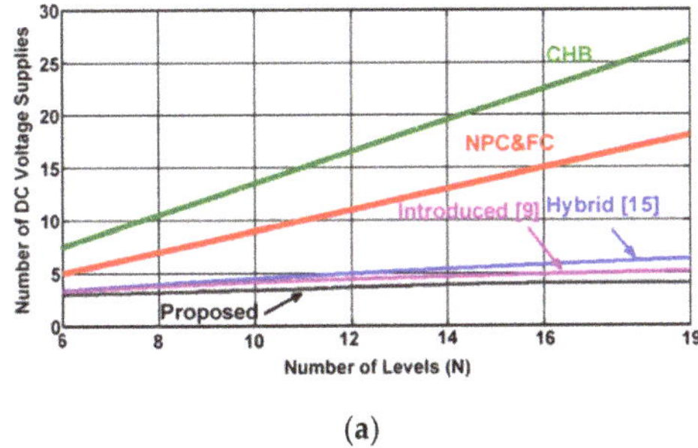

(a)

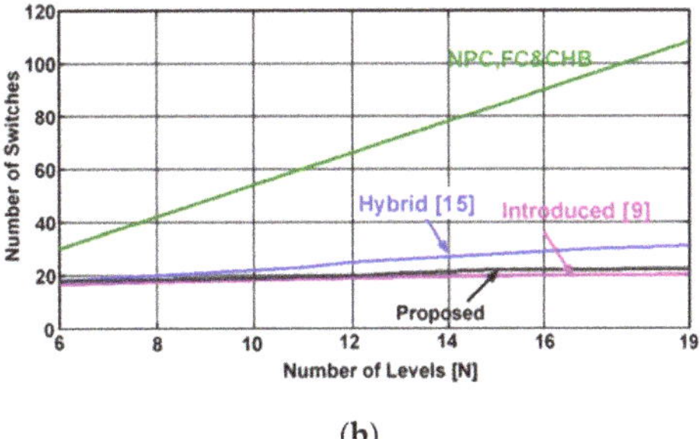

(b)

Figure 7. Comparison in terms of: (**a**) DC supplies, (**b**) switches.

Table 4. Voltage ratings of proposed topology switches.

Switches	Main Bridge	Bidirectional	Half-Bridge	Full-Bridge
Voltage rating	$(N-1)\,Vdc$	$(N-2)\,Vdc$	$3^n\,(Vdc)$	$3^{(n-1)}\,(Vdc)$

Table 5. Voltage rating of other multilevel inverter topologies.

Voltage Ratings	NPC	FC	CHB	[9]	[15]
Switch	Vdc	Vdc	Vdc	$(N-1)\,Vdc$	Vdc
Diode	Vdc	0	0	0	0
Capacitor	0	Vdc	0	0	0

5. Experimental Validation and Results

5.1. Experimental Setup

An inverter prototype is built and tested by applying $f = 50$ Hz and $Ma = (1.3, 1.15, 0.98$ and $0.8)$ for validating the compatibility of the studied modulation method and the performance of the proposed inverter, as shown in Figure 8. Three isolated DC voltage supplies having different magnitudes ($Vdc = 20$ V, $3Vdc = 60$ V and $2Vdc = 40$ V) are used to form the multilevel DC-link. A three-phase load (237 Ω–0.5 H) is used as the output. The inverter control algorithm based on the suggested modulation technique is developed using Simulink/ MATLAB software and TMS320F28335 DSP controller. Figure 9 illustrates the proposed MLI's control diagram.

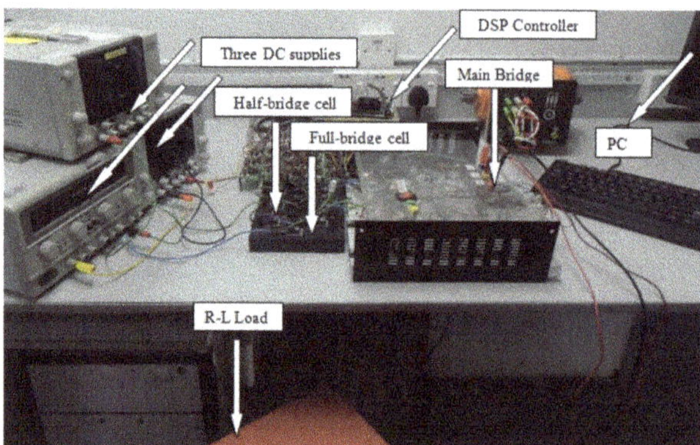

Figure 8. Laboratory setup of the hardware prototype.

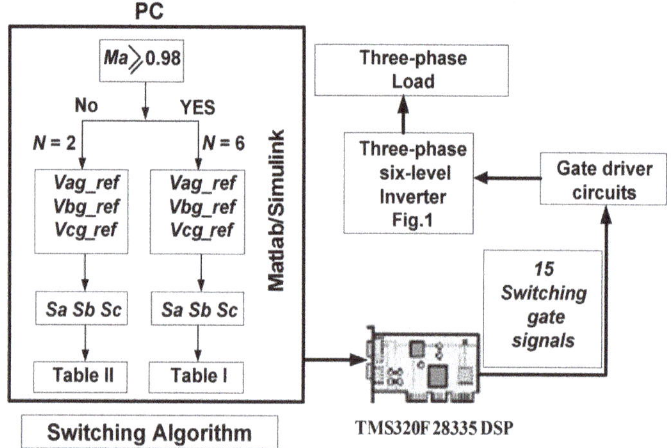

Figure 9. Control block diagram.

5.2. Experimental Results

The inverter line to ground, line to line, line to neutral, line to center and center to ground outputs are shown in Figure 10. For $Ma = 1.3$, the waveforms for Vag, Vbg and Vcg have six symmetric voltage levels with the highest value of 100 V, as shown in Figure 10a. It should be addressed that the experimental results are presented following the RGB color format, which means all the waveforms of phase a are represented by red color, while the waveforms of phase b and phase c are represented by a green and blue color, respectively.

The waveforms of Vab, Vbc and Vca are composed of 11 voltage steps with a peak value of 100 V, as depicted by Figure 10b. In Figure 10c, the inverter line to neutral VaN reached its peak value of 63 V with 16 voltage steps, while the waveform of Vao attained the peak voltage of 80 V producing nine voltage levels as shown in Figure 10d. From Figure 10e, four different voltage values are taken from inverter mid-point to ground Vog, with the peak value of 80 V. Then, within a full cycle of Vag, a repetition occurs three times. In Figure 10f,g, the proposed MLI's line to line output voltage at modulation index, $Ma = 1.15$ and 0.98 are depicted, respectively. The performance of the proposed MLI becomes the same as that of a traditional two-level inverter when $Ma = 0.8$, as shown in Figure 10h. The harmonic spectrum for the waveforms of inverter line to line voltage and load current at $Ma = 1$ are depicted in Figure 11.

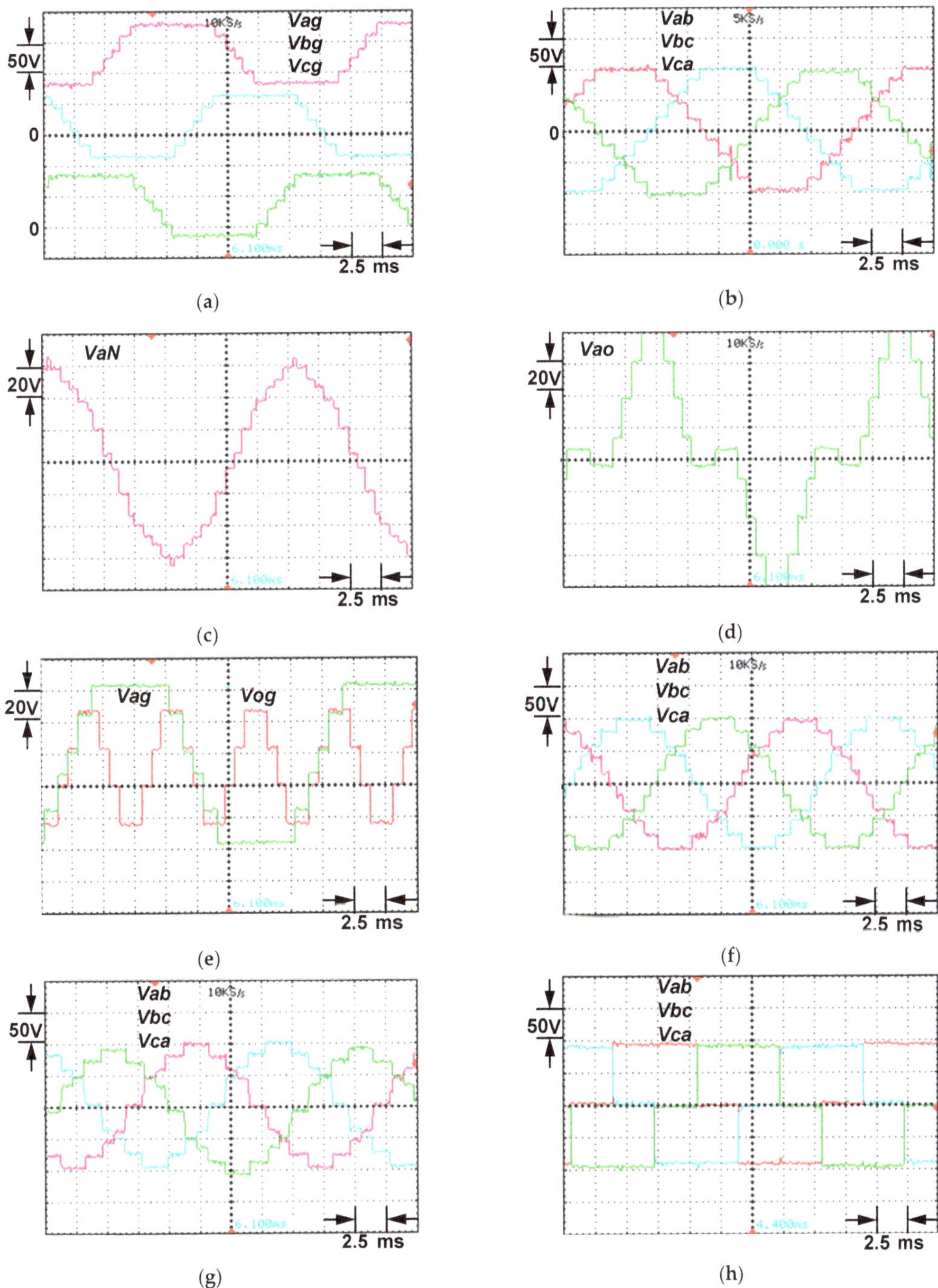

Figure 10. For *Ma* =1.3, the MLIs are: (**a**) line to ground voltages, (**b**) line to line voltages, (**c**) line to neutral voltage, (**d**) line to mid-point voltage and (**e**) line to ground and mid-point ground voltages, (**f**) For *Ma* =1.15, line to line voltages, (**g**) For *Ma* = 0.98, line to line voltages, (**h**) For *Ma* = 0.8, line to line voltages.

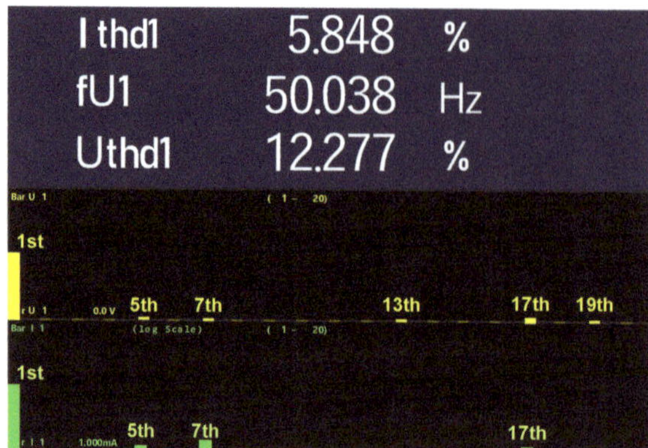

Figure 11. Harmonic spectrum of load current (yellow) and line to line voltage (green).

According to Figure 11, both spectrums do not contain any triplen harmonics such as 3rd, 9th and 15th harmonics. Furthermore, all even harmonics are eliminated because the MLI's output voltage and load current have obtained symmetry. %THD for the load current is only 5.8%, while the THD for MLI's line to line voltage is 12.3%.

6. Conclusions

A novel three-phase six-level multilevel inverter structure was developed and is outlined in this manuscript. The designed MLI possesses some advantageous characteristics topology such as it requires the least amount of power components and DC supplies, the prototype can be built-in cost-effective manner and it can be controlled using a simple modulation technique. A new approach based on the nearest vector modulation technique was easily implemented and the switching gate signals were successfully generated. Furthermore, a three-phase *N*-level structure based on the proposed configuration was suggested and discussed. The proposed MLI validity and suggested modulation technique compatibility was validated by employing both experimental and simulation approaches. The simple and pragmatic design of the proposed MLI would make it a highly suitable candidate for medium voltage applications in distributed renewable energy applications and grid-connected systems. The low current THD and higher efficiency generated by the MLI will allow it to be utilized in various industrial applications. The concluding remarks of the manuscript can be listed as follows:

1. The proposed MLI utilized a reduced number of power electric components compared to classical and other recently developed MLIs.
2. The modularity of the MLI makes it a suitable candidate for high voltage operations.
3. The operating efficiency of the MLI is 97.59% and effectively follows the IEEE 1547 standard, making it suitable to be applied as an interconnected device in grid systems.
4. The low current THD of 5.8% would allow this inverter to be utilized in the grid's current and voltage compensation systems, including shunt active power filters (SAPFs), voltage restorers and unified power quality conditioners (UPQCs).

Author Contributions: Conceptualization, S.T.M. and A.M.; methodology, S.T.M. and A.M.; formal analysis, S.T.M., A.M. and K.H.; investigation, S.T.M., K.H. and N.Z.Y.; resources, S.T.M., A.M. and N.Z.Y.; data curation, S.T.M., A.M., K.H. and F.M.; writing—original draft preparation, S.T.M., A.M. and K.H.; writing—review and editing, M.S.H.L., S.H.M.A., A.H. and M.M.O.; supervision, M.S.H.L.; project administration, M.S.H.L.; funding acquisition, M.S.H.L. All authors have read and agreed to the published version of the manuscript.

Funding: This work was supported by Universiti Kebangsaan Malaysia under Grant Code GP-2021-K023221.

Institutional Review Board Statement: Not applicable.

Informed Consent Statement: Not applicable.

Data Availability Statement: Not applicable.

Conflicts of Interest: The authors declare no conflict of interest.

References

1. Shayestegan, M.; Shakeri, M.; Abunima, H.; Reza, S.S.; Akhtaruzzaman, M.; Bais, B.; Mat, S.; Sopian, K.; Amin, N. An overview on prospects of new generation single-phase transformerless inverters for grid-connected photovoltaic (PV) systems. *Renew. Sustain. Energy Rev.* **2018**, *82*, 515–530. [CrossRef]
2. Yuan, X. A set of multilevel modular medium-voltage high power converters for 10-MW wind turbines. *IEEE Trans. Sustain. Energy* **2014**, *5*, 524–534. [CrossRef]
3. Meraj, S.T.; Hasan, M.K.; Islam, J.; El-Ebiary, Y.A.B.; Nebhen, J.; Hossain, M.; Alam, K.; Vo, N. A diamond shaped multilevel inverter with dual mode of operation. *IEEE Access* **2021**, *9*, 59873–59887. [CrossRef]
4. Hamidi, M.N.; Ishak, D.; Zainuri, M.A.A.M.; Ooi, C.A. An asymmetrical multilevel inverter with optimum number of components based on new basic structure for photovoltaic renewable energy system. *Sol. Energy* **2020**, *204*, 13–25. [CrossRef]
5. Hannan, M.A.; Ali, J.A.; Mohamed, A.; Uddin, M.N. A random forest regression based space vector PWM inverter controller for the induction motor drive. *IEEE Trans. Ind. Electron.* **2016**, *64*, 2689–2699. [CrossRef]
6. Law, K.H. An effective voltage controller for quasi-Z-source inverter-based STATCOM with constant DC-link voltage. *IEEE Trans. Power Electron.* **2018**, *33*, 8137–8150. [CrossRef]
7. Jibhakate, C.; Chaudhari, M.; Renge, M. A reduced switch AC–AC converter with the application of D-STATCOM and induction motor drive. *Electronics* **2018**, *7*, 110. [CrossRef]
8. Lee, J.; Kang, D.; Lee, J. A study on the improved capacitor voltage balancing method for modular multilevel converter based on hardware-in-the-loop simulation. *Electronics* **2019**, *8*, 1070. [CrossRef]
9. Katebi, R.; He, J.; Weise, N. An advanced three-level active neutral-point-clamped converter with improved fault-tolerant capabilities. *IEEE Trans. Power Electron.* **2018**, *33*, 6897–6909. [CrossRef]
10. Nair, V.; Rahul, A.; Kaarthik, S.; Kshirsagar, A.; Gopakumar, K. Generation of higher number of voltage levels by stacking inverters of lower multilevel structures with low voltage devices for drives. *IEEE Trans. Power Electron.* **2017**, *32*, 52–59. [CrossRef]
11. Samadaei, E.; Kaviani, M.; Iranian, M.; Pouresmaeil, E. The P-type module with virtual DC links to increase levels in multilevel inverters. *Electronics* **2019**, *8*, 1460. [CrossRef]
12. Meraj, S.T.; Yahaya, N.Z.; Hasan, K.; Masaoud, A. A hybrid T-type (HT-type) multilevel inverter with reduced components. *Ain Shams Eng. J.* **2021**, *12*, 1959–1971. [CrossRef]
13. Mahato, B.; Mittal, S.; Majumdar, S.; Jana, K.; Nayak, P.K. Multilevel inverter with optimal reduction of power semi-conductor switches. In *Renewable Energy and Its Innovative Technologies*; Springer: Singapore, 2018; pp. 31–50. [CrossRef]
14. Zeng, J.; Lin, W.; Cen, D.; Liu, J. Novel K-type multilevel inverter with reduced components and self-balance. *IEEE J. Emerg. Sel. Top. Power Electron.* **2020**, *8*, 4343–4354. [CrossRef]
15. Gautam, S.P.; Kumar, L.; Gupta, S. Hybrid topology of symmetrical multilevel inverter using less number of devices. *IET Power Electron.* **2015**, *8*, 2125–2135. [CrossRef]
16. Vijayarajan, P.; Shunmugalatha, A.; Sathik, J. A new hybrid multilevel inverter topology for medium and high voltage applications. *Appl. Math. Inf. Sci.* **2017**, *11*, 497–508. [CrossRef]
17. Babaei, E.; Laali, S.; Alilu, S. Cascaded multilevel inverter with series connection of novel h-bridge basic units. *IEEE Trans. Ind. Electron.* **2014**, *61*, 6664–6671. [CrossRef]
18. Masaoud, A.; Mekhilef, S.; Ping, H.W.; Wong, K.I. A simplified structure for three-phase 4-level inverter employing fundamental frequency switching technique. *IET Power Electron.* **2017**, *10*, 1870–1877. [CrossRef]
19. Raushan, R.; Mahato, B.; Jana, K.C. Comprehensive analysis of a novel three-phase multilevel inverter with minimum number of switches. *IET Power Electron.* **2016**, *9*, 1600–1607. [CrossRef]
20. Hota, A.; Jain, S.; Agarwal, V. An optimized three-phase multilevel inverter topology with separate level and phase sequence generation part. *IEEE Trans. Power Electron.* **2017**, *32*, 7414–7418. [CrossRef]
21. Norambuena, M.; Kouro, S.; Dieckerhoff, S.; Rodriguez, J. Reduced multilevel converter: A novel multilevel converter with a reduced number of active switches. *IEEE Trans. Ind. Electron.* **2018**, *65*, 3636–3645. [CrossRef]
22. Thiyagarajan, V.; Somasundaram, P.; Kumar, K.R. Simulation and analysis of novel extendable multilevel inverter topology. *J. Circuits Syst. Comput.* **2019**, *28*, 1950089. [CrossRef]
23. Hamidi, M.N.; Ishak, D.; Zainuri, M.A.A.M.; Ooi, C.A. Multilevel inverter with improved basic unit structure for symmetric and asymmetric source configuration. *IET Power Electron.* **2020**, *13*, 1445–1455. [CrossRef]
24. Meraj, S.T.; Yahaya, N.Z.; Hasan, K.; Masaoud, A. Single phase 21 level hybrid multilevel inverter with reduced power components employing low frequency modulation technique. *Int. J. Power Electron. Drive Syst.* **2020**, *11*, 810–822. [CrossRef]
25. Rodriguez, J.; Lai, J.-S.; Peng, F.Z. Multilevel inverters: A survey of topologies, controls, and applications. *IEEE Trans. Ind. Electron.* **2002**, *49*, 724–738. [CrossRef]

26. Ahmadi, D.; Wang, J. Online selective harmonic compensation and power generation with distributed energy resources. *IEEE Trans. Power Electron.* **2013**, *29*, 3738–3747. [CrossRef]
27. Islam, J.; Meraj, S.T.; Masaoud, A.; Mahmud, A.; Nazir, A.; Kabir, M.A.; Hossain, M.; Mumtaz, F. Opposition-based quantum bat algorithm to eliminate lower-order harmonics of multilevel inverters. *IEEE Access* **2021**, *9*, 103610–103626. [CrossRef]
28. Oskouei, A.B.; Dehghanzadeh, A.R. Generalized space vector controls for MLZSI. *Ain Shams Eng. J.* **2015**, *6*, 1161–1169. [CrossRef]
29. Sayago, J.A.; Bruckner, T.; Bernet, S. How to select the system voltage of mv drives—A comparison of semiconductor expenses. *IEEE Trans. Ind. Electron.* **2008**, *55*, 3381–3390. [CrossRef]
30. Luo, J.; Lin, K.; Li, J.; Xue, Y.; Zhang, X.-P. Cost analysis and comparison between modular multilevel converter (MMC) and modular multilevel matrix converter (M3C) for offshore wind power transmission. In Proceedings of the 15th IET International Conference on AC and DC Power Transmission (ACDC 2019), Coventry, UK, 5–7 February 2019.
31. Wang, Z.; Li, G.; Tseng, M.-L.; Wong, W.-P.; Liu, B. Distributed systematic grid-connected inverter using IGBT junction temperature predictive control method: An optimization approach. *Symmetry* **2020**, *12*, 825. [CrossRef]

Article

Noncascading Quadratic Buck-Boost Converter for Photovoltaic Applications

Rodrigo Loera-Palomo [1,*], Jorge A. Morales-Saldaña [2], Michel Rivero [3], Carlos Álvarez-Macías [4*] and Cesar A. Hernández-Jacobo [4]

1 CONACYT–TecNM/Instituto Tecnológico de La Laguna, Torreón 27000, Mexico
2 Facultad de Ingeniería, UASLP, San Luis Potosí 78290, Mexico; jmorales@uaslp.mx
3 Instituto de Investigaciones en Materiales, Unidad Morelia, UNAM, Morelia 58190, Mexico; mrivero@materiales.unam.mx
4 TecNM/Instituto Tecnológico de La Laguna, Torreón 27000, Mexico; calvarezm@correo.itlalaguna.edu.mx (C.Á.-M.); cahj_22@hotmail.com (C.A.H.-J.)
* Correspondence: rloerapa@conacyt.mx

Abstract: The development of switching converters to perform with the power processing of photovoltaic (PV) applications has been a topic receiving growing interest in recent years. This work presents a nonisolated buck-boost converter with a quadratic voltage conversion gain based on the I–IIA noncascading structure. The converter has a reduced component count and it is formed by a pair of L–C networks and two active switches, which are operated synchronously to achieve a wide conversion ratio and a quadratic dependence with the duty ratio. Additionally, the analysis using different sources and loads demonstrates the differences in the behavior of the converter, as well as the pertinence of including PV devices (current sources) into the analysis of new switching converter topologies for PV applications. In this work, the voltage conversion ratio, steady-state operating conditions and semiconductor stresses of the proposed converter are discussed in the context of PV applications. The operation of the converter in a PV scenario is verified by experimental results.

Keywords: switching DC-DC converters; quadratic converter; PV systems

Citation: Loera-Palomo, R.; Morales-Saldaña, J.A.; Rivero, M.; Álvarez-Macías, C.; Hernández-Jacobo, C.A. Noncascading Quadratic Buck-Boost Converter for Photovoltaic Applications. *Micromachines* **2021**, *12*, 984. https://doi.org/10.3390/mi12080984

Academic Editor: Francisco J. Perez-Pinal

Received: 13 July 2021
Accepted: 16 August 2021
Published: 19 August 2021

1. Introduction

In recent decades, the development of power supply systems has exhibited significant growth due to their use in portable systems and equipment, ranging from power supply for the Internet of things (IoT) applications to the energy management of renewable energy sources. In this scenario, the current trend in power electronics systems has required the incorporation of new standards and specifications that must be met, such as higher efficiency, conversion ratios, and power density, amongst others [1]. In these systems, the switching converter is the key of every modern power supply system [2], since it has energy-processing functions aided by suitable control schemes. In this sense, the development of new topologies of switching converters is a topic of interest, as are new structures and/or schemes for the interconnection of converters with high efficiencies. There is also an interest in the application to semiconductor elements with low voltage and/or current stress.

In recent years, a significant number of topologies have been developed that satisfy high transformation ratios, based on isolated and nonisolated schemes, where the implementation is based primarily on the particular application. For example, the main aim in alternating current (AC) power supply systems based on switching converters is to achieve lower total harmonic distortion (THD), a higher power factor, and regulation in the output voltages, as well as higher efficiency [3,4]. In these applications, it is desirable for converters to exhibit galvanic isolation between the output and input of the power system. In the case of direct current (DC) power systems, there are two relevant trends: one is related to the power supply of portable and electronics equipment; the other is related to energy processing and control from renewable sources [5] to develop power systems with high transformation ratios and high efficiency [6,7].

In the emerging renewable energy market, it is a common practice to use switching converters under maximum power point tracking (MPPT) control schemes to obtain the maximum power available in photovoltaic (PV) and/or wind systems. In addition, switching converters must have high conversion ratios, since the voltage levels provided by photovoltaic systems, fuel cells, and/or batteries is a few tens of volts [8,9]. This condition requires systems that allow the processing of energy at the levels of hundreds of volts without or with galvanic isolation capacity.

Currently, switching converters with the capability of increasing or reducing the input voltage for different applications have gained attention. Al–Saffar et al. [4] presented a converter topology based on the interconnection of buck-boost and quadratic buck topologies to obtain high power factor and low output voltage. An approximation for supply systems for light-emitting diode (LED) applications with a high power factor is presented in [10] by Alonso et al. using a quadratic buck-boost converter. The synthesis of a switching converter based on quadratic buck and basic boost topologies was introduced by Nousiainen and Suntio [11,12], where the converter was applied on a photovoltaic system to adapt the energy injected to an inverter. The development of switching converters has led to converters with floating loads [13,14], which restrict their possible applications. In addition, some topologies of circuit synthesis combine the properties of quadratic converters with basic topologies [4,10,11,15,16]. Nowadays, emergent topologies address actual challenges faced by power supply systems for renewable energies, for example, continuous input and output currents, high voltage conversion ratios, common ground path between the input and output ports [14,17–21], and bidirectional power flow properties with quadratic voltage conversion ratios [15,22].

In he technical literature, several solutions exist to obtain switching converters with high voltage conversion ratios. The cascade connection of basic converters is a solution to this problem. Carbajal-Gutiérrez, et al. [23] presented a modeling approach for a quadratic buck converter, where the converter has a high gain to step down the output voltage. In [24], Morales-Saldaña et al. discuss the implementation of a multiloop control scheme in a quadratic boost converter. In [25], Loera-Palomo et al. present a family of quadratic step-down converters based on the noncascade connection of basic switching converters; in [26], a set of converters with a quadratic step-up voltage conversion ratio is introduced. Other quadratic or high-gain converters are presented in [27–30]. In the context of quadratic converters, in recent years, converters with high buck-boost conversion ratios have been appearing [7,16–18,31].

This paper presents a switching converter topology on the basis of the reduced redundant power processing concept, which is based on the interconnection of basic converters in a noncascade structure. The obtained conversion ratio is characterized by a step-up and step-down of the output voltage by a quadratic factor. The proposed converter has a low component count, noninverting output voltage, a common node between the input and output ports, and continuous input current. Additionally, the converter is analyzed with different types of sources and loads where the impact of the PV system in the operation of the converter is investigated .

The remainder of this paper is organized as follows: In Section 2, the proposed converter with wide step-up/down conversion ratio is presented. In addition, a steady-state analysis is given under three scenarios: (1) voltage source–resistive load, (2) PV source–resistive load, and (3) PV source–clamped voltage, to provide insight into the operating conditions of the converter. In Section 3, a representative example is presented to illustrate the main differences in the operating point in steady-state and voltage/current stresses on semiconductor devices. The experimental results of the converter in a PV application are presented in Section 4. Finally, the paper concludes with some remarks in Section 5.

2. Proposed Converter

The proposed converter is derived from the noncascading I–IIA structure, which was introduced by Tse et al. in [32,33]. The structure, shown in Figure 1, relates two unidirectional ports (source and load), a bidirectional port (storage element, C_1), and two general blocks (A and B), which are formed with basic switching DC-DC cells.

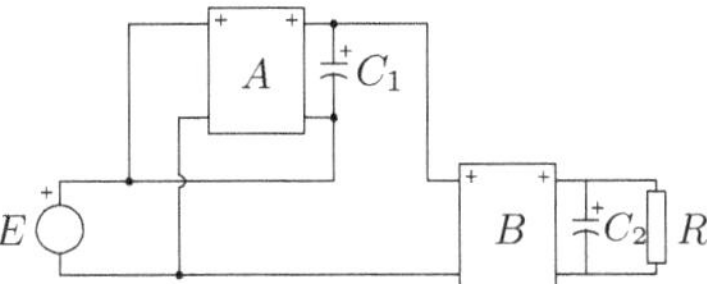

Figure 1. Noncascading I–IIA structure based on the R^2P^2 concept.

The development of a high-gain switching converter is based on the implementation of two basic switching cells. In block A, a boost cell is implemented with inverted output terminals, as shown in Figure 2a. This boost cell changes the voltage polarity in the storage element C_1, which assures a negative voltage in the input port of block B. In turn, a buck-boost cell, as depicted in Figure 2b, provides a positive voltage in the output port of the I–IIA structure.

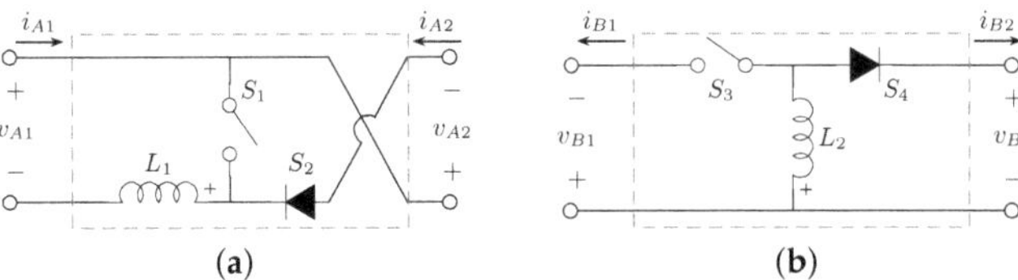

Figure 2. Nonisolated basic switching cells: (**a**) boost; (**b**) buck-boost.

The proposed quadratic buck-boost converter is shown in e Figure 3. This topology uses two active switches, S_1 and S_3; two passive switches, S_2 and S_4; two inductors, L_1 and L_2; a storage element, C_1; and an output capacitor, C_2. Additionally, E represents a voltage source and the load is modeled by a resistance R.

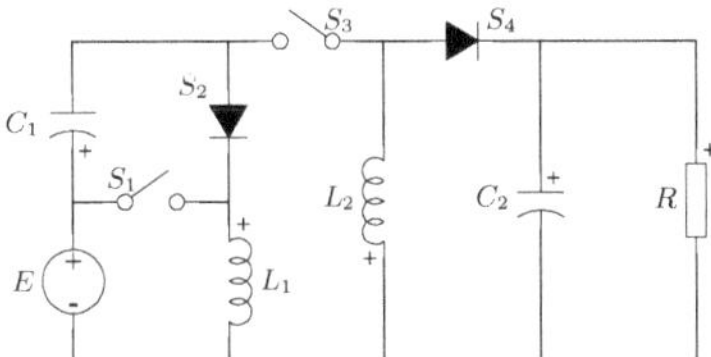

Figure 3. Proposed quadratic buck–boost converter.

Notably, other topologies with high gain or quadratic voltage conversion ratio have been developed under this kind of noncascading structure for PV applications, as discussed in [34–36]. Nevertheless, different procedures exist to develop topologies with specific high voltage gain, as stated in [37].

2.1. Steady-State Analysis: Voltage Source

In this section, we widen the steady-state analysis of the proposed converter, where the input port of the converter is connected to a voltage supply. In next subsection, the contribution of the PV module and converter is discussed.

The operating modes of the converter are defined by the conditions of the active switches, where the following assumptions are stated:

- The converter operates in continuous conduction mode (CCM).
- The active switches S_1 and S_3 operate synchronized to achieve a high voltage gain.
- The operation of the active switches is described by the switching function q, and for the passive switches (S_2 and S_4) by $\overline{q} = (1 - q)$. They are provided in Equation (1).

$$q = \begin{cases} 1 & \rightarrow \quad 0 < t \leq t_{on}, \\ 0 & \rightarrow \quad t_{on} < t \leq T_s, \end{cases} \qquad \overline{q} = \begin{cases} 0 & \rightarrow \quad 0 < t \leq t_{on}, \\ 1 & \rightarrow \quad t_{on} < t \leq T_s, \end{cases} \tag{1}$$

where t_{on} is the conducting time of the active switches and T_s is the switching period.

The operation modes of the converter are described by the on- and off-state of the active switches. During the switch on-state (Figure 4a), inductor L_1 is connected to the source voltage E, whereas inductor L_2 exhibits voltage $v_{C1} - E$. In this state, capacitor C_1 transfers energy to inductor L_2, whereas capacitor C_2 supports the power demanded by load R. In this operating mode, the differential equations that describe the behavior of the system are:

$$\begin{aligned} L_1 \frac{di_{L1}}{dt} &= E, \\ L_2 \frac{di_{L2}}{dt} &= v_{C1} - E, \\ C_1 \frac{dv_{C1}}{dt} &= -i_{L2}, \\ C_2 \frac{dv_{C2}}{dt} &= -\frac{v_{C2}}{R}. \end{aligned} \tag{2}$$

When the switches are in the turn-off state (Figure 4b), inductor L_1 transfers energy to capacitor C_1 through diode S_2, whereas inductor L_2 supplies energy to conjunct R–C_2 via diode S_4. The corresponding differential equations are given by:

$$\begin{aligned} L_1 \frac{di_{L1}}{dt} &= E - v_{C1}, \\ L_2 \frac{di_{L2}}{dt} &= -v_{C2}, \\ C_1 \frac{dv_{C1}}{dt} &= i_{L1}, \\ C_2 \frac{dv_{C2}}{dt} &= i_{L2} - \frac{v_{C2}}{R}. \end{aligned} \tag{3}$$

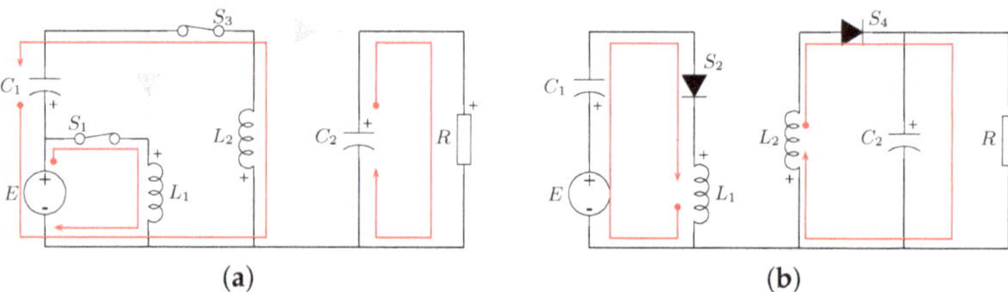

Figure 4. Operating modes of the proposed converter: (**a**) function $q = 1$; (**b**) function $q = 0$.

The set of Equations (2) and (3) defines the voltages at the terminals of the inductors and the currents through the capacitors in one switching period. For these voltages and currents, the principles of inductor volt-second balance and capacitor-charge balance describe a steady-state operation of the converter, which are given by Equations (4) and (5).

$$\langle v_L(t)\rangle = 0 = \frac{1}{T_s}\int_0^{T_s} v_L(t)dt, \tag{4}$$

$$\langle i_C(t)\rangle = 0 = \frac{1}{T_s}\int_0^{T_s} i_C(t)dt. \tag{5}$$

Using these principles allows us to express the averaged inductor currents and averaged capacitor voltages in the steady-state condition, with the result in Equations (6)–(9).

$$I_{L1} = \frac{ED^3}{(1-D)^4 R}, \tag{6}$$

$$I_{L2} = \frac{ED^2}{(1-D)^3 R}, \tag{7}$$

$$V_{C1} = \frac{E}{(1-D)}, \tag{8}$$

$$V_{C2} = \frac{ED^2}{(1-D)^2}, \tag{9}$$

where D is the nominal duty ratio of the converter that defines the on-state time of the active switches by $t_{on} = DT_s$. Then, the voltage gain of the converter has a quadratic dependence on the duty ratio, which implies a wide output step-up or step-down voltage conversion characteristic. Therefore, the voltage gain is expressed in Equation (10).

$$M = \frac{V_{C2}}{E} = \frac{D^2}{(1-D)^2}. \tag{10}$$

It is important to notice that the voltage conversion ratio and the steady-state operating condition are valid only during CCM operation of the converter. In addition, the conversion ratio is positive and the converter presents a common ground between the input and output ports. In this operating mode, two passive networks are formed (Figure 4), in contrast to the discontinuous conduction mode (DCM), where three passive networks are formed, losing the quadratic voltage gain characteristic of the converter.

The boundary conditions between the CCM and DCM operation of the converter are given by the dimensionless parameters as,

$$k_1 = \frac{2L_1}{RT}, \qquad k_2 = \frac{2L_2}{RT}, \tag{11}$$

for inductor L_1 and L_2, respectively, for the limit parameter in the following form,

$$k_{crit(1)} = \frac{(1-D)^4}{D^2}, \qquad k_{crit(2)} = (1-D)^2. \tag{12}$$

Then, the converter operates in CCM if it satisfies the criteria in Equation (13).

$$k_1 > k_{crit(1)}, \qquad k_2 > k_{crit(2)}. \tag{13}$$

Figure 5 graphs the limit parameter for each switching cell in the converter, as well as the operative regions depending on the parameters k_1 and k_2. In CCM operation, it is easy to show that $k_2 > 1$ satisfies the condition for the switching cell S_3, S_4, and L_2. However, the front-end switching cell (S_1, S_2, and L_1) has an operational limit given by $D' < D < 1$, where D' is the duty ratio that satisfies $k_1 = k_{crit(1)}$. Therefore, an increase in k_1 corresponds to an increase in the operational limit of the converter in CCM (reduced value of D').

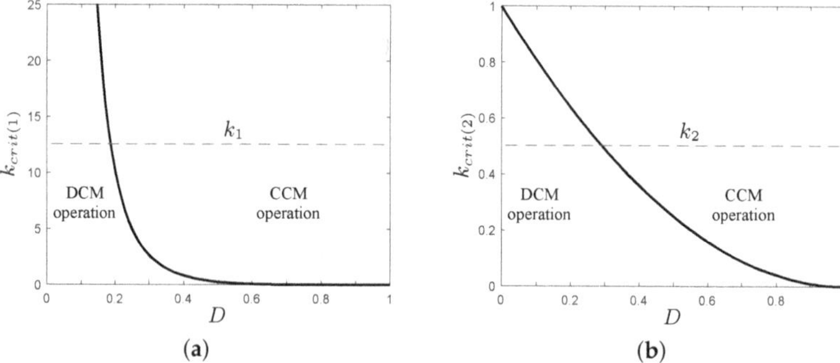

Figure 5. Boundary operation of the proposed converter under parameter variations: (**a**) parameter $k_{crit(1)}$; (**b**) parameter $k_{crit(2)}$.

A complete operation of the converter in DCM requires both switching cells to have the same operational limit given by $0 < D < D'$. The duty ratio D' sets the values of k_1 and k_2 through the equalities given by Equations (14).

$$k_1 = k_{crit(1)}(D'), \qquad k_2 = k_{crit(2)}(D'). \tag{14}$$

Then, the complete voltage conversion ratio of the proposed converter is given by Equation (15). In CCM operation, the voltage gain of the converter depends on the quadratic of the duty ratio, whereas in DCM, it depends on the duty ratio and parameter k_1.

$$M = \frac{V_{C2}}{E} = \begin{cases} \dfrac{D^2}{(1-D)^2} & \rightarrow \; k_j > k_{crit(j)} \text{ with } j = 1,2, \\ \dfrac{D}{\sqrt{k_1}} & \rightarrow \; k_j < k_{crit(j)} \text{ with } j = 1,2. \end{cases} \tag{15}$$

In CCM operation, it is assumed that current ripples on the inductors are small, that is, $\Delta I_{L1} \leq \alpha_1 I_{L1}$ and $\Delta I_{L2} \leq \alpha_2 I_{L2}$, where $0 < \alpha_j \leq 1$. In this sense, the inductances required to keep the ripples values below a given threshold are provided by Equation (16).

$$L_1 = \frac{(1-D)^4 RT}{\alpha_1 D^2}, \qquad L_2 = \frac{(1-D)^2 RT}{\alpha_2}. \tag{16}$$

2.2. Steady-State Analysis: Current Source

The consideration of a current source in the converter is related to PV applications, since a PV module/array is modeled as a current source. As a consequence, the voltage source of the converter in Figure 3 is changed by a PV module and a coupling capacitor C_i. In this system, the switching function in Equation (1) describes the operation of the active and passive switches. Figure 6 shows the resulting networks by the operation of the converter according to this switching function.

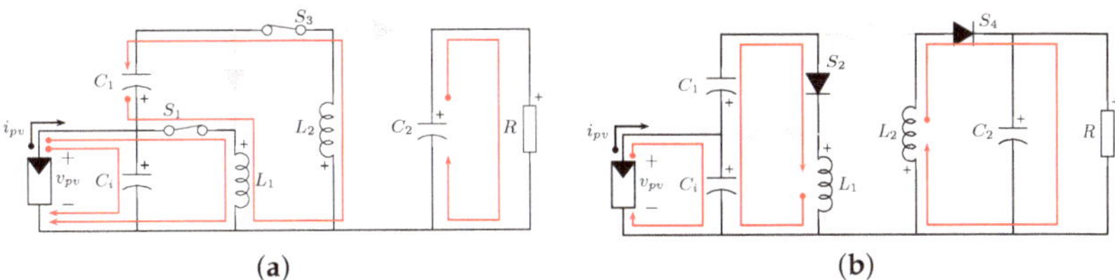

Figure 6. Operating modes of the proposed converter using a PV module. (**a**) Function $q = 1$. (**b**) Function $q = 0$.

In PV applications, during the switch on-state (Figure 6a), the inductor L_1 is connected to coupling capacitor C_i and exhibits voltage v_{Ci}, whereas inductor L_2 exhibits voltage $v_{C1} - v_{Ci}$. In this scenario, both inductors are in charge mode. The PV module transfers energy to capacitor C_i; in turn, capacitor C_1 transfers energy to inductor L_2 and C_2 supports the load power demand. Equations (17) describe the behavior of the system in this state.

$$\begin{aligned} L_1\frac{di_{L1}}{dt} &= v_{Ci}, \\ L_2\frac{di_{L2}}{dt} &= v_{C1} - v_{Ci}, \\ C_i\frac{dv_{Ci}}{dt} &= i_{pv} - i_{L1} + i_{L2}, \\ C_1\frac{dv_{C1}}{dt} &= -i_{L2}, \\ C_2\frac{dv_{C2}}{dt} &= -\frac{v_{C2}}{R}. \end{aligned} \tag{17}$$

Conversely, in the turn-off state of the active switches (Figure 6b), inductor L_1 and capacitor C_i transfer energy to capacitor C_1, while inductor L_2 supplies energy to conjunct R–C_2. The differential equations in this condition are given by Equations (18).

$$\begin{aligned} L_1\frac{di_{L1}}{dt} &= v_{Ci} - v_{C1}, \\ L_2\frac{di_{L2}}{dt} &= -v_{C2}, \\ C_i\frac{dv_{Ci}}{dt} &= i_{pv} - i_{L1}, \\ C_1\frac{dv_{C1}}{dt} &= i_{L1}, \\ C_2\frac{dv_{C2}}{dt} &= i_{L2} - \frac{v_{C2}}{R}. \end{aligned} \tag{18}$$

The above expressions define the voltages at the terminals of the inductors and the currents through the capacitors in one switching period. The steady-state condition of the converter can by obtained by using the principles of volt-second and charge balance, which result in Equations (19)–(23).

$$I_{L1} = \frac{I_{pv}}{D}, \tag{19}$$

$$I_{L2} = \frac{(1-D)I_{pv}}{D^2}, \tag{20}$$

$$V_{Ci} = \frac{(1-D)^4 I_{pv} R}{D^4}, \tag{21}$$

$$V_{C1} = \frac{(1-D)^3 I_{pv} R}{D^4}, \tag{22}$$

$$V_{C2} = \frac{(1-D)^2 I_{pv} R}{D^2}. \tag{23}$$

The voltage conversion gain of the converter can be obtained using Expressions (21) and (23), where the relation between the output voltage and the coupling capacitor voltage maintain a quadratic dependence, given by Equation (24).

$$M = \frac{V_{C2}}{V_{Ci}} = \frac{D^2}{(1-D)^2}. \tag{24}$$

The power generated by the PV module/array depends on weather conditions, that is, incident solar irradiance and temperature. In addition, the voltage at the terminals permits the adjustment of the power supplied by the PV module. Since the maximum power of a PV module is related to a specific voltage, Equation (21) shows that the voltage in the PV module can be selected by using the duty ratio signal of the converter, that is

$$D = \frac{[I_{pv} R]^{\frac{1}{4}}}{[I_{pv} R]^{\frac{1}{4}} + [V_{Ci}]^{\frac{1}{4}}}. \tag{25}$$

Finally, current I_o injected to load (R) is given by Equation (26). This implies that the output voltage of converter (V_{C2}) is not fixed and it depends on the duty ratio, the generated current by the PV module, and the value of load R.

$$I_o = \frac{(1-D)^2 I_{pv}}{D^2} = \frac{I_{pv}}{M}. \tag{26}$$

The inequality (13), which is the boundary condition between the CCM and DCM operation of the converter, holds independently of the PV application. Additionaly, the inductance values are given by Expressions (16).

PV systems can be applied in different ways; however, there are some applications in which the output voltage of the power converter is clamped. Figure 7a shows the proposed converter topology, where the output power condition is given in terms of the output current and output voltage. Figure 7b shows the implementation of the PV system in a DC microgrid, where the output voltage of the converter corresponds to the bus voltage of the microgrid. Finally, Figure 7c shows a grid-connected application, where the output voltage of the converter is clamped or regulated by the inverter (DC-AC converter). In this kind of scenario, the operating point of the converter and the semiconductor stresses are different than those presented by a pure resistive load.

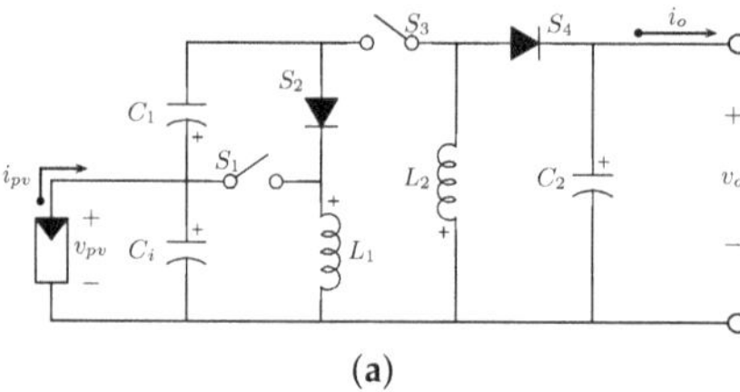

(a)

Figure 7. *Cont.*

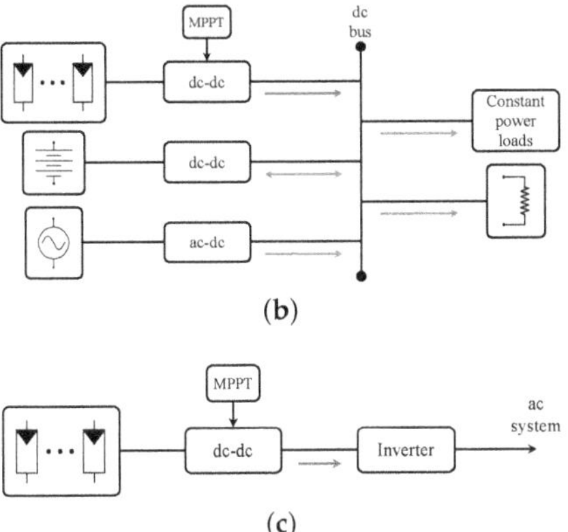

Figure 7. Structure of PV applications. (**a**) Proposed converter. (**b**) DC network application. (**c**) Grid-connected PV system.

A similar analysis applied to the converter topology in Figure 7a shows that the steady-state condition of the converter results in Equations (27)–(31).

$$I_{L1} = \frac{I_{pv}}{D}, \tag{27}$$

$$I_{L2} = \frac{(1-D)I_{pv}}{D^2}, \tag{28}$$

$$V_{Ci} = \frac{(1-D)^2 v_o}{D^2}, \tag{29}$$

$$V_{C1} = \frac{V_{Ci}}{(1-D)}, \tag{30}$$

$$V_{C2} = v_o. \tag{31}$$

Equation (29) shows that the voltage conversion ratio ($M = V_{C2}/V_{Ci}$) of the converter maintains a quadratic dependence on the duty ratio. Additionally, to achieve a specific voltage at the terminals of the PV module requires the duty ratio to be

$$D = \frac{\sqrt{V_{C2}}}{\sqrt{V_{C2}} + \sqrt{V_{Ci}}}. \tag{32}$$

In addition, the current injected to the next system I_o corresponds to the expression given in Equation (26).

2.3. Semiconductor Stress

An aspect of interest in designing renewable-energy-based power systems is the stress level of the semiconductor elements. As such, the expressions for the current and voltage are derived.

For the proposed converter, the expressions of voltage and current stress on the semiconductor devices are presented in Table 1 for the three applications analyzed in this work. V_{s1} (I_{s1}) and V_{s3} (I_{s3}) are the voltage (current) stress on the active switches S_1 and S_3, respectively; whereas V_{s2} (I_{s2}) and V_{s4} (I_{s4}) are the voltage (current) stress on the passive switches (diodes) S_2 and S_4, respectively.

It is important to notice that the voltage and current stress on semiconductor devices depend on the duty ratio of the converter. However, depending on the external elements (sources and loads) connected to the converter, the voltage and current stress might differ. The main difference in the stress on a semiconductor is related to the PV application, where a PV module has a particular behavior depending on the weather conditions and the operation point.

Table 1. Voltage and current stress on a semiconductor devices.

Semiconductor General Expression	$V_{S1}=V_{S2}$ V_{C1}	$V_{S3}=V_{S4}$ $V_{Ci}-V_{C1}-V_{C2}$	$I_{S1}=I_{S2}$ I_{L1}	$I_{S3}=I_{S4}$ I_{L2}
Voltage converter appl	$\frac{E}{(1-D)}$	$\frac{DE}{(1-D)^2}$	$\frac{D^3E}{(1-D)^4R}$	$\frac{D^2E}{(1-D)^3R}$
PV appl–resistive load	$\frac{(1-D)^3 I_{pv} R}{D^4}$	$\frac{(1-D)^2 I_{pv} R}{D^3}$	$\frac{I_{pv}}{D}$	$\frac{(1-D)I_{pv}}{D^2}$
PV appl–clamped voltage	$\frac{V_{Ci}}{(1-D)}$	$\frac{v_o}{D}$	$\frac{I_{pv}}{D}$	$\frac{(1-D)I_{pv}}{D^2}$

2.4. Dynamic Model

A state-space averaged model of converters is a key aspect in the control of power electronic systems. In this study, averaged models were developed based on common techniques to complement the study of the proposed converter under previously mentioned applications. The resulting model, when a voltage source is used in the input port, is given by Equations (33).

$$
\begin{aligned}
L_1\frac{di_{L1}}{dt} &= E-(1-d)v_{C1}-R_{p1}i_{L1},\\
L_2\frac{di_{L2}}{dt} &= d(v_{C1}-E)-(1-d)v_{C2}-R_{p2}i_{L2},\\
C_1\frac{dv_{C1}}{dt} &= (1-d)i_{L1}-di_{L2},\\
C_2\frac{dv_{C2}}{dt} &= (1-d)i_{L2}-\frac{v_{C2}}{R}.
\end{aligned} \tag{33}
$$

The averaged model, when a PV module with a coupling capacitor is used in the input port and a resistive load in the output port, results in

$$
\begin{aligned}
L_1\frac{di_{L1}}{dt} &= v_{Ci}-(1-d)v_{C1}-R_{p1}i_{L1},\\
L_2\frac{di_{L2}}{dt} &= d(v_{C1}-v_{Ci})-(1-d)v_{C2}-R_{p2}i_{L2},\\
C_1\frac{dv_{C1}}{dt} &= (1-d)i_{L1}-di_{L2},\\
C_2\frac{dv_{C2}}{dt} &= (1-d)i_{L2}-\frac{v_{C2}}{R},\\
C_i\frac{dv_{Ci}}{dt} &= i_{pv}-i_{L1}+di_{L2}.
\end{aligned} \tag{34}
$$

Finally, the averaged model of the converter in a PV application connected to a DC voltage system corresponds to

$$
\begin{aligned}
L_1\frac{di_{L1}}{dt} &= v_{Ci}-(1-d)v_{C1}-R_{p1}i_{L1},\\
L_2\frac{di_{L2}}{dt} &= d(v_{C1}-v_{Ci})-(1-d)v_{C2}-R_{p2}i_{L2},\\
C_1\frac{dv_{C1}}{dt} &= (1-d)i_{L1}-di_{L2},\\
C_2\frac{dv_{C2}}{dt} &= (1-d)i_{L2}-i_o,\\
C_i\frac{dv_{Ci}}{dt} &= i_{pv}-i_{L1}+di_{L2}.
\end{aligned} \tag{35}
$$

In the above models, $d(t)$ represents the averaged duty ratio; R_{pn}, with $n = 1,2$, represents the parasitic resistances in the converter in Equation (36).

$$R_{pn} = R_{Ln} + R_{Mn}d + R_{Dn}(1-d), \tag{36}$$

where R_{Ln} is the resistance in the inductor, and R_{Mn} and R_{Dn} are the on-resistances of the active and passive switches, respectively.

2.5. Comparison

In the literature, several solutions of quadratic buck-boost converters for photovoltaic applications have been published. The properties of these converters are summarized in Table 2, where the proposed converted is included for the sake of comparison. Note that all converters present a quadratic gain.

Table 2. Main characteristics of existing quadratic buck-boost converters.

Topology	[7]	[17]	[16]	[31]	[18]		**Proposed**	
Source Load	Voltage Resistive	Voltage Resistive	Voltage Resistive	Voltage Resistive	Voltage Resistive	Voltage Resistive	PV Resistive	PV Clamped voltage
Gain, $\frac{V_o}{E}$	$\frac{D^2}{(1-D)^2}$	$-\frac{D(2-D)}{(1-D)^2}$	$\frac{D^2}{(1-D)^2}$	$\frac{D^2}{(1-D)^2}$	$\frac{D^2}{(1-D)^2}$		$\frac{D^2}{(1-D)^2}$	
Inductors	3	2	3	2	3		2	
Capacitors	3	2	3	2	3		2	
Diodes	2	3	5	2	2		2	
Switches	2	1	1	2	2		2	
Volt. stress switches	$S_1 : \frac{E}{(1-D)^2}$ $S_2 : \frac{DE}{(1-D)^2}$	$S_1 : \frac{E}{(1-D)^2}$	$S_1 : \frac{E}{(1-D)^2}$	$S_1 : \frac{E}{(1-D)}$ $S_2 : \frac{DE}{(1-D)^2}$	$S_1 : \frac{E}{(1-D)}$ $S_2 : \frac{DE}{(1-D)^2}$	$S_1 : \frac{E}{(1-D)}$ $S_3 : \frac{DE}{(1-D)}$	$S_1 : \frac{(1-D)^3 I_{pv} R}{D^4}$ $S_3 : \frac{(1-D)^2 I_{pv} R}{D^3}$	$S_1 : \frac{V_{pv}}{(1-D)}$ $S_3 : \frac{V_o}{D}$
Volt. stress diodes	$D_1 : \frac{E}{(1-D)}$ $D_2 : \frac{DE}{(1-D)^2}$	$D_1 : \frac{E}{(1-D)}$ $D_2 : \frac{DE}{(1-D)^2}$ $D_3 : \frac{E}{(1-D)^2}$	$D_{1,4} : \frac{E}{(1-D)}$ $D_{2,5} : \frac{DE}{(1-D)^2}$ $D_3 : \frac{E}{(1-D)^2}$	$D_1 : \frac{E}{(1-D)}$ $D_2 : \frac{DE}{(1-D)^2}$	$D_1 : \frac{E}{(1-D)}$ $D_2 : \frac{DE}{(1-D)^2}$	$S_2 : \frac{E}{(1-D)}$ $S_4 : \frac{DE}{(1-D)}$	$S_2 : \frac{(1-D)^3 I_{pv} R}{D^4}$ $S_4 : \frac{(1-D)^2 I_{pv} R}{D^3}$	$S_2 : \frac{V_{pv}}{(1-D)}$ $S_4 : \frac{V_o}{D}$
Output polarity	Positive	Negative	Negative	Positive	Positive		Positive	
Continuous I_1	Yes	No	Yes	No	Yes		Yes	
Continuous I_o	Yes	No	Yes	No	Yes		No	
Common ground	Yes	Yes	Yes	Yes	Yes		Yes	
PV appl	Yes	Yes	No	No	Yes		Yes	

The topologies presented in [7,16,18] share a common characteristic, continuous input and output current, but this feature is penalized by the component number, since three inductors and three capacitors are required; additionally, [16] increased the semiconductors by two with respect to the other solutions. The proposed converter partially fulfills this requirement, where an output filter can be added to ensure a continuous output current, increasing the number of components.

In semiconductor devices, the voltage stress in previous topologies and in the proposed converter is similar when a voltage source and a resistive load are used. An interesting point regarding the previous topologies is their use or applicability in PV applications. Those works considered a voltage source and a resistive load that do not represent the operation of the converter in a PV application. Conversely, in this paper, we show that a PV source with a resistive load or clamped output voltage changes the voltage stress on semiconductor devices.

3. Numerical Analysis and Discussion

In this section, a numerical analysis is described to better demonstrate the performance of the proposed converter under different operating conditions, as well as to evaluate the electrical stress on the semiconductor devices. In a PV application scenario, we used a photovoltaic module LUXEN LNSA-160P (Luxen Solar Energy Co., Suzhou, China). The PV module (listed in Table 3) was characterized under the following conditions:

814 W/m^2 solar irradiance, ambient temperature 29.2 °C, and operating temperature of 63.13 °C.

Table 3. Specifications of the LUXEN LNSA-160P PV module.

Characteristic	Value	Characteristic	Value
Power in MPP, $P_{pv,mpp}$	103.9 W	Open-circuit voltage, V_{oc}	18.86 V
Voltage in MPP, $V_{pv,mpp}$	14.01 V	Short-circuit current, I_{sc}	8.190 A
Current in MPP, $I_{pv,mpp}$	7.413 A		

Now, we turn our attention to the three studied cases (voltage converter application, PV application with resistive load, and PV application with clamped output voltage) under equal nominal operating conditions, but taking into account the maximum power point (MPP) of the PV module. The specifications of the proposed converter are: nominal supplied voltage $E = 14.01$ V, nominal PV current $I_{pv} = 7.413$ A, output port voltage of 56 V, nominal power of 103.9 W, and resistive load $R = 30.183\ \Omega$. Using the aforementioned specifications allowed us to determine the nominal operating point of the converter in each application, described in Table 4.

Table 4. Nominal operating point of the proposed converter.

	D	I_{L1}	I_{L2}	V_{C1}	V_{C2}	$E//V_{Ci}$
Voltage appl	0.6666	11.125	5.565	42.020	56	14.01
PV resistive	0.6666	11.121	5.563	42.017	56	14.01
PV clamped volt	0.6666	11.121	5.562	42.020	56	14.01

From the previous analysis, we observed that the proposed converter presents a buck-boost voltage conversion ratio that depends on the square of the duty ratio. Figure 8 plots the converter gain as a function of the duty ratio for the three application cases. In this plot, the voltage/PV applications with resistive load present the same voltage gain;,whereas the PV application with clamped voltage has a constant gain for $D \leq 0.633$, and then it exhibits the same gain. This behavior is due to the operating characteristics of the PV module, where the voltage at the terminals has a limited range, given by $0 < V_{pv} < V_{oc}$, with V_{oc} being the maximum voltage in the PV module (zero power supply). Expression (32) shows that the point of constant gain can change depending on the characteristics/technology of the PV module and the selected output voltage.

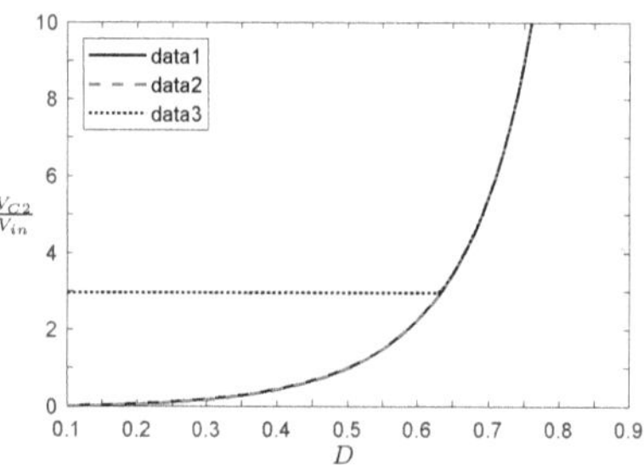

Figure 8. Voltage conversion gain (data 1: voltage converter appl; data 2: PV appl, resistive load; data 3: PV appl, clamped voltage).

The converter presents the same voltage gain (or partially) in different application cases; however, the voltages and currents in the converter are different due to the PV module's behavior. Figure 9 shows the voltages in the capacitors as a function of the duty ratio. In a voltage converter application (data 1, solid line), it is evident that supply voltage is constant (ideal voltage source); for this reason, when the duty cycle increases, the voltage in capacitors C_1 and C_2 (output port) increases too. A PV application with a resistive load (data 2, dashed line) behaves differently. In this scenario, for low duty ratios, the voltage at the terminals of the PV module is equal to V_{oc}, and when the duty cycle increases, the voltage V_{pv} reduces. This finding implies that the voltage in the PV module can be controlled by the duty cycle signal, as well as the power delivered by the PV module. Additionally, the output voltage increases until the maximum PV power (nominal operating point) is reached; after this point, the output voltage decreases as the voltage and current in the PV module tend to zero. Finally, in the PV application with clamped output voltage (data 3, dotted line), V_{pv} is constant (equal to V_{oc}) at low duty cycles since the voltage gain of the converter is lower than the relation V_{C2}/V_{oc}. When the duty cycle increases and reaches an appropriate voltage converter gain ($M(D) = V_{C2}/V_{oc}$), a further increase in the duty ratio implies a reduction in V_{pv}, where the voltage in the PV module can be controlled by the converter. Then, the operative region of the converter depends on the gain of the converter, clamped output voltage, and voltage range ($0 < V_{pv} < V_{oc}$) of the PV module.

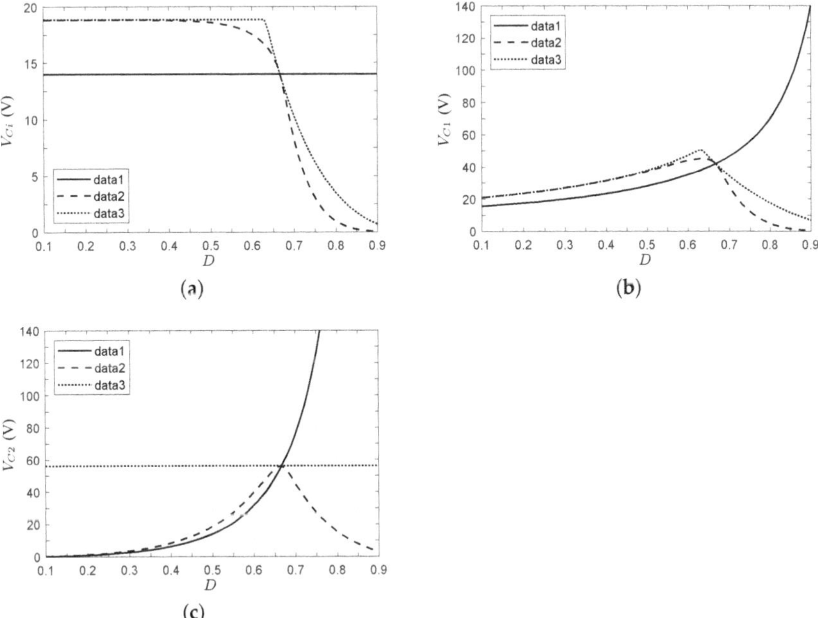

Figure 9. Capacitor voltages in the steady state of the proposed converter (data 1: voltage converter appl; data 2: PV appl, resistive load; data 3: PV appl, clamped voltage). (**a**) Input port. (**b**) Capacitor C_1. (**c**) Output port.

Figure 10 shows the inductor currents of the converter for the three cases analyzed. Here, it can be observed that the inductor currents in the voltage converter application increase with duty cycle, and these currents are higher above the nominal operating point due to the increase in the power demand. In the case of a PV with a resistive load, the inductor currents increase until the nominal operating point is reached; after, the current decreases as a function of the duty cycle and the current I_{pv}, which tends to the maximum current I_{sc}. In the case where the output voltage is clamped, the inductor currents are zero until the converter reaches an appropriate voltage gain, after which the currents increase in value.

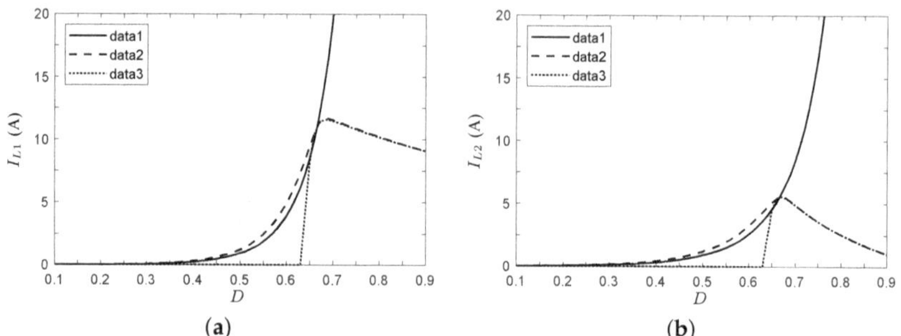

Figure 10. Inductor currents in the steady state of the proposed converter (data 1: voltage converter appl; data 2: PV appl, resistive load; data 3: PV appl, clamped voltage). (**a**) Inductor L_1. (**b**) Inductor L_2.

Now, considering the voltage and current stress on semiconductor devices, and based on the expressions in Table 1, it can be shown that the voltage and current stress on active switch S_1 and passive switch S_2 are related to voltage V_{C1} and current I_{L1}; the current stress on semiconductors S_3 and S_4 is related to current I_{L2}. For semiconductors S_1 and S_2, the voltage stress is equal ($V_{S1} = V_{S2} = V_{C1}$), as shown in Figure 9b. In this plot, the PV application presents a greater voltage stress than the voltage source application, which can be explained by the voltage of the PV module being near the V_{oc} value. After this value, the voltage stress reduces as the voltage of the PV module tends to zero. In turn, the current stresses ($I_{S1} = I_{S2} = I_{L1}$ and $I_{S3} = I_{S4} = I_{L2}$) for the clamped voltage application are lower than for the voltage source application, since the PV module current approaches zero when the PV module voltage is near the V_{oc} value (Figure 10). After the maximum power point of the PV module, the current stress is limited since the PV module current is practically constant. Finally, the voltage stress on semiconductors S_3 and S_4 is plotted in Figure 11. Here, it can be observed that the voltage stress is greater for the output voltage clamped case than for the PV application with load resistance case.

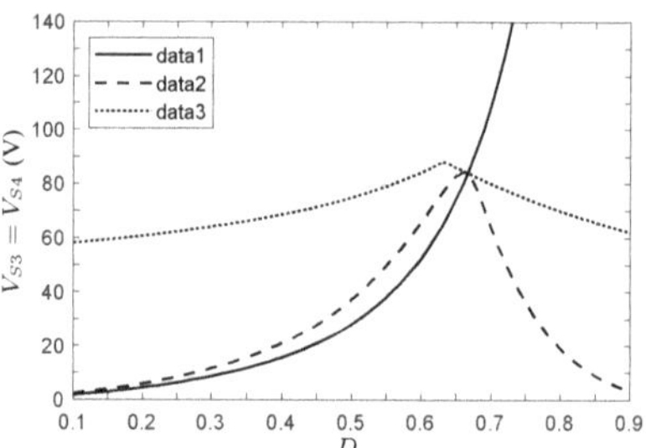

Figure 11. Voltage stress on semiconductors S_3 and S_4 (data 1: voltage converter appl; data 2: PV appl-, resistive load; data 3: PV appl, clamped voltage).

4. Experimental Results

To verify the behavior of the proposed converter, we conducted an experimental text. The test circuit is shown in Figure 12. The PV module LUXEN LNSA-160P has specifications at standard test conditions (STCs) of 160 W, 18.30 V, and 8.75 A at the maximum power point. The source v_o fixes the voltage to a desired value. Additionally, since the voltage source v_o does not accept input currents, a load R was added to dissipate the power generated by the PV module. Finally, a switching frequency of 50 kHz was used for the proposed converter.

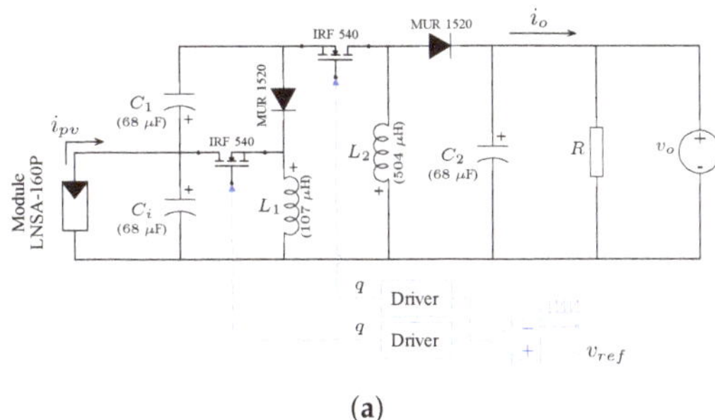

(**a**)

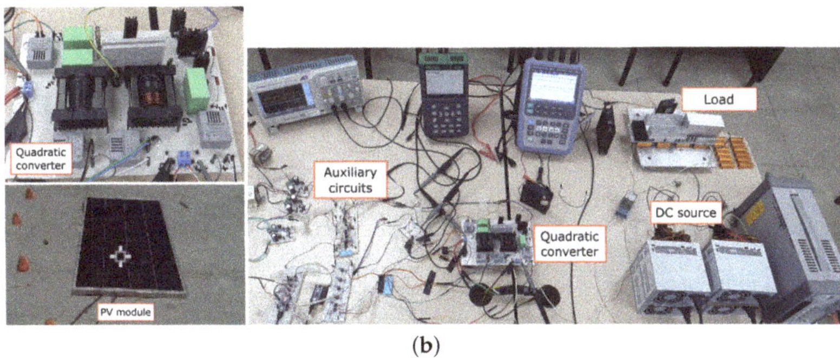

(**b**)

Figure 12. Experimental prototype of step-up/down quadratic converter. (**a**) Circuit diagram. (**b**) Experimental setup.

The resulting waveforms of currents and voltages from the operation of the proposed converter are shown in Figures 13–15. This test was performed with a clamped output voltage of $v_o = 56$ V, an incident solar radiation of 912 W/m^2, and a PV module temperature of 67 °C. Figure 13 shows the voltage at the terminals of each switching devices, where it can be observed that active switches operate synchronously, whereas passive switches (diodes) have a complementary operation. In all cases, well-defined transitions are observed.

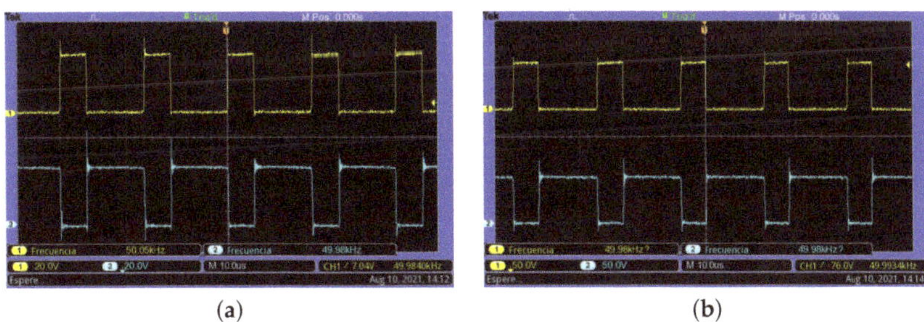

(**a**) (**b**)

Figure 13. Switching device voltages. (**a**) Active switch S_1 (**Top**) and diode S_2 (**Bottom**). (**b**) Active switch S_3 (**Top**) and diode S_4 (**Bottom**).

Figure 14 shows the waveforms of the current of the PV module and the inductor currents. Here, it can be observed that both inductor currents increase linearly when the active switches are turned on, and the current decreases when the switches are turned off. The measured average current in each element is $I_{pv} = 7.622$ A, $I_{L1} = 10.8$ A ,and $I_{L2} = 4.53$ A, whose values are consistent with those obtained from Expressions (27) and (28).

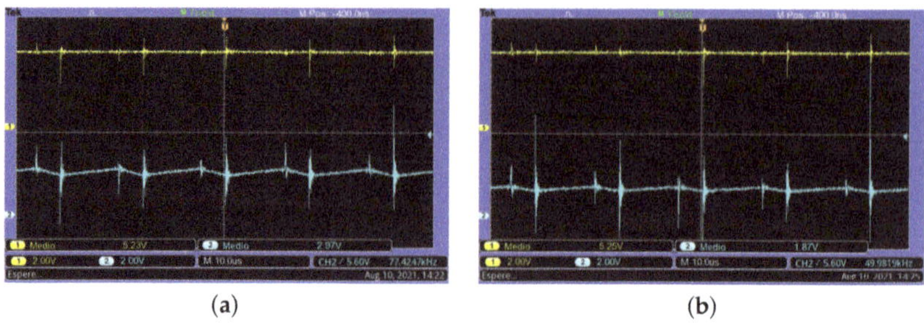

(**a**) (**b**)

Figure 14. Inductor and PV module currents. (**a**) PV module (**Top**) and inductor L_1 (**Bottom**). (**b**) PV module (**Top**) and inductor L_2 (**Bottom**).

Finally, the waveforms of the voltages in the switching converter capacitor C_i (input port) and capacitor C_1 are shown in Figure 15. In this condition, the voltage at the terminals of the PV module is $V_{pv} = 13$ V, which is lower than the open-circuit voltage $V_{oc} = 19.18$ V, whereas the voltage in the capacitor C_1 is $V_{C1} = 37$ V.

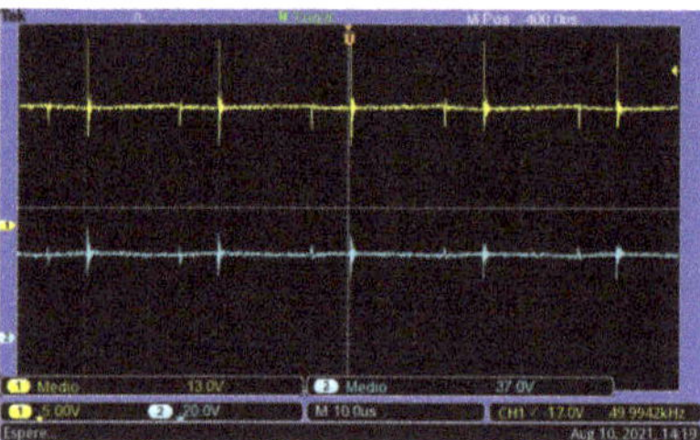

Figure 15. Capacitor voltages: capacitor C_i (**Top**) and capacitor C_1 (**Bottom**).

The experimental voltage conversion ratio ($M = V_o/V_{Ci}$) of the proposed converter is shown in Figure 16. In the test, the output voltage was maintained constant (test 1: V_o = 56 V, and test 2: V_o = 48 V), whereas the duty ratio was changed step-by-step from 0.5 to 0.9. At the start of the test, the conversion ratio was constant since the voltage at terminals of the PV module is equal to its open-circuit voltage ($V_{pv} = 19.9$ V). When the duty ratio increases, the converter forces the reduction in the PV module voltage, increasing the conversion ratio of the converter. The maximum conversion ratio at $D = 0.9$ is $M = 25.64$ for $V_o = 56$ V, and $M = 21.87$ for $V_o = 48$ V.

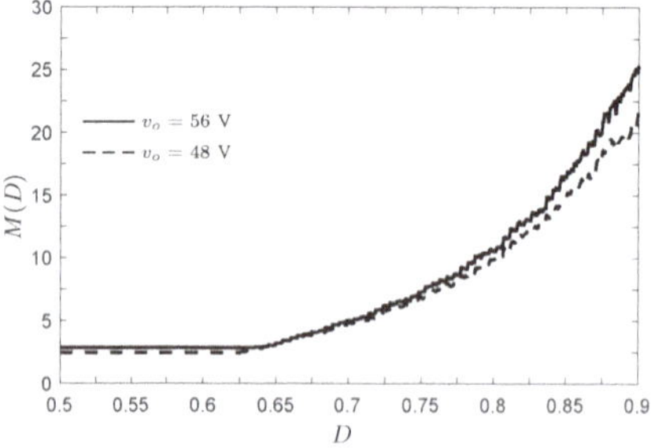

Figure 16. Experimental voltage conversion ratio using $v_o = 56$ V (solid line) and $v_o = 48$ V (dashed line).

Figure 17 shows the main voltages and currents of the converter when the duty cycle is varied from 0.6 to 0.9. In Figure 17a, the PV module voltage is reduced when the duty ratio increases; therefore, the PV module voltage can be controlled through the duty ratio, as can the power delivered by the PV module. It can be observed in Figure 17b that the PV module current increases until it reaches a constant current region. Here, a further increase in the duty ratio implies that voltages V_{pv} and V_{C1} tend to zero, whereas the current I_{pv} tends to the short-circuit current, which is consistent with the operation of the PV module.

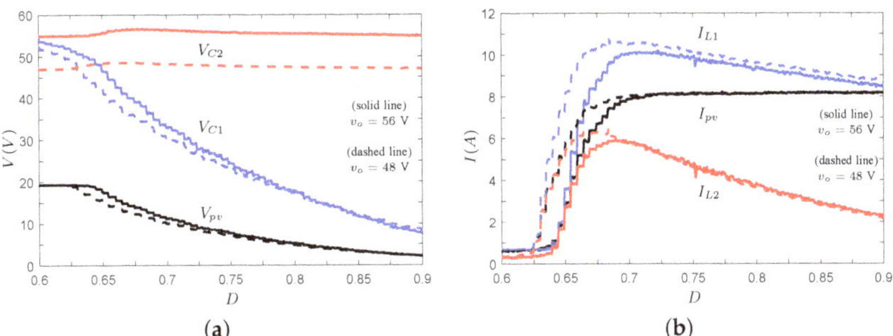

Figure 17. Currents and voltages in the converter. (**a**) Voltage at the terminal of the PV module: capacitor C_1 and capacitor C_2. (**b**) Current generated by the PV module: inductor L_1 and inductor L_2.

Figure 18 shows the power–voltage (P–V) and current–voltage (I–V) curves of the PV module, which were built with the measured input current and voltage of the converter. In Figure 18a, it can be observed that an increase in the duty ratio increases the current I_{pv}, whereas the voltage V_{pv} decreases. This behavior continues until the PV module enters the constant current region. Figure 18b presents the behavior of the power delivered in the PV module when the duty ratio is varied. In this test, ia maximum power of 95.61 W is obtained at $D = 0.68$, where $I_{pv} = 7.054$ A and $V_{pv} = 13.55$ V.

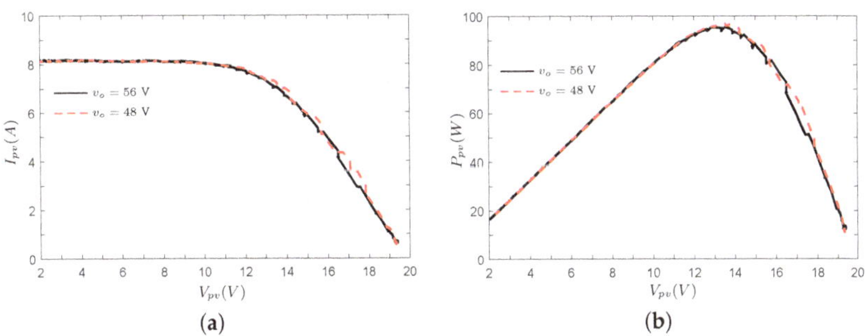

Figure 18. Experimental results in the converter's input port. (**a**) I–V curve. (**b**) P–V curve.

Then, the proposed converter was tested in a scenario to determine the maximum power of the PV module. In this test, the duty cycle of the converter was determined by a perturb and observe (P&O) MPPT algorithm. Figure 19 shows the principal measurements that confirm the operation of the converter in a day with partially cloudy conditions. Global solar irradiance shows a significant variability due to clouds, which is strongly correlated with the current and power developed. The voltage at the terminals of the PV module varies according to the perturbations introduced by the P&O algorithm. In the test, the efficiency was over 80%. Notably, efficiency can be increased by adopting high-quality components such as SiC semiconductors [38] and techniques such as synchronous rectification, amongst others; however, in this work, experimental evaluation was performed to support the theoretical results of the conversion ratio, operation, and its potential use in PV applications.

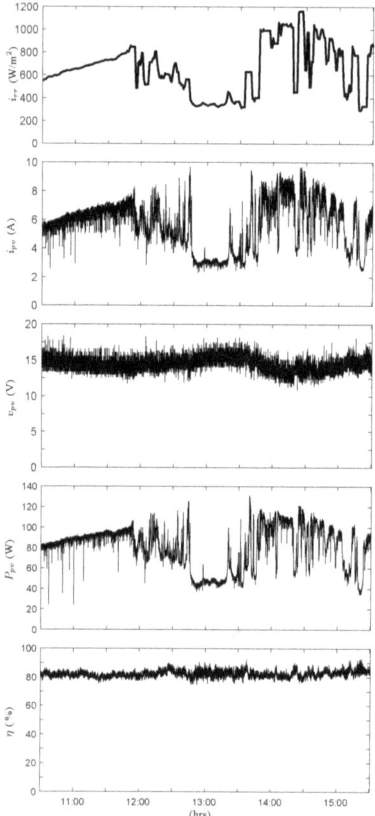

Figure 19. Time responses of the switching converter with an MPPT perturb and observe algorithm. (**Top**) to (**bottom**): Measured global solar irradiance, generated current of the PV module, voltage at the terminals, power developed by the PV module, and efficiency (partially cloudy day).

Finally, the large-signal model of the converter (35) was implemented in MATLAB/Simulink (MathWorks Inc., R2021a, Natick, MA, USA), where transient responses were obtained when steps in the duty ratio were applied. Simulations and experimental results are shown in Figure 20 for capacitor voltages. In this test, the duty ratio was changed from 0.64 to 0.71 and back. As can be observed, the simulated responses predicted the dynamic response of the converter prototype, confirming the validity of the developed models.

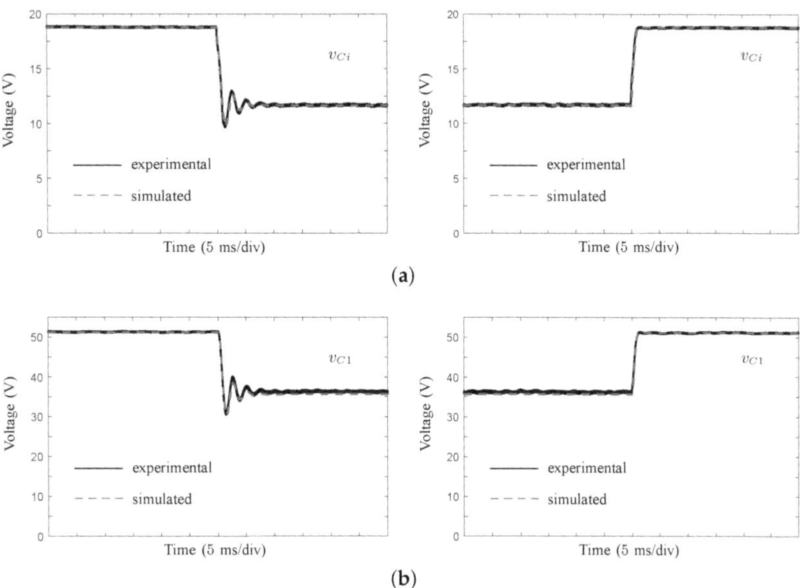

Figure 20. Simulation and experimental results of the transient responses of the converter under duty ratio steps. (**a**) Voltage in capacitor C_i (time: 5 ms/div). (**b**) Voltage in capacitor C_1 (time: 5 ms/div).

5. Conclusions

This paper presented a quadratic buck-boost converter based on the interconnection of two basic switching cells in a noncascading structure, which is suitable for PV applications since the converter properly links a PV device with a DC power system. The proposed converter provides a wide voltage conversion ratio, a non-inverting output voltage, and a common node between input and output ports. The steady-state analysis demonstrated that the operation in a PV application differs when it is used in a voltage source application. Additionally, the voltage stress on semiconductors is more stringent in a PV application, whereas the current stress is low under this condition.

The experimental results showed an adequate operation of the converter under real conditions, and that the duty ratio controls the voltage at the terminals of the PV module, adjusting the generated power. The converter presents a high voltage conversion ratio, which allows wide input voltage variations with a clamped output voltage, desirable in a grid-connected applications through an inverter or microgrid applications.

Author Contributions: Conceptualization, R.L.-P. and J.A.M.-S.; methodology, R.L.-P., J.A.M.S., and M.R.; software, C.A.H.-J.; validation, R.L.-P., C.A.H.-J., and C.Á.-M.; formal analysis, R.L.-P. and C.A.H.-J.; investigation, J.A.M.-S.; data curation, R.L.-P. and C.A.H.-J.; writing—original draft preparation, R.L.-P.; writing—review and editing, R.L.-P., J.A.M.-S., M.R., and C.Á.-M.; visualization, M.R. and C.Á.-M.; supervision, R.L.-P. and J.A.M.-S.; project administration, R.L.-P. All authors have read and agreed to the published version of the manuscript.

Funding: This research received no external funding.

Institutional Review Board Statement: Not applicable.

Informed Consent Statement: Not applicable.

Data Availability Statement: Data are contained within the article.

Acknowledgments: This research was supported by CONACYT, México, under project 1982 of Cátedras CONACYT; and by PRODEP through ITLAG-CA-10.

Conflicts of Interest: The authors declare no conflict of interest.

Abbreviations

The following abbreviations are used in this manuscript:

PV	Photovoltaic
AC	Alternating current
DC	Direct current
THD	Total harmonic distortion
MPPT	Maximum power point tracking
MPP	Maximum power point
LED	Light-emitting diode
CCM	Continuous conduction mode
DCM	Discontinuous conduction mode
P&O	Perturb and observe

References

1. López-Santos, O.; Martínez-Salamero, L.; García, G.; Valderrama-Blavi, H.; Mercuri, D.O. Efficiency analysis of a sliding-mode controlled quadratic boost converter. *IET Power Electron.* **2013**, *6*, 364–373. doi: 10.1049/iet-pel.2012.0417. [CrossRef]
2. Alonge, F.; Pucci, M.; Rabbeni, R.; Vitale, G. Dynamic modelling of a quadratic DC/DC single-switch boost converter. *Electr. Power Syst. Res.* **2017**, *152*, 130–139. doi: 10.1016/j.epsr.2017.07.008. [CrossRef]
3. Yan, T.; Xu, J.; Dong, Z.; Shu, L.; Yang, P. Quadratic boost PFC converter with fast dynamic response and low output voltage ripple. In Proceedings of the 2013 International Conference on Communications, Circuits and Systems (ICCCAS), Chengdu, China, 15–17 November 2013. doi: 10.1109/ICCCAS.2013.6765367. [CrossRef]
4. Al–Saffar, M.A.; Ismail, E.H.; Sabzali, A.J. Integrated buck-boost-quadratic buck PFC rectifier for universal input applications. *IEEE Trans. Power Electron.* **2009**, *24*, 2886–2896. doi: 10.1109/TPEL.2009.2023323. [CrossRef]
5. Ozdemir, S.; Altin, N.; Sefa, I. Fuzzy logic based MPPT controller for high conversion ratio quadratic boost converter. *Int. J. Hydrogen Energy* **2017**, *42*, 17748–17759. doi: 10.1016/j.ijhydene.2017.02.191. [CrossRef]
6. García-Vite, P.M.; soriano-Rangel, C.A.; Rosas-Caro, J.C.; Mancilla-David, F. A DC-DC converter with quadratic gain and input current ripple cancelation at a selectable duty cycle. *Renew. Energy* **2017**, *101*, 431–436. doi: 10.1016/j.renene.2016.09.010. [CrossRef]
7. Sarikhani, A.; Allahverdinejad, B.; Hamzeh, M.; Afjei, E. A continuous input and output current quadratic buck-boost converter with positive output voltage for photovoltaic applications. *Sol. Energy* **2019**, *188*, 19–27. doi: 10.1016/j.solener.2019.05.025. [CrossRef]
8. Lotfi–Nejad, M.; Poorali, B.; Adib, E.; Motie-Birjandi, A.A. New cascade boost converter with reduced losses. *IET Power Electron.* **2016**, *9*, 1213–1219. doi: 10.1049/iet-pel.2015.0240. [CrossRef]
9. Baekhoej-Kjaer, S.; Pedersen, J.K.; Blaabjerg, F. A review of single-phase grid-connected inverters for photovoltaic modules. *IEEE Trans. Ind. Appl.* **2005**, *41*, 1213–1219. doi: 10.1109/TIA.2005.853371. [CrossRef]
10. Alonso, J.M.; Viña, J.; Gacio, D.; Campa, L.; Osorio, R. Analysis and design of the quadratic buck-boost converter as a high-power-factor driver for power-LED lamps. In Proceedings of the IEEE 36th Annual Conference on Industrial Electronics Society, Glendale, AZ, USA, 7–10 November 2010. doi: 10.1109/IECON.2010.5675145. [CrossRef]
11. Nousiainen, L.; Suntio, T. Dynamic characteristics of current-fed semi-quadratic buck-boost converter in photovoltaic applications. In Proceedings of the IEEE Energy Conversion Congress and Exposition, Phoenix, AZ, USA, 17–22 September 2011. doi: 10.1109/ECCE.2011.6063886. [CrossRef]
12. Nousiainen, L.; Suntio, T. Dual-mode current-fed semi-quadratic buck-boost converter for transformerless modular photovoltaic applications. In Proceedings of the 14th European Conference on Power Electronics ans Applications, Birmingham, UK, 30 August 2011–1 September 2011.
13. Mostaan, A.; Gorji, S.A.; Soltani, M.; Ektesabi, M. A novel quadratic buck-boost DC-DC converter without floating gate-driver. In Proceedings of the IEEE 2nd Annual Southern Power Electronics Conference (SPEC), Auckland, New Zealand, 5–8 December 2016. doi: 10.1109/SPEC.2016.7846053. [CrossRef]
14. García-Vite, P.M.; Rosas-Caro, J.C.; Martínez-Salazar, A.L.; Chavez, J.J.; Valderrábano-González, A.; Sánchez-Huerta, V.M. Quadratic buck-boost converter with reduced input current ripple and wide conversion range. *IET Power Electron.* **2019**, *12*, 3977–3986. doi: 10.1049/iet-pel.2018.5616. [CrossRef]
15. Baby, B.; Abraham, T.M. Bidirectional buck-boost quadratic converter using fuzzy controller for distributed generation systems. In Proceedings of the International Conference on Current Trends towards Converging Technologies, Coimbatore, India, 1–3 March 2018. doi: 10.1109/ICCTCT.2018.8551025. [CrossRef]
16. Zhang, N.; Zhang, G.; See, K.W.; Zhang, B. A single-switch quadratic buck-boost converter with continuous input port current and continuous output port current. *IEEE Trans. Power Electron.* **2018**, *33*, 4157–4166. doi: 10.1109/TPEL.2017.2717462. [CrossRef]
17. Gorji, S.A.; Mostaan, A.; My, H.T.; Ektesabi, M. Non-isolated buck-boost DC-DC converter with quadratic voltage gain ratio. *IET Power Electron.* **2019**, *12*, 1425–1433. doi: 10.1049/iet-pel.2018.5703. [CrossRef]

18. Gholizadeh, H.; Sarikhani, A.; Hamzeh, M. A transformerless quadratic buck-boost converter suitable for renewable applications. In Proceedings of the International Power Electronics, Drive Systems and Technologies Conference (PEDSTC), Shiraz, Iran, 12–14 February 2019. doi: 10.1109/PEDSTC.2019.8697232. [CrossRef]
19. Mayo-Maldonado, J.C.; Valdez-Resendiz, J.E.; Garcia-Vite, P.M.; Rosas-Caro, J.C.; Rivera-Espinosa, M.R.; Valderrábano-González, A. Quadratic buck-boost converter with zero output voltage ripple at a selectable operating point. *IEEE Tans. Ind. Appl.* **2019**, *55*, 2813–2822. doi: 10.1109/TIA.2018.2889421. [CrossRef]
20. Rosas-Caro, J.C.; Valdez-Resendiz, J.E.; Mayo-Maldonado, J.C.; Alejo-Reyes, A.; Valderrábano-González, A. Quadratic buck-boost converter with positive output voltage and minimum ripple point design. *IET Power Electron.* **2018**, *11*, 1306–1313. doi: 10.1049/iet-pel.2017.0090. [CrossRef]
21. Rosas-Caro, J.C.; Sánchez, V.M.; Valdez-Resendiz, J.E.; Mayo-Maldonado, J.C.; Beltran-Carbajal, F.; Valderrábano-González, A. Quadratic buck-boost converter with positive output voltage and continuous input current. In Proceedings of the International Conference on Electronics, Communications and Computers, Cholula Puebla, Mexico, 21–23 February 2018. doi: 10.1109/CONIELECOMP.2018.8327191. [CrossRef]
22. Dubey, N.; Kumar-Sharma, A. Analysis of bi-directional DC-DC buck-boost quadratic converter for energy storage devices. In Proceedings of the International Conference on Communication and Electronics Systems, Coimbatore, India, 15–16 November 2019. doi: 10.1109/ICCES45898.2019.9002112. [CrossRef]
23. Carbajal-Gutiérrez, E.E.; Morales-Saldanña, J.A.; Leyva-Ramos, J. Modeling of a single-switch quadratic buck converter. *IEEE Tran. Aerosp. Electron. Syst.* **2005**, *41*, 1450–1456. doi: 10.1109/TAES.2005.1561895. [CrossRef]
24. Morales-Saldanña, J.A.; Galarza-Quirino, R.; Leyva-Ramos, J.; Carbajal-Gutiérrez, E.E. Multiloop controller design for a quadratic boost converter. *IET Power Electron.* **2007**, *1*, 362–367. doi: 10.1049/iet-epa:20060426. [CrossRef]
25. Loera-Palomo, R.; Morales-Saldanña, J.A.; Palacios-Hernández, E. Quadratic step-down DC-DC converters based on reduced redundant power processing approach. *IET Power Electron.* **2013**, *6*, 136–145. doi: 10.1049/iet-pel.2012.0122. [CrossRef]
26. Loera-Palomo, R.; Morales-Saldanña, J.A. Family of quadratic step-up DC-DC converters based on noncascading structures. *IET Power Electron.* **2015**, *8*, 793–801. doi: 10.1049/iet-pel.2013.0879. [CrossRef]
27. Moschopoulos, G. Quadratic power conversion for industrial applications. In Proceedings of the 2010 Twenty-Fifth Annual IEEE Applied Power Electronics Conference and Exposition (APEC'10), Orlando, FL, USA, 21–25 February 2010. doi: 10.1109/APEC.2010.5433400. [CrossRef]
28. Ismail, E.H.; Al-Saffar, M.A.; Sabzali, A.J.; Fardoun, A.A. A family of single-switch PWM converters with high step-up conversion ratio. *IEEE Trans. Circuits Syst.* **2008**, *55*, 1159–1171. doi: 10.1109/TCSI.2008.916427. [CrossRef]
29. Yuan-Mao, Y.; Cheng, K.W.E. Quadratic boost converter with low buffer capacitor stress. *IET Power Electron.* **2014**, *7*, 1162–1170. doi: 10.1049/iet-pel.2013.0205. [CrossRef]
30. Hwu, K.I.; Peng, T.J. High-voltage-boosting converter with charge pump capacitor and coupling inductor combined with buck-boost converter. *IET Power Electron.* **2014**, *7*, 177–188. doi: 10.1049/iet-pel.2013.0229. [CrossRef]
31. Miao, S.; Wang, F.; Ma, X. A new transformerless buck-boost converter with positive output voltage. *IEEE Trans. Ind. Electron.* **2016**, *63*, 2965–2975. doi: 10.1109/TIE.2016.2518118. [CrossRef]
32. Tse, C.K.; Chow, M.H.L. Theoretical study of switching power converters with power factor correction and output regulation. *IEEE Trans. Circuits Syst. I Fundam. Theory Appl.* **2000**, *47*, 1047–1055. doi: 10.1109/81.855460. [CrossRef]
33. Tse, C.K.; Chow, M.H.L.; Cheung, M.K.H. A family of PFC voltage regulator configurations with reduced redundant power processing. *IEEE Trans. Power Electron.* **2001**, *16*, 794–802. doi: 10.1109/63.974377. [CrossRef]
34. Herrera-Carrillo, N.E.; Loera-Palomo, R.; Rivero-Corona, M.A.; Álvarez-Macías, C. Noncascading quadratic boost converter for PV applications. In Proceedings of the IEEE International Autumn Meeting on Power, Electronics and Computing, Ixtapa, Guerrero, Mexico, 14–16 November 2018. doi: 10.1109/ROPEC.2018.8661478. [CrossRef]
35. Hernández-Jacobo, C.A.; Loera-Palomo, R.; Sellschopp-Sánchez, F.S.; Álvarez-Macías, C.; Rivero-Corona, M.A. Quadratic buck/boost converter in series connection for photovoltaic applications. In Proceedings of the IEEE International Autumn Meeting on Power, Electronics and Computing, Ixtapa, Guerrero, Mexico, 4–6 November 2020. doi: 10.1109/ROPEC50909.2020.9258687. [CrossRef]
36. de la Rosa-Romo, D.; Loera-Palomo, R.; Rivero, M.; Sellschopp-Sánchez, F.S. Averaged current mode control for maximum power point tracking in high-gain photovoltaic applications. *J. Power Electron.* **2020**, *20*, 1650–1661. doi: 10.1007/s43236-020-00144-1. [CrossRef]
37. Panigrahi, R.; Mishra, S.K.; Joshi, A. Synthesizing a family of converters for a specified conversion ratio using flux balance principle. *IEEE Trans. Ind. Electron.* **2020**, *68*, 3854–3864. doi: 10.1109/TIE.2020.2984450. [CrossRef]
38. Hernando, M.M.; Fernández, A.; García, J.; Lamar, D.G.; Rascón, M. Comparing Si and SiC diode performance in commercial AC-to-DC rectifiers with power-factor correction. *IEEE Trans. Ind. Electron.* **2006**, *53*, 705–707. doi: 10.1109/PESC.2003.1217755. [CrossRef]

Article

High Step-Up Converter Based on Non-Series Energy Transfer Structure for Renewable Power Applications

Luis Humberto Diaz-Saldierna * and **Jesus Leyva-Ramos**

Instituto Potosino de Investigación Científica y Tecnológica, Camino a la Presa San José No. 2055, San Luis Potosí 78216, Mexico; jleyva@ipicyt.edu.mx
* Correspondence: ldiaz@ipicyt.edu.mx; Tel.: +52-444-834-2000

Abstract: In this paper, a high step-up boost converter with a non-isolated configuration is proposed. This configuration has a quadratic voltage gain, suitable for processing energy from alternative sources. It consists of two boost converters, including a transfer capacitor connected in a non-series power transfer structure between input and output. High power efficiencies are achieved with this arrangement. Additionally, the converter has a common ground and non-pulsating input current. Design conditions and power efficiency analysis are developed. Bilinear and linear models are derived for control purposes. Experimental verification with a laboratory prototype of 500 W is provided. The proposed configuration and similar quadratic configurations are compared experimentally using the same number of components to demonstrate the power efficiency improvement. The resulting power efficiency of the prototype was above 95% at nominal load.

Keywords: renewable energy sources; dc-dc power electronic converters; energy efficiency

Citation: Diaz-Saldierna, L.H.; Leyva-Ramos, J. High Step-Up Converter Based on Non-Series Energy Transfer Structure for Renewable Power Applications. *Micromachines* **2021**, *20*, 689. https://doi.org/10.3390/mi12060689

Academic Editors: Francisco J. Perez-Pinal and Young-Ho Cho

Received: 26 May 2021
Accepted: 10 June 2021
Published: 13 June 2021

1. Introduction

Nowadays, the even increase in energy consumption and the worldwide concern about environmental issues have led to increase the power generation through renewable sources, like photovoltaic and fuel-cell sources [1,2]. The output voltage of these alternative sources is low and unregulated [3,4]; therefore, an interface with DC-DC power converters is needed to obtain a high and regulated output voltage. [5]. On the other hand, these energy sources can suffer permanent damage if high ripples appear in the demanded current [6,7]. According to the above, DC-DC switching power converters for renewable applications should have high-voltage gains, low-input current ripples and high power efficiencies [8].

The conventional boost converter has been proposed for renewable applications [9,10]; however, larger duty cycles are required to achieve high voltage gains. The above produces large voltage spikes, diode reverse recovery time problems, and high conduction losses on the active switch due to the intrinsic resistances [11,12].

In the open literature, several configurations are available to further extend the voltage gain without large duty cycles. A boost converter with a quadratic conversion ratio is discussed in [13]. This configuration is based on stackable switching cells with a modular structure that can be extended to boost the output voltage. This converter exhibits a pulsating input current; therefore, a coupled capacitor is needed for renewable energy applications [14]. A quadratic boost converter with two conventional boost converters connected in series (QBC-SC) is proposed in [15]. The power flows in cascade from the input source to the first converter; then, the power flows to the second converter and later to the load.

The quadratic relationship of the conversion ratio provides a high-voltage gain for this configuration. A quadratic boost converter with a single switch (QBC-SS) is discussed in [16]. It is similar to the QBC-SC, the power flows in series through the two converters and load; however, it includes only one switch. A QBC-SS that includes a voltage doubler to achieve an extra voltage gain (QBC-TR) is proposed in [17]. It includes a voltage doubler to achieve an extra voltage gain. The voltage doubler consists of two capacitors, two diodes, and a coupled inductor that provides an additional gain respect to the quadratic conversion ratio. A current-fed boost converter with a quadratic voltage gain (QBC-CF) is proposed in [18]. This converter has two operation modes. Energy from the input source is stored in two inductors; after that, the energy is transfer to two output capacitors. The quadratic gain is obtained by a series arrangement of capacitors, where the output voltage is the sum of both capacitor voltages. The number of diodes and the way to charge both inductors have an impact on the power efficiency. A boost converter with an active switched inductor-capacitor (LC) network is proposed in [19]. The configuration consists of two inductors that are charged in parallel across two active switches, one diode, and one capacitor. This configuration uses a non-common grounding. The energy stored in both inductors is transferred in series to the output capacitor and the load. The structure of this configuration allows an enlargement of the conversion ratio respect to the gain of the quadratic boost converter. On the other hand, a coupling capacitor needs to be connected to the input of the converter because the input current has a pulsating waveform. Two boost configurations based on the quadratic and Cuk converters are shown in [20]. These configurations offer an extra voltage gain by adding a factor increment to the quadratic voltage gain. They are classified into two types, depending on the voltage gain of each configuration. Both converters have non-common grounding. An isolated boost converter that includes forward and flyback configurations is proposed in [21]. This converter is based on a quadratic configuration with an extended voltage gain because of the turns ratio of the transformer; however, the input current has a pulsating waveform.

The quadratic boost converter is a feasible option for processing energy generated by renewable sources, because it has a higher voltage conversion ratio. Quadratic conversion ratios can achieve higher voltage gains with duty cycles less than 0.7. This configuration generates a non-pulsating input current, which is critical for renewable sources as fuel and photovoltaic cells. The circuit of the conventional quadratic boost converter (QBC-SC) is depicted in Figure 1.

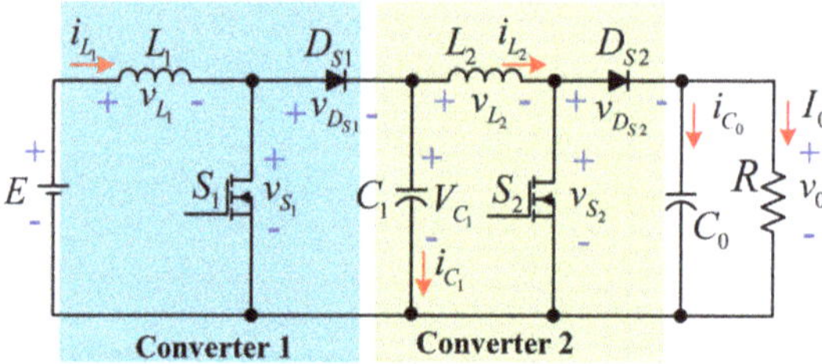

Figure 1. Quadratic boost converter with series power transfer.

A major drawback of the conventional quadratic configuration is the series power transfer between the input source and the load, resulting in a low overall efficiency. The above implies a reprocessing of the energy generated by the source; thus, a slow energy propagation between the input of the converter and the load is obtained. On the other hand, energy reprocessing yields an increment on the power losses due to the parasitics of the converter elements. This issue is described in Figure 2, where the power is transferred throughout the two boost converters of the QBC-SC. As can be seen, the power flow is done in series between the input source and the load, where the capacitor C_1 acts as the input source of the second converter.

A boost converter with a quadratic voltage gain, common ground, and non-series power transfer for better efficiency is proposed in this work. This configuration consists of two conventional boost converters with a transfer capacitor between both converters. This capacitor allows a parallel energy transfer between both converters and the output. The main advantage is an increase in power efficiency due to the non-series power transfer.

Moreover, this configuration can be used for processing energy from alternative sources without the need to add a filter for reducing the input current ripple.

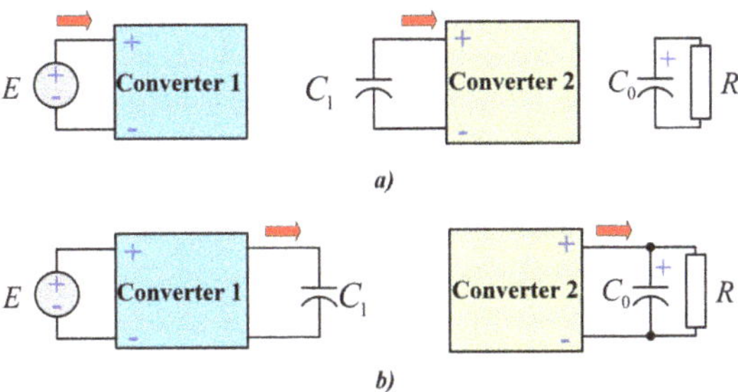

Figure 2. Power flows of the QBC-SC: (**a**) Equivalent scheme when both inductor are storing energy, and (**b**) equivalent scheme when both inductor are delivering energy.

The organization of this work is as follows. A description and operation of the proposed configuration are shown in Section 2. The state-steady conditions and the converter design are developed in Section 3. The power efficiency of the proposed configuration is analyzed in Section 4. Bilinear and linear models of the proposed converter are developed in Section 5. Design criteria of the proposed converter are shown in Section 6. Experimental verification with a 500 W laboratory prototype is provided in Section 7. Final remarks are given in Section 8.

2. Converter Operation

The proposed configuration is depicted in Figure 3. The input voltage is represented by E, the currents and voltages through L_1 and L_2 are denoted by i_{L_1}, v_{L_1}, i_{L_2}, and v_{L_2}, respectively. The voltages and currents through C_p, and C_0 are denoted by v_{C_p}, i_{C_p}, v_0, and i_{C_0}, respectively. On the other hand, the voltages through S_1, D_{S1}, S_2 and D_{S2} are denoted by v_{S1}, $v_{D_{S1}}$, v_{S2}, and $v_{D_{S2}}$. Finally, the current through the load R is denoted by I_0.

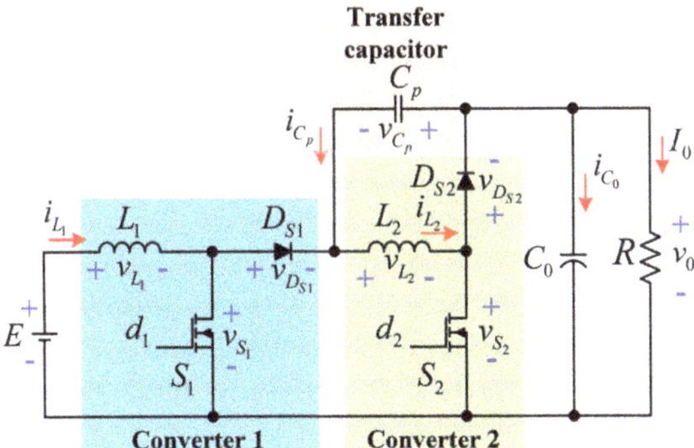

Figure 3. Quadratic boost converter with non-series power transfer.

The analysis is carried out considering the following points:

- The study is developed for continuous conduction mode operation (CCM).
- The passive elements (L_1, L_2, C_p, C_0), the active switches (S_1, S_2), and the passive switches (D_{S1}, D_{S2}) are considered as an ideal components.
- The active switches operate under the same duty cycle D. Thus, $D_1 = D_2 = D$, $f_s = 1/T_s$ is the switching frequency, and T_s is the period of the PWM signal that drives the active switches.

- For steady-state operation, $T_s = t_{ON} + t_{OFF}$, where $t_{ON} = DT_s$ and $t_{OFF} = (1 - D)T_s$.

Due to the above assumptions, the proposed configuration operates in two modes. The power transfer from the input to the output of the proposed configuration is shown in Figure 4. As can be seen, a non-series power transfer between the input source and the load occurs due to the transfer capacitor C_p.

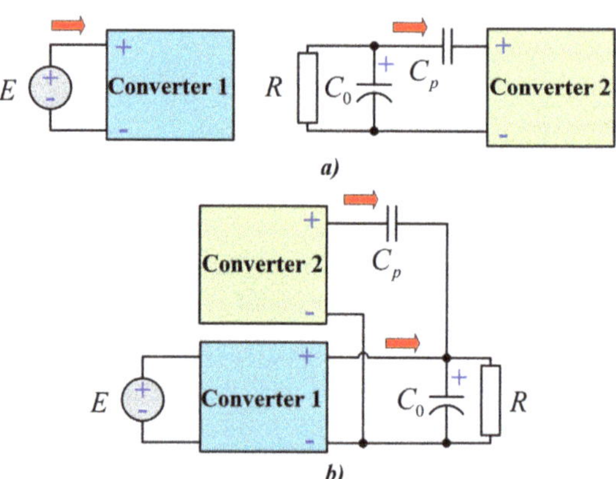

Figure 4. Power flows of the proposed configuration: (**a**) Equivalent scheme when both inductor are storing energy, and (**b**) equivalent scheme when both inductors are delivering energy.

The two operating modes of the converter are described as follows:

Mode I [t_{ON}]: Both active switches (S_1, S_2) are turned ON. The diodes D_{S1} and D_{S2} are not conducting; then, the input source E delivers energy to the inductor L_1. The inductor L_2 is charged through capacitor C_p by the energy stored in the capacitor C_0, which also supplies energy to the load. The circuit that describes this operating mode is exhibited in Figure 5.

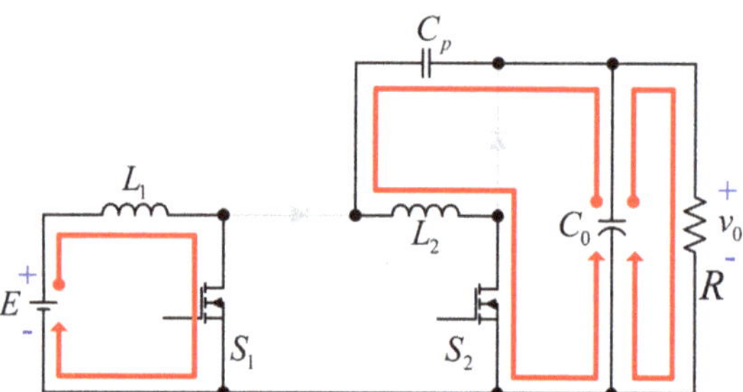

Figure 5. Circuit for the intervale t_{ON}.

Mode II [t_{OFF}]: Both active switches (S_1, S_2) are turned OFF. For this interval, the output capacitor C_0 and the load R are supplied by the energy stored in inductor L_1, through capacitor C_p. The inductor L_2 delivers energy to the output (C_0 and R) through the diode D_{S_2}. The circuit that describes this operating mode is exhibited in Figure 6.

The analysis, modeling, and the expressions in steady-state are easy to develop in two operating modes. On the other hand, the capacitor C_p acts as a power transfer element between the first converter and the output capacitor C_0. Due to this arrangement, the cascade power transfer between the first and second converter is avoided; thus, there is an improvement in the converter efficiency.

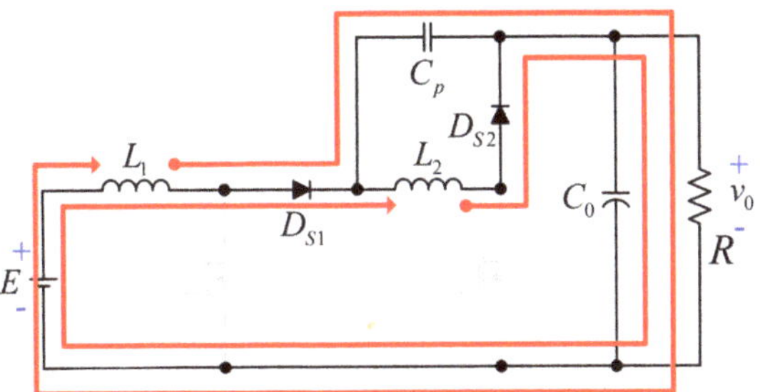

Figure 6. Circuit for the interval t_{OFF}.

3. Converter Analysis

Figure 7 shows the theoretical waveforms of the proposed configuration. The input converter current i_{L_1} has a non-pulsating waveform, and its ripple amplitude depends on the inductance value.

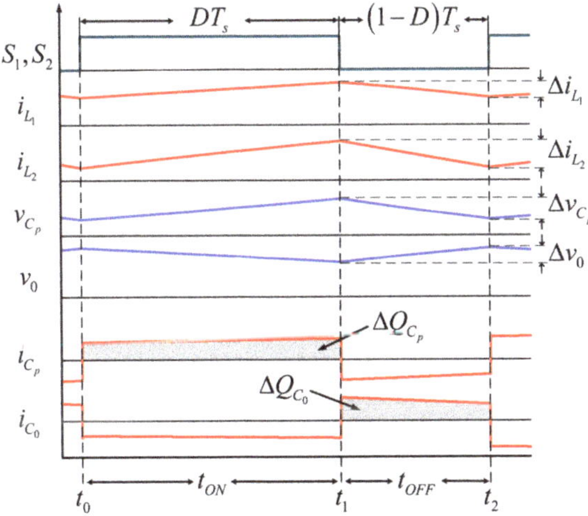

Figure 7. Current and voltage waveforms of the converter in the time T_s.

3.1. Converter Voltage Gain

Applying the volt-second balance principle to L_1 and L_2 results in:

$$\begin{aligned} EDT_s + (E + V_{C_p} - V_0)(1-D)T_s = 0 \\ (-V_{C_p} + V_0)DT_s - V_{C_p}(1-D)T_s = 0 \end{aligned} \tag{1}$$

By using the last equations, the expression for V_{C_p} can be derived:

$$V_{C_p} = V_0 - \frac{E}{(1-D)} = DV_0 \tag{2}$$

From the above expression, the voltage gain M is:

$$M = \frac{V_0}{E} = \frac{1}{(1-D)^2} \tag{3}$$

According to Equation (3), the voltage gain M has a quadratic conversion ratio.

3.2. Steady-State Conditions

For calculating the inductances L_1 and L_2, the period when the active switches are turned ON is analyzed, see Figure 5. For large enough inductances, the currents rise linearly from their minimum value to their maximum value during t_{ON}; then, the voltages in the inductor terminals of L_1 and the inductor terminals of L_2 are, respectively:

$$
\begin{aligned}
V_{L_1} &= L_1 \frac{\Delta I_{L_1}}{t_{ON}} = E \\
V_{L_2} &= L_2 \frac{\Delta I_{L2}}{t_{ON}} = -V_{C_p} + V_0
\end{aligned} \tag{4}
$$

Substituting $t_{ON} = DT_s$, $V_{C_p} = DV_0$, and $V_0 = E/(1-D)^2$ into Equation (4), the expressions for the inductor current ripples are:

$$
\begin{aligned}
\Delta i_{L_1} &= \frac{ED}{L_1 f_s} \\
\Delta i_{L_2} &= \frac{ED}{(1-D)L_2 f_s}
\end{aligned} \tag{5}
$$

Considering the equality between the input and output power ($EI_{L_1} = V_0^2/R$), the average current through L_1 can be computed. On the other hand, the average current of L_2 can be computed using the relationship $I_{L_2} = (1-D)I_{L_1}$; then, the corresponding expressions for I_{L_1} and I_{L_2} are:

$$
\begin{aligned}
I_{L_1} &= \frac{E}{R(1-D)^4} \\
I_{L_2} &= \frac{E}{R(1-D)^3}
\end{aligned} \tag{6}
$$

The maximum and minimum current values reached by the first and second inductors can be obtained as follows:

$$
\begin{aligned}
I_{L_{1MAX}} &= I_{L_1} + \frac{\Delta I_{L_1}}{2}, \; I_{L_{1MIN}} = I_{L_1} - \frac{\Delta I_{L_1}}{2} \\
I_{L_{2MAX}} &= I_{L_2} + \frac{\Delta I_{L2}}{2}, \; I_{L_{2MIN}} = I_{L_2} - \frac{\Delta I_{L_2}}{2}
\end{aligned} \tag{7}
$$

In switching converters, it is essential to find out the critical inductance values for operation in CCM [22]. The critical inductance values of the proposed configuration can be computed using the relationships:

$$
\begin{aligned}
0 &= I_{L_1} - \frac{\Delta I_{L_1}}{2} \\
0 &= I_{L_2} - \frac{\Delta I_{L_2}}{2}
\end{aligned} \tag{8}
$$

then, the inductances for the CCM operation are:

$$
\begin{aligned}
L_1 &> \frac{DR(1-D)^4}{2f_s} \\
L_2 &> \frac{DR(1-D)^2}{2f_s}
\end{aligned} \tag{9}
$$

The charge variation in a capacitor is defined as $\Delta Q_C = C\Delta v_C$, where ΔQ_C depicted the area under the current curve when the capacitor stores energy, C denotes the capacitance, and Δv_C is the voltage ripple (see Figure 7). For t_{ON}, the capacitor C_p is charged by the current i_{L_2} ($i_{L_2} = i_{C_p}$). On the other hand, the capacitor C_0 is charged by the current of L_1 at time t_{OFF}. In this interval, the current through C_0 is $i_{C_0} = i_{L_1} - I_0$, that is:

$$\begin{aligned} \Delta Q_{C_p} &= \int_{t_0}^{t_1} i_{L_2}(t)dt = C_p \Delta v_{C_p} \\ \Delta Q_{C_0} &= \int_{t_1}^{t_2} [i_{L_1}(t) - I_0]dt = C_0 \Delta v_{C_0} \end{aligned} \tag{10}$$

where $t_{ON} = t_1 - t_0$ and $t_{OFF} = t_2 - t_1$. Calculating ΔQ_C for C_p and C_0, the following expressions for the capacitor ripples are obtained:

$$\begin{aligned} \Delta v_{C_p} &= \frac{V_{C_p}}{R(1-D)^3 C_p f_s} \\ \Delta v_0 &= \frac{V_0 D(2-D)}{R(1-D) C_0 f_s} \end{aligned} \tag{11}$$

According to Equations (5) and (11), the inductor current and capacitor voltage ripples can be chosen for a specific amplitude. Selecting a ripple value is essential, especially in the first inductor current (Δi_{L_1}). The above is due to the requirements of renewable sources as in fuel and photovoltaic cells, which do not allow high ripples on the demanded currents. Large magnitude of current ripple in high frequency (>10 kHz) cause degradation of the catalyst of fuel cell plates. Moreover, fuel-cell currents should not have a high pulsating either negative profile.

3.3. Stress on Semiconductor Devices

The current stress values on active and passive switches are computed using the Equations (5)–(7). The current stress on S_1, D_{S1}, S_2, and D_{S2} are:

$$\begin{aligned} I_{S_1} = I_{D_{S1}} &= \frac{E[2L_1 f_s + DR(1-D)^4)]}{2R(1-D)^4 L_1 f_s} \\ I_{S_2} = I_{D_{S2}} &= \frac{E[2L_2 f_s + DR(1-D)^2)]}{2R(1-D)^3 L_2 f_s} \end{aligned} \tag{12}$$

Using the Figure 5, the voltage stress values on active and passive switches are computed as:

$$\begin{aligned} V_{S_1} &= -V_{D_{S1}} = V_0(1-D) \\ V_{S_2} &= -V_{D_{S2}} = V_0 \end{aligned} \tag{13}$$

The Equation (13) shows that the voltage stress on S_1 and D_{S1} increases when the duty cycle is reduced. The stress on S_2 and D_{S2} does not depend on the duty cycle.

3.4. Comparison with Other Quadratic Converters

A comparison of the proposed converter with other quadratic configurations described in the Introduction section is now given. The number of components and voltage gains of each converter are shown in Table 1. It can be noticed that, QBC-CS, QBC-SS and the proposed converter consist of only eight components, while QBC-CF includes nine components, and QBC-TR twelve components. Moreover, the voltage gains have the same relationship, with the exception of QBC-TR, which depends on the turns ratio of the transformer $n = N_2/N_1$.

Table 1. Comparison of the proposed converter with other quadratic configurations.

Components	Proposed Converter	QBC-CS Ref. [15]	QBC-SS Ref. [16]	QBC-CF Ref. [17]	QBC-TR Ref. [18]
Capacitors	2	2	2	4	2
Inductors	2	2	2	1	2
Diodes	2	2	3	5	4
Switches	2	2	1	1	1
Transformers	0	0	0	1	0
Gain (CCM)	$\frac{1}{(1-D)^2}$	$\frac{1}{(1-D)^2}$	$\frac{1}{(1-D)^2}$	$\frac{1+n}{(1-D)^2}$	$\frac{1}{(1-D)^2}$

4. Study of the Power Efficiency

Now, the analysis and expressions for computing the power losses of each element for the proposed converter are provided. According to [23], the corresponding circuit including the parasitics of all elements of the configuration is exhibited in Figure 8. The resistances R_{L_1}, R_{L_2}, R_{C_p}, R_{C_0}, R_{D1}, R_{D2}, R_{ON1}, and R_{ON2} are the equivalent series resistance (ESR) of L_1, L_2, C_p, C_0, D_{S_1}, D_{S_2}, S_1, and S_2, respectively. The voltages $V_{F_{DS1}}$ and $V_{F_{DS2}}$ are the forward voltage drops of D_{S_1}, and D_{S_2}. The gate voltage for S_1 and S_2 is V_g. In the interval t_{ON}, both active switches are conducting. The respective circuit for this interval is described in Figure 9. On the other hand, the respective circuit for the interval t_{OFF} (both active switches are not conducting) is depicted in Figure 10.

Using the volt-second balance principle over L_1 and L_2, two expressions for V_{C_p} are derived. From these two expressions, it is possible to find the relationship for computing the duty cycle (D_{loss}) and the losses of each component of the converter. To precisely quantify the losses associated with the parasitics of the converter, it is necessary to recalculate the duty cycle ($D_{loss} > D$). The resulting equation for D_{loss} has a fourth-order behavior:

$$D_{loss}^4 + aD_{loss}^3 + bD_{loss}^2 + cD_{loss} + d = 0 \tag{14}$$

where:

$$a = -\frac{4RV_0 + RV_{F_{DS1}} + 4RV_{F_{DS2}} + R_{D2} - R_{ON2}V_0}{RV_0 + RV_{F_{DS2}}}$$

$$b = \frac{6RV_0 + 3RV_{F_{DS1}} + 6RV_{F_{DS2}} + 3R_{D2}V_0 + R_{L_2}V_0 - 2R_{ON2}V_0 + ER}{RV_0 + RV_{F_{DS2}}}$$

$$c = \frac{(R_{ON1} + R_{ON2} - 2R_{L_2} - 4R - R_{D1} - 3R_{D2})V_0 + 2ER - 3RV_{F_{DS1}} - 4RV_{F_{DS2}}}{RV_0 + RV_{F_{DS2}}}$$

$$d = \frac{RV_0 + RV_{F_{DS1}} + R_{D1}V0 + RV_{F_{DS2}} + R_{D2}V_0 + R_{L_1}V_0 + R_{L_2}V_0 - ER}{RV_0 + RV_{F_{DS2}}}$$

to obtain the duty cycle D_{loss}, it is necessary to solve Equation (14) and choose the adequate root.

The expressions for the power losses of each element are shown in Table 2 (conduction and switching losses are included). The diode junction capacitances of D_{S1} and D_{S2} are C_{j_DS1} and C_{j_DS2}, respectively. The time intervals are $t_{rr1} = t_{r1} + t_{ON1}$, $t_{rr2} = t_{r2} + t_{ON2}$, $t_{ff1} = t_{f1} + t_{OFF1}$, and $t_{ff2} = t_{f2} + t_{OFF2}$, where t_{r1}, t_{ON1}, t_{r2}, t_{ON2}, t_{f1}, t_{OFF1}, t_{f2}, and t_{OFF2} are the rise time, turn ON delay time, fall time, and turn OFF delay time of S_1, and S_2, respectively. The input capacitances of S_1 and S_2 are C_{iss1} and C_{iss2}, respectively. The RMS current values can be obtained as:

$$I^2_{L1_{RMS}} = \frac{V_0^2}{R^2(1-D_{loss})^4}, \quad I^2_{L2_{RMS}} = \frac{V_0^2}{R^2(1-D_{loss})^2}$$

$$I^2_{S1_{RMS}} = I^2_{L1_{RMS}} D_{loss}, \quad I^2_{D1_{RMS}} = I^2_{L1_{RMS}}(1-D_{loss})$$

$$I^2_{S2_{RMS}} = I^2_{L2_{RMS}} D_{loss}, \quad I^2_{D2_{RMS}} = I^2_{L2_{RMS}}(1-D_{loss})$$

$$I^2_{CP_{RMS}} = \frac{V_0 D_{loss}(D^2_{loss} - D_{loss} + 1)}{R^2(1-D_{loss})^6}$$

$$I^2_{C0_{RMS}} = \frac{V_0^2 D_{loss}(2-D_{loss})^2}{R^2(1-D_{loss})^3}$$

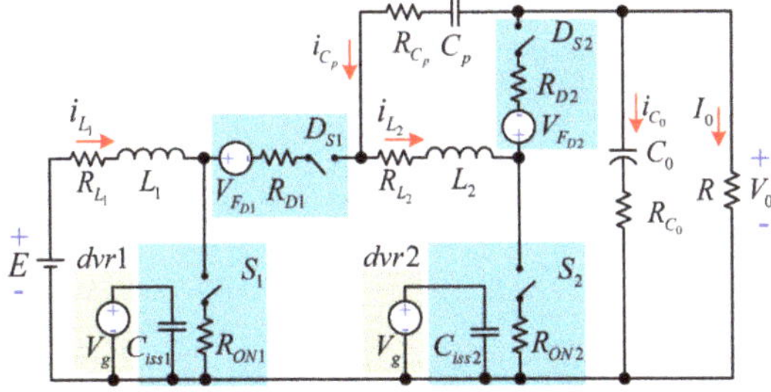

Figure 8. Equivalent circuit with the parasitics of elements.

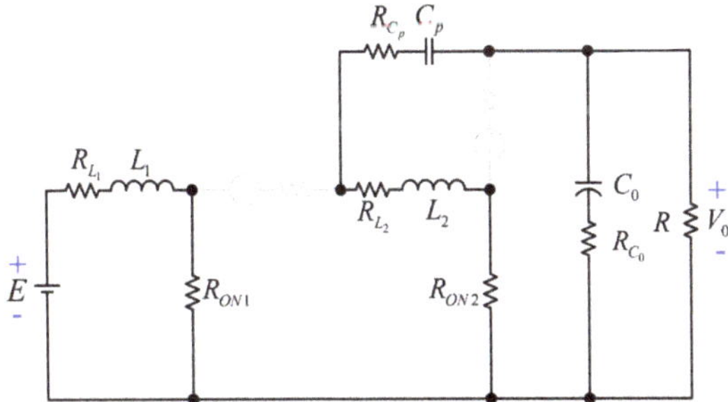
Figure 9. Equivalent circuit for the interval t_{ON}.

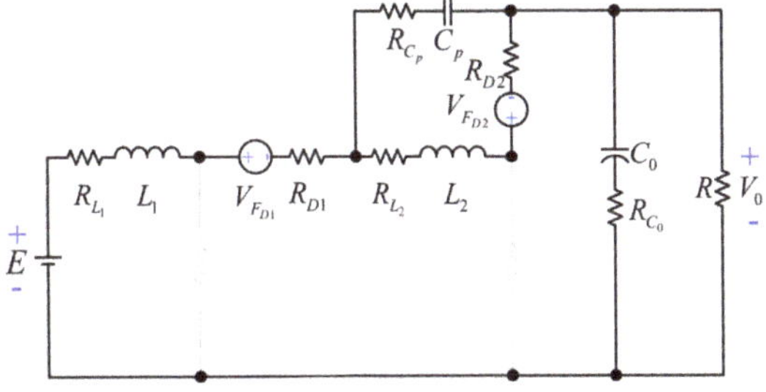
Figure 10. Equivalent circuit for the interval t_{OFF}.

Table 2. Power losses calculation.

Inductors	$P_{loss\_L1} = I^2_{L1_RMS} R_{L1}$ $P_{loss\_L2} = I^2_{L2_{RMS}} R_{L2}$
Capacitors	$P_{loss\_CP} = I^2_{CP_{RMS}} R_{C_P}$ $P_{loss\_C0} = I^2_{C0_{RMS}} R_{C_0}$
Diodes	$P_{loss\_D_1} = I^2_{D1_{RMS}} R_{D_1} + I_{L_1} V_{F_{D1}} + \frac{1}{2} C_{j\_D1} V_{S_1}$ $P_{loss\_D_2} = I^2_{D2_{RMS}} R_{D_2} + I_{L_2} V_{F_{D2}} + \frac{1}{2} C_{j\_D2} V_{S_2}$
Switches	$P_{loss\_S_1} = I^2_{S1_{RMS}} R_{ON1} + \frac{1}{2}(t_{rr2} + t_{ff2}) I_{L_2} V_{S_2}$ $P_{loss\_S_2} = I^2_{S2_{RMS}} R_{ON2} + \frac{1}{2}(t_{rr2} + t_{ff2}) I_{L_2} V_{S_2}$
Drivers	$P_{loss\_dvr1} = \frac{1}{2} C_{iss1} V_g^2 f_s$ $P_{loss\_dvr2} = \frac{1}{2} C_{iss2} V_g^2 f_s$

The total power loss is:

$$\begin{aligned} P_{T_{loss}} = & P_{loss\_L1} + P_{loss\_L2} + P_{loss\_CP} + P_{loss\_C0} + P_{loss\_D_1} \\ & + P_{loss\_D_2} + P_{loss\_S_1} + P_{loss\_S_2} + P_{loss\_dvr1} + P_{loss\_S_2}. \end{aligned} \tag{15}$$

The expression for computing the power efficiency is:

$$\eta = \frac{P_0}{P_0 + P_{T_{loss}}} \tag{16}$$

where P_0 is defined as $P_0 = V_0^2/R$.

5. Modeling of the Converter

The average modeling is commonly used to describe the behavior of power electronic circuits [24]. The PWM-switch model is a useful technique for describing the electrical behavior of DC-DC power converters.

Nonlinear Average Model of the Converter

The differential equations for each mode can be derived from Figures 5 and 6, respectively. The differential equations for mode I and mode II are :

$$\begin{aligned} \dot{i}_{L_1} &= \frac{1}{L_1}\Big(E\Big) \\ \dot{i}_{L_2} &= \frac{1}{L_2}\Big(-v_{C_p} + v_0\Big) \\ \dot{v}_{C_p} &= \frac{1}{v_{C_p}}\Big(i_{L_2}\Big) \\ \dot{v}_{C_0} &= \frac{1}{C_0}\Big(-i_{L_2} - \frac{v_0}{R}\Big) \end{aligned} \tag{17}$$

$$\begin{aligned} \dot{i}_{L_1} &= \frac{1}{L_1}\Big(v_{C_p} - v_0 + E\Big) \\ \dot{i}_{L_2} &= \frac{1}{L_2}\Big(-v_{C_p} + v_0\Big) \\ \dot{v}_{C_p} &= \frac{1}{v_{C_p}}\Big(i_{L_2}\Big) \\ \dot{v}_{C_0} &= \frac{1}{C_0}\Big(-i_{L_2} - \frac{v_0}{R}\Big) \end{aligned} \tag{18}$$

By using the switching function as a weighting factor, the average non-linear model can be derived as:

$$\begin{bmatrix} \dot{i}_{L_1} \\ \dot{i}_{L_2} \\ \dot{v}_{C_p} \\ \dot{v}_{C_0} \end{bmatrix} = \begin{bmatrix} 0 & 0 & \frac{(1-d)}{L_1} & -\frac{(1-d)}{L_1} \\ 0 & 0 & -\frac{1}{L_2} & \frac{d}{L_2} \\ -\frac{(1-d)}{C_p} & \frac{1}{C_p} & 0 & 0 \\ \frac{(1-d)}{C_0} & -\frac{d}{C_0} & 0 & -\frac{1}{RC_0} \end{bmatrix} \begin{bmatrix} i_{L_1} \\ i_{L_2} \\ v_{C_p} \\ v_{C_0} \end{bmatrix} + \begin{bmatrix} \frac{1}{L_1} \\ 0 \\ 0 \\ 0 \end{bmatrix} e \tag{19}$$

The above description can be generalized as:

$$\dot{x}(t) = A(d)x + B(d)e \tag{20}$$

where the state vector is $x(t) = [i_{L_1}, i_{L_2}, v_{C_p}, v_0]^\top \in \mathbb{R}^4_+$, the control signal $d \in (0,\ 1)$, and the input voltage $e \in \mathbb{R}$. The model described in (19) is bilinear, since the signal d is multiplying to all state variables.

Linealization of non-linear systems is a useful technique for analyzing and controlling complex high-order dynamical systems. This process describes the converter behaviour to small perturbations around an operation point, where the perturbations are applied to the input signals [25].

According to above, each state variable and the input signal are the sum of DC and AC components, which can be decomposed as:

$$\begin{aligned} i_{L_1} &= I_{L_1} + \tilde{i}_{L_1} \\ i_{L_2} &= I_{L_2} + \tilde{i}_{L_2} \\ v_{C_p} &= V_{C_p} + \tilde{v}_{C_1} \\ v_{C_0} &= V_{C_0} + \tilde{v}_{C_0} \\ d &= D + \tilde{d} \\ e &= E + \tilde{e} \end{aligned} \tag{21}$$

where $I_{L_1}, I_{L_2}, V_{C_p}, V_{C_0}, D, E$ represent the DC components, and $\tilde{i}_{L_1}, \tilde{i}_{L_2}, \tilde{v}_{C_p}, \tilde{v}_{C_0}, \tilde{d}, \tilde{e}$ are the AC components. In steady state, the AC components are equal to zero.

Linearizing around an equilibrium point

$$\mathcal{E} := (I_{L_1}, I_{L_2}, V_{C_p}, V_{C_0}) \in \mathbb{R}^4_+ \tag{22}$$

The linear representation of the systems shown in (19) can be rewritten as:

$$\dot{\tilde{x}} = A\tilde{x} + B\tilde{u} \tag{23}$$

where $\tilde{x} \in \mathbb{R}^4_+$ is the state vector, and $\tilde{u} = [\tilde{d}, \tilde{e}]^\top \in \mathbb{R}^2_+$ is the vector with inputs. A is a constant matrix in R^{4X4}, and B is a constant matrix in R^{4X2}. The average linear time-invariant model is:

$$
\begin{bmatrix} \dot{\tilde{i}}_{L_1} \\ \dot{\tilde{i}}_{L_2} \\ \dot{\tilde{v}}_{C_p} \\ \dot{\tilde{v}}_{C_0} \end{bmatrix} = \begin{bmatrix} 0 & 0 & \frac{(1-D)}{L_1} & -\frac{(1-D)}{L_1} \\ 0 & 0 & -\frac{1}{L_2} & \frac{D}{L_2} \\ -\frac{(1-D)}{C_p} & \frac{1}{C_p} & 0 & 0 \\ \frac{(1-D)}{C_0} & -\frac{d}{C_0} & 0 & -\frac{1}{RC_0} \end{bmatrix} \begin{bmatrix} \tilde{i}_{L_1} \\ \tilde{i}_{L_2} \\ \tilde{v}_{C_p} \\ \tilde{v}_{C_0} \end{bmatrix} + \begin{bmatrix} \frac{E}{L_1(1-D)} & \frac{1}{L_1} \\ \frac{E}{L_2(1-D)^2} & 0 \\ \frac{E}{RC_p(1-D)^4} & 0 \\ -\frac{E(2-D)}{RC_0(1-D)^4} & 0 \end{bmatrix} \begin{bmatrix} \tilde{d} \\ \tilde{e} \end{bmatrix} \quad (24)
$$

6. Converter Design

Now, the design of a 500 W power converter using the procedure shown in Section 3 is now described. The input voltage E is 30 V, the output voltage V_0 is set to 220 V, and the load R is 96.8 Ω. A switching frequency of $f_s = 100$ kHz is selected with an ideal duty cycle of $D = 0.63$. The selection criterion of the switching frequency was made based on the sizing of the passive elements, which increase in value with low switching frequencies. Due to the above, the power efficiency decreases when high inductance and capacitance values are selected. On the other hand, generally the value of the switching frequency used for medium and high power DC-DC converters ranges from 25 kHz to 100 kHz. The corresponding parameters for the designed converter are shown in Table 3.

Table 3. Parameters of the proposed converter.

$L_1 = 90\ \mu$H	$L_2 = 330\ \mu$H	$C_p = 20\ \mu$F	$C_0 = 20\ \mu$F
$I_{L_1} = 16.6$ A	$I_{L_2} = 6.1$ A	$V_{C_p} = 138$ V	$V_0 = 220$ V
$\Delta i_{L_1} = 2$ A	$\Delta i_{L_2} = 1.5$ A	$\Delta v_{C_p} = 2$ V	$\Delta v_0 = 2.6$ V
$r_I = 12.6\%$	$r_I = 25.2\%$	$r_V = 1.4\%$	$r_V = 1.2\%$

where $r_I = 100 \times (\Delta I_L / I_L)$ and $r_V = 100 \times (\Delta V_C / V_C)$.

The theoretical power efficiency at the nominal load of the proposed configuration is 95.4%, which was obtained using (14)–(16). The parameters of Table 2 were computed using the datasheet of the semiconductor devices used used to build the prototypes. The ESR values of the two inductors and two capacitors were measured using the meter model LCR-821 from GW Insteak. The theoretical duty cycle, including the power losses, is $D_{loss} = 0.635$.

Figure 11 shows a comparison between the ideal gain obtained with expression (3) and the gain given in expression (14), where all losses are included. It can be observed that both plots are similar until the duty cycle reaches 0.8; after that, the non-idealities produce a difference as shown in the plot.

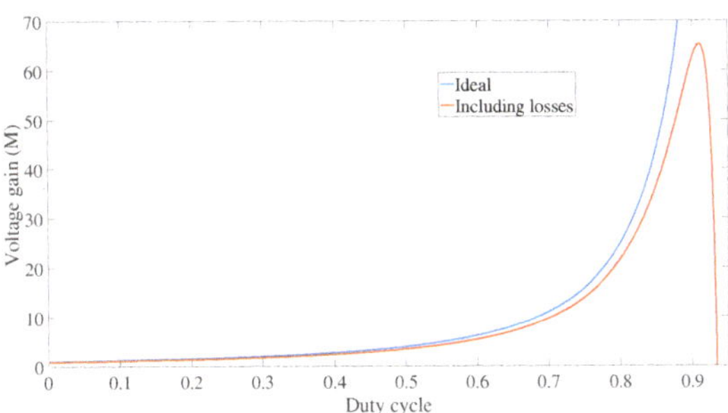

Figure 11. Comparison between the ideal gain and the gain including the power losses.

7. Experimental Verification

Experimental verification with a 500 W laboratory prototype is provided in this section. The prototype was designed according to the parameters given in Table 3, and it is shown in Figure 12. The prototype for QBC-SC and QBC-SS converter is shown in Figure 13. Both converters were built in the same prototype, only one switch was replaced by a diode and vice versa.

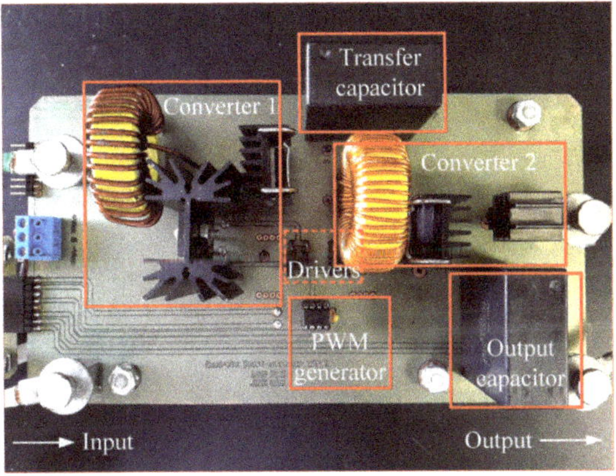

Figure 12. Laboratory prototype of the proposed converter.

The experimental set-up of the converter is described in Figure 14. The relationship of the voltage gain (V_0/E) is 220 V/30 V, and the peak value of the ramp signal is 5 V. Additionally, QBC-SC (Figure 15) and QBC-SS (Figure 16) prototypes were built to compare the experimental power efficiencies of the three configurations.

For a fair comparison, all converters were built using the same components. For the QBC-SS, the active switch S_1 was replaced by the Schottky diode DSA90C200HB.

The parameters of the QBC-SC and the QBC-SS are shown in Table 4. The parameters were obtained by using the expressions developed in [26]. The theoretical and experimental plots for the comparison of efficiencies are shown in Figure 17. The teoretical plot was obtained using the Equations (14)–(16), while the experimental plots of the three converter were obtained for a voltage gain of 220 V/30 V, and varying the load from 484 Ω (100 W) to 96.8 Ω (500 W), with steps of 100 W. The voltage and current values were obtained using the osciloscopie model MDO3034 and the current probe model TCP303 from Tektronix. As can be seen, the theoretical and experimental plots for the proposed converter are similar, only with small variations. The experimental efficiency of the proposed converter ranges from 97.8% (20% of nominal load) to 95.1% (nominal load). It can be noticed that the proposed configuration offers an improvement in the power efficiency due to the non-series power transfer. One part of the energy processed by the first converter flows directly to the output through the transfer capacitor without being reprocessed by the second converter. A pie chart of breakdown losses for nominal load is shown in Figure 18. It is clear that the biggest losses are in the diodes and inductors, while the switch losses are not significant due to the low ESR value of both switches.

Figure 13. Laboratory prototype of the QBC-SC and QBC-SS converters.

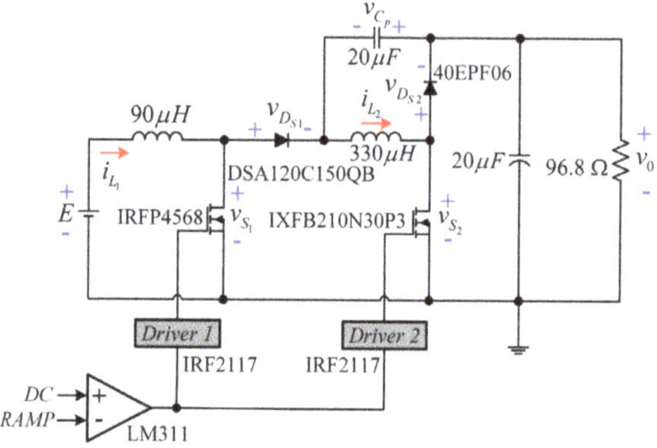

Figure 14. Experimental schematic for the proposed converter.

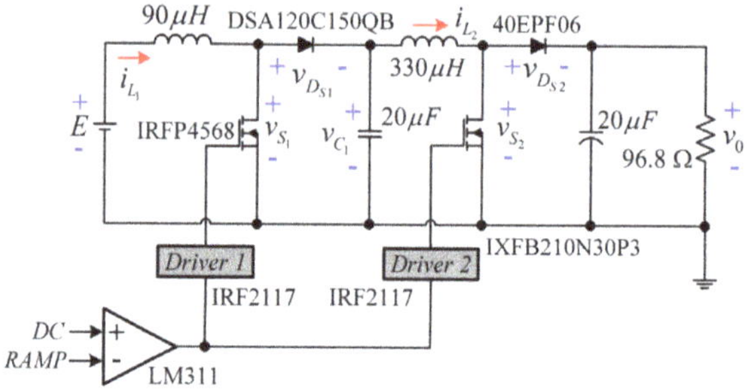

Figure 15. Circuit of the QBC-SC prototype for the experimental efficiency comparison with the proposed converter.

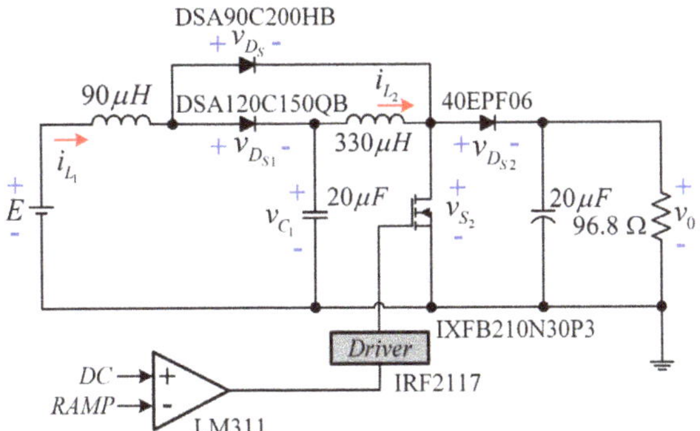

Figure 16. Circuit of the QBC-SS prototype for the experimental efficiency comparison with the proposed converter.

Table 4. Parameters of the QBC-SC and QBC-SS converters.

$L_1 = 90$ μH	$L_2 = 330$ μH	$C_1 = 20$ μF	$C_0 = 20$ μF
$I_{L_1} = 16.6$ A	$I_{L_2} = 6.1$ A	$V_{C_p} = 81$ V	$V_0 = 220$ V
$\Delta i_{L_1} = 2$ A	$\Delta i_{L_2} = 1.5$ A	$\Delta v_{C_p} = 2$ V	$\Delta v_0 = 0.7$ V
$r_I = 12.6\%$	$r_I = 25.2\%$	$r_V = 2.4\%$	$r_V = 0.3\%$

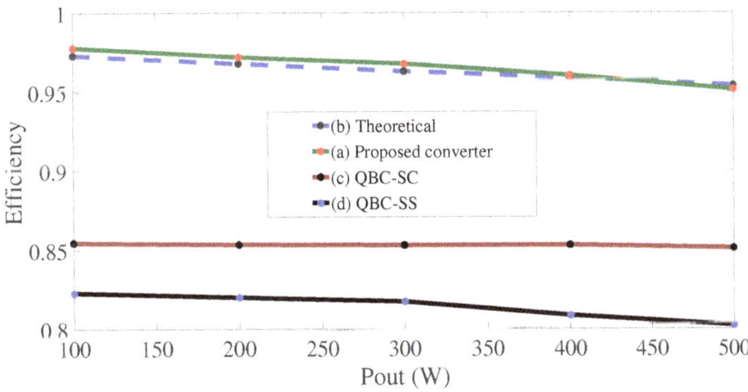

Figure 17. Comparison of power efficiencies. (From top to bottom) (**a**) Proposed converter, (**b**) theoretical efficiency, (**c**) QBC-SC, and (**d**) QBC-SS.

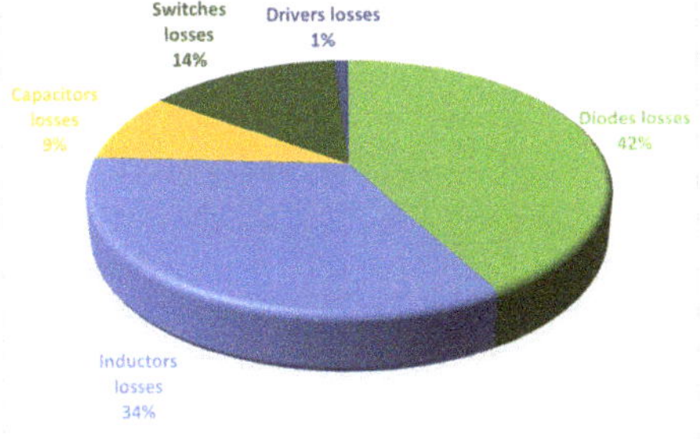

Figure 18. Pie chart of loss breakdown at nominal load.

The inductor and output currents of the prototype are exhibited in Figure 19. The average value of i_{L_1} is 17.5 A, the average value of i_{L_2} is 6.3 A, and for I_0 is 2.7 A. In Figure 20, the input and capacitors voltages are exhibited The value of E is 30 V, the average transfer capacitor voltage V_{C_p} is 139 V, while the average output capacitor voltage V_0 is 220 V.

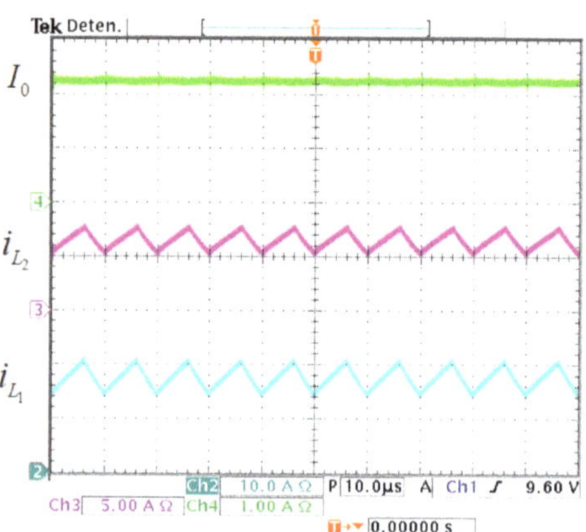

Figure 19. Current waveforms of the prototype. (From top to bottom) Load current I_0 (1 A/div), second inductor current i_{L_2} (5 A/div), and first inductor current i_{L_1} (10 A/div) (10 µs/div).

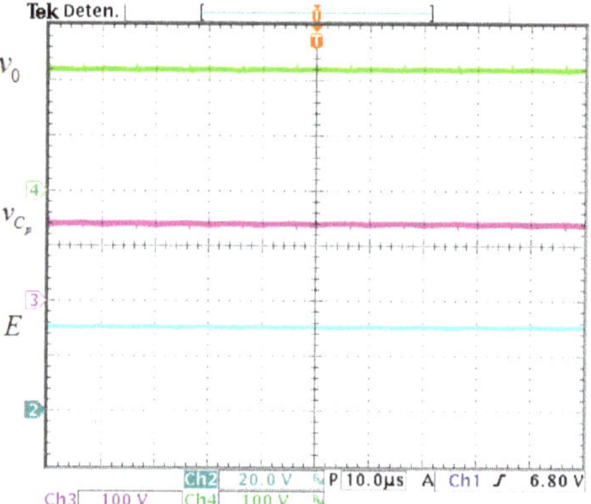

Figure 20. Voltage waveforms of the prototype. (From top to bottom) Output capacitor voltage v_0 (100 V/div), transfer capacitor voltage v_{C_p} (100 V/div), and input voltage E (20 V/div) (10 µs/div).

The voltage waveforms in the active switch S_1 and the diode D_{S1} are exhibited in Figure 21. The voltage stress on S_1 and D_{S1} is 98 V. The voltage waveforms in the active switch S_2 and the diode D_{S2} are exhibited in Figure 22, where the voltage stress on S_2 and D_{S2} is 222 V.

Voltage and current ripples of the prototype are exhibited in Figure 23. The value of Δv_0 is 2.8 V (1.30%), the value of Δv_{C_p} is 2.2 V (1.60%), the value of Δi_{L_2} is 1.7 A (27%), and the value of Δi_{L_1} is 3 A (17.1%).

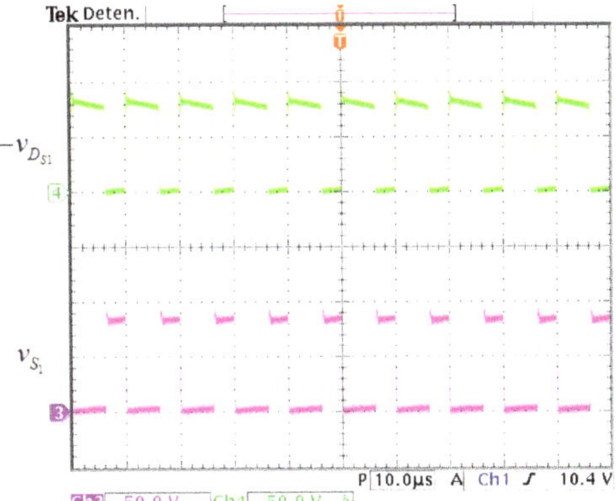

Figure 21. Voltage waveforms in the input of converter. (From top to bottom) Voltage waveforms in D_{S1} (50 V/div) and S_1 (50 V/div) (10 µs/div).

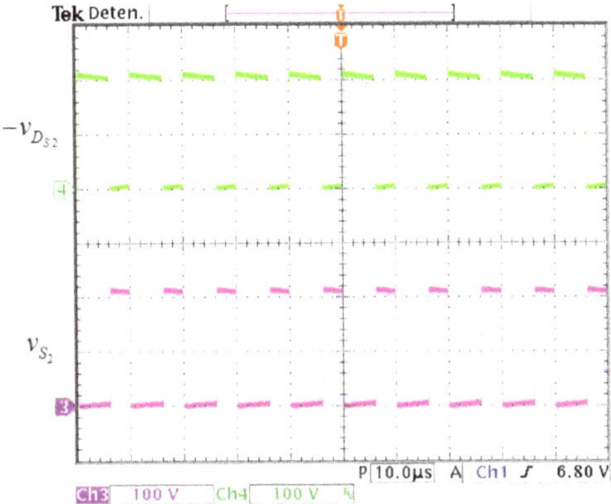

Figure 22. Voltage waveforms in the output of converter. (From top to bottom) Voltage waveforms in D_{S2} (100 V/div) and S_2 (100 V/div) (10 µs/div).

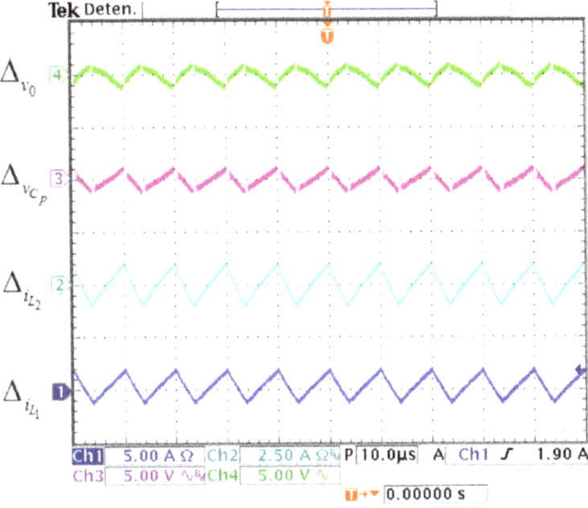

Figure 23. Voltage and current ripples of the prototype. (From top to bottom) Output capacitor ripple Δv_0 (5 V/div), transfer capacitor ripple Δv_{C_p} (5 V/div), second inductor ripple Δi_{L_2} (2.5 A/div) and first inductor ripple Δi_{L_1} (5 A/div) (10 µs/div).

8. Concluding Remarks

A step-up configuration with a quadratic conversion ratio and a non-series power transfer is proposed in this work. It consists of two conventional boost converters, with two active switches operating with the same duty cycle. This configuration uses a transfer capacitor to avoid reprocessing power between both converters. The proposed configuration can be used for processing energy from renewable generation systems, due to the wide voltage conversion ratio, non-pulsating input current, and higher efficiency compared to other quadratic configurations. The steady-state values and power efficiency analysis are derived for design purposes. A power converter with 220 V output voltage @ 500 W is designed using the procedure given in this work. Experimental validation with a laboratory prototype is exhibited to prove the theoretical results given within. By comparing the experimental power efficiency with the QBC-SC, and the QBC-SS shows that the proposed configuration offers an improvement in the power efficiency. The average linear time-invariant model is derived. This model is a useful tool for control purposes due to the transfer functions of all state variables can be obtained. An appropriate control technique for controlling the proposed converter is the average current-mode control, which consists of two feedback loops (inner and outer). The inner loop uses the inductor current, while the outer loop uses the output voltage for regulation purposes. This configuration can be constructed using a single active switch; however, there is a reduction in the power efficiency.

Author Contributions: Conceptualization, L.H.D.-S. and J.L.-R.; methodology, L.H.D.-S.; software, L.H.D.-S.; validation, L.H.D.-S. and J.L.-R.; formal analysis, J.L.-R.; investigation, L.H.D.-S. and J.L.-R.; writing—original draft preparation, L.H.D.-S.; writing—review and editing, J.L.-R.; visualization,L.H.D.-S.; supervision, J.L.-R.; project administration, J.L.-R.; funding acquisition, J.L.-R. All authors have read and agreed to the published version of the manuscript.

Funding: This research received funding from the Consejo Nacional de Ciencia y Tecnología (CONACyT).

Conflicts of Interest: The authors declare no conflict of interest.

References

1. Singh, K.A.; Chaudhary, K. Design and development of a new three-phase AC-DC single-stage wind energy conversion system. *IET Power Electron.* **2021**, *14*, 302–312. [CrossRef]
2. Saganeiti, L.; Pilogallo, A.; Faruolo, G.; Scorza, F.; Murgante, B. Territorial Fragmentation and Renewable Energy Source Plants: Which Relationship? *Sustainability* **2020**, *12*, 1828. [CrossRef]
3. Mohan Appikonda, M.; Kaliaperumal, D. Signal flow graph model and control of dual input boost converter with voltage multiplier cell. *Int. J. Electron. Commun.* **2020**, *125*, 153345. [CrossRef]
4. Luta, D.N.; Raji, A.K. Fuzzy Rule-Based and Particle Swarm Optimisation MPPT Techniques for a Fuel Cell Stack. *Energies* **2019**, *12*, 936. [CrossRef]
5. Tseng, K.; Cheng, C.; Chen, C. High Step-Up Interleaved Boost Converter for Distributed Generation Using Renewable and Alternative Power Sources. *IEEE J. Emerg. Sel. Top. Power Electron.* **2017**, *5*, 713–722. [CrossRef]
6. Hoang, K.D.; Lee, H. Accurate Power Sharing with Balanced Battery State of Charge in Distributed DC Microgrid. *IEEE Trans. Ind. Electron.* **2019**, *66*, 1883–1893. [CrossRef]
7. Kong, X.; Khambadkone, A.M. Analysis and Implementation of a High Efficiency, Interleaved Current-Fed Full Bridge Converter for Fuel Cell System. *IEEE Trans. Power Electron.* **2007**, *22*, 543–550. [CrossRef]
8. Hasanpour, S.; Baghramian, A.; Mojallali, H. A Modified SEPIC-Based High Step-Up DC–DC Converter with Quasi-Resonant Operation for Renewable Energy Applications. *IEEE Trans. Ind. Electron.* **2019**, *66*, 3539–3549. [CrossRef]
9. Parada Salado, J.G.; Herrera Ramirez, C.A.; Soriano Sanchez, A.G.; Rodriguez Licea, M.A. Nonlinear Stabilization Controller for the Boost Converter with a Constant Power Load in Both Continuous and Discontinuous Conduction Modes. *Micromachines* **2021**, *12*, 512. [CrossRef]
10. Diaz-Saldierna, L.H.; Leyva-Ramos, J.; Langarica-Cordoba, D.; Morales-Saldaña, J.A. Control strategy of switching regulators for fuel-cell power applications. *IET Renew. Power Gener.* **2017**, *11*, 799–805. [CrossRef]
11. Hosseini, S.M.; Ghazi, R.; Nikbahar, A.; Eydi, M. A new enhanced-boost switched-capacitor quasi Z-source network. *IET Power Electron.* **2021**, *12*, 412–421. [CrossRef]
12. Shreelakshmi, M.P.; Moumita, D.; Vivek, A. Design and Development of a Novel High Voltage Gain, High-Efficiency Bidirectional DC–DC Converter for Storage Interface. *IEEE Trans. Ind. Electron.* **2019**, *66*, 4490–4501.
13. Valdez-Resendiz, J.E.; Rosas-Caro, J.C.; Mayo-Maldonado, J.C.; Llamas-Terres, A. Quadratic boost converter based on stackable switching stages. *IET Power Electron.* **2018**, *11*, 1373–1381. [CrossRef]

14. Langarica-Córdoba, D.; Diaz-Saldierna, L.H.; Leyva-Ramos, J. Fuel-cell energy processing using a quadratic boost converter for high conversion ratios. In Proceedings of the 2015 IEEE 6th International Symposium on Power Electronics for Distributed Generation Systems (PEDG), Aachen, Germany, 22–25 June 2015.
15. Morales-Saldana, J.A.; Carbajal-Gutierrez, E.E.; Leyva-Ramos, J. Modeling of switch-mode dc-dc cascade converters. *IEEE Trans. Aerosp. Electron. Syst.* **2002**, *38*, 295–299. [CrossRef]
16. Luo, F.L.; Ye, H. Positive output cascade boost converters. *IEE Proc. Electr. Power Appl.* **2004**, *151*, 590–606. [CrossRef]
17. Wang, Y.; Qiu, Y; Bian, Q.; Guan, Y.; Xu, D. A Single Switch Quadratic Boost High Step Up DC–DC Converter. *IEEE Trans. Ind. Electron.* **2019**, *66*, 4387–4397. [CrossRef]
18. Marsala, G.; Pucci, M.; Rabbeni, R.; Vitale, G. Analysis and design of a dc-dc converter with high boosting and reduced current ripple for PEM FC. In Proceedings of the 2011 IEEE Vehicle Power and Propulsion Conference, Chicago, IL, USA, 6–9 September 2011.
19. Gu, Y.; Chen, Y.; Zhang, B.; Qiu, D.; Xie, F. High Step-Up DC–DC Converter With Active Switched LC-Network for Photovoltaic Systems. *IEEE Trans. Energy Convers.* **2019**, *34*, 321–329. [CrossRef]
20. Pires, B.F.; Cordeiro, A.; Foito, D.; Silva, J.F. High Step-Up DC–DC Converter for Fuel Cell Vehicles Based on Merged Quadratic Boost–Ćuk. *IEEE Trans. Veh. Technol.* **2019**, *68*, 7521–7530. [CrossRef]
21. Tseng, K.-C.; Huang, H.-S.; Cheng, C.-A. Integrated boost-forward-flyback converter with high step-up for green energy power-conversion applications. *IET Power Electron.* **2021**, *14*, 27–37. [CrossRef]
22. Babaei, E.; Varesi, K.; Vosoughi, N. Calculation of critical inductance in n-input buck dc–dc converter. *IET Power Electron.* **2016**, *19*, 2434–2444. [CrossRef]
23. López-Santos, O.; Martínez-Salamero, L.; García, G.; Valderrama-Blavi, H.; Mercuri, D.O. Efficiency analysis of a sliding-mode controlled quadratic boost converter. *IET Power Electron.* **2013**, *6*, 364–373. [CrossRef]
24. Van Dijk, E.; Spruijt, J.N.; O'Sullivan, D.M.; Klaassens, J.B. PWM-switch modeling of DC-DC converters. *IEEE Trans. Power Electron.* **1995**, *10*, 659–665. [CrossRef]
25. Carbajal-Gutierrez, E.E.; Morales-Saldana, J.A.; Leyva-Ramos, J. Modeling of a single-switch quadratic buck converter. *IEEE Trans. Aerosp. Electron. Syst.* **2005**, *41*, 1450–1456. [CrossRef]
26. Ortiz-Lopez, M.G.; Leyva-Ramos, J.; Carbajal-Gutierrez, E.E.; Morales-Saldana, J.A. Modelling and analysis of switch-mode cascade converters with a single active switch. *IET Power Electron.* **2008**, *1*, 478–487. [CrossRef]

Article

New Equivalent Electrical Model of a Fuel Cell and Comparative Study of Several Existing Models with Experimental Data from the PEMFC Nexa 1200 W

Fatima Zahra Belhaj [1,*], Hassan El Fadil [1,*], Zakariae El Idrissi [1,*], Abdessamad Intidam [1], Mohamed Koundi [1] and Fouad Giri [2]

1 ISA Laboratory ENSA, Ibn Tofail University, Kénitra 14000, Morocco; intidam.abdessamad@gmail.com (A.I.); mohamed.koundi@uit.ac.ma (M.K.)
2 LAC Lab, University of Caen Normandie (UNICAEN), 14000 Caen, France; fouad.giri@unicaen.fr
* Correspondence: fatimazahra.belhaj@uit.ac.ma (F.Z.B.); hassan.elfadil@uit.ac.ma (H.E.F.); zakariae.elidrissi@uit.ac.ma (Z.E.I.)

Abstract: The present work investigates different models of polymer electrolyte membrane fuel cell. More specifically, three models are studied: a nonlinear state-space model, a generic dynamic model integrated into MATLAB/Simulink, and an equivalent RC electrical circuit. A new equivalent electrical RL model is proposed, and the methodology for determining its parameters is also given. An experimental test bench, based on a 1200-W commercial PEMFC, is built to compare the static and dynamic behaviour of the existing models and the proposed RL model with the experimental data. The comparative analysis highlights the advantages and drawbacks of each of these models. The major advantages of the proposed RL model lie in both its simplicity and its ability to provide a similar transitory behaviour compared to the commercially manufactured PEMFC employed in this research.

Keywords: fuel cell; new equivalent electrical model of fuel cell; PEMFC NEXA 1200; comparative study; experimental validation

Citation: Belhaj, F.Z.; El Fadil, H.; El Idrissi, Z.; Intidam, A.; Koundi, M.; Giri, F. New Equivalent Electrical Model of a Fuel Cell and Comparative Study of Several Existing Models with Experimental Data from the PEMFC Nexa 1200 W. *Micromachines* **2021**, *12*, 1047. https://doi.org/10.3390/mi12091047

Academic Editor: Francisco J. Perez-Pinal

Received: 15 May 2021
Accepted: 21 July 2021
Published: 30 August 2021

1. Introduction

Today, protecting our planet is a major issue that involves several policies pertaining to transport and energy. Respecting the Kyoto Protocol and Paris Agreement on the reduction of greenhouse gas emissions, and keeping a global temperature rise this century below two degrees Celsius, requires drastic measures in favour of energy savings and the development of renewable energy. Indeed, given the increase in the global population, attention is being paid to the fact that energy supplies are necessarily limited, and that the risk of one day being faced with an energy shortage may become a reality. The transport sector today is seriously threatened because it is, on the one hand, extremely dependent on oil and, on the other hand, it is partly responsible for greenhouse gas emissions. In this respect, the use of fuel cells (FCs) in a traction system of electric vehicles is a hopeful solution, because it ultimately promises zero pollution [1]. In addition, the hydrogen sector has the advantage of being able to reduce the dependence of the transport sector on fossil fuels. Fuel cell electric vehicles (FCEVs) are classified as zero-emission vehicles (ZEVs) because they only release water. Therefore, hydrogen fuel cells have been targeted for their potential to contribute to decarbonization in the transportation sector [2,3]. The first FCEVs, which use polymer electrolyte membrane fuel cells (PEMFCs), were introduced in 2013 [4,5]. The advantages of these vehicles relative to current battery electric vehicles (BEVs) include higher driving ranges (over 500 km) and faster refuelling (3–5 min to refill the hydrogen storage tank).Therefore, the PEMFC is the essential choice for developing distributed generation power systems, hybrid electric vehicles, and other emerging fuel cell applications. It is therefore important for electrical and automation engineers and researchers to understand the dynamic behaviour of the PEM fuel cell for its successful use in different applications. In the literature, many research works are attempts to develop models of the PEM fuel cell.

In the beginning, electrochemistry-based models of the PEM fuel cell were introduced [6,7]. Then, dynamic models started to emerge [8–12]. In [13] a dynamic model of PEMFC, using the exact linearization approach, was presented. Although these models provide a certain understanding of the PEMFC, they remain insufficient to design adequate controllers for PEM fuel cell systems. It is for this reason that state-space models were introduced in some works [14,15]. However, state-space models are further complicated because they are highly nonlinear, and involve a large number of state variables and parameters. Then, some works have attempted to develop equivalent electric models, because they are still simple and easy to understand and to implement [16–25].

The present work investigates different classes of models proposed in the literature. More specifically, three models are presented: a nonlinear state-space model, a generic dynamic model integrated into MATLAB/Simulink, and an equivalent electric RC circuit. Using the dynamic behaviour of a 1200-W commercialized PEMFC, a new equivalent electric model is proposed. A comparative study between the proposed model and the previous models is conducted, showing the pros and cons of each model.

The rest of the paper is organized as follows: In Section 2, the electrochemical principle of a PEMFC is presented. Section 2.1 is devoted to the presentation of the nonlinear state-space model. A generic dynamical model integrated into MATLAB/Simulink is illustrated in Section 2.2. In Section 2.3, an equivalent electrical RC circuit is presented. The proposed equivalent electrical RL circuit is shown in Section 2.4. Section 3. is devoted to the experimental behaviour of a 1200-W commercialized PEMFC. The comparative study between different models is conducted in Section 4. A conclusion and a reference list end the paper.

2. Theoretical Principle

A fuel cell (FC) is an electrochemical energy generator used to directly transform the chemical energy of a fuel (hydrogen, hydrocarbons, alcohols, etc.) into electrical energy. Figure 1 shows a schematic of a hydrogen PEMFC. The FC core consists of three elements, including two electrodes—an oxidizing anode (electron emitter), and a reducing cathode (electron collector)—separated by an electrolyte. The FC is supplied by an injection of hydrogen at the anode and air at the cathode. Continuous electrical energy is then available across the FC.

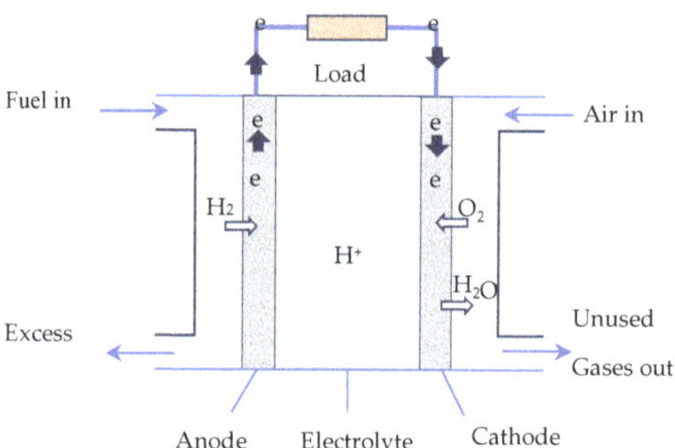

Figure 1. Schematic diagram of a PEMFC.

In the core of a hydrogen fuel cell of the PEMFC type, two electrochemical reactions occur successively [26,27]:

At the anode: catalytic oxidation of the hydrogen, which dissociates from its electrons:

$$H_2 \rightarrow 2H^+ + 2e- \quad (1)$$

At the cathode: catalytic reduction of the oxygen, which captures the H^+ ions that have passed through the electrolyte membrane, as well as the electrons arriving from the external circuit. The reaction produces heat and water:

$$\frac{1}{2}O_2 + 2H^+ + 2e- \rightarrow H_2O + Q(heat) \quad (2)$$

To evaluate the PEMFC's performance, and for control purposes, several mathematical models of PEMFC have been developed in the literature. They can be classified into three main categories:

Static models representing the input–output behaviour of the FC—in particular, the nonlinear current–voltage characteristic (see Figure 2). The output voltage of the fuel cell is dependent on the thermodynamically predicted fuel cell voltage output, and three major losses: activation losses (due to the electrochemical reaction), ohmic losses (due to the ionic electronic condition), and concentration losses (due to mass transport).

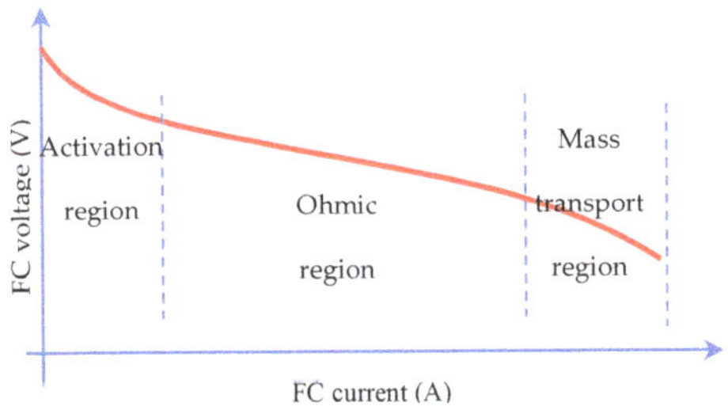

Figure 2. Nonlinear *i–v* characteristic of the fuel cell.

Nonlinear state-space representing the internal behaviour of the fuel cell and equivalent electrical circuits.

In this paper a comparison between four models is investigated: a nonlinear state-space model, a generic model from MATLAB Toolbox, an equivalent electrical circuit RC, and a new proposed equivalent electrical circuit RL. The resulting models will be compared to the experimental results using a 1.2-kW fuel cell module from Ballard (the Nexa 1200).

2.1. Nonlinear State-Space Model of the PEMFC (NLM)

It has been suggested in many studies [12,15,16,28] that a nonlinear state-space model of the PEM fuel cell could be represented by Equations (3)–(8). In this model, the open-circuit output voltage of the PEM fuel cell, mass balance and thermodynamic energy balance, irreversible voltage losses, and the formation of the charged double layer in the PEM fuel cell is modelled:

$$\dot{x}_1 = -\theta_1 x_1 + \theta_1 u_{T_R} - L(x) I_{fc} \quad (3)$$

$$\dot{x}_2 = 2\theta_2 x_1 u_{P_A} - 2\theta_2 x_1 x_2 - \theta_3 x_1 I_{fc} \quad (4)$$

$$\dot{x}_3 = 2\theta_4 x_1 u_{P_C} - 2\theta_4 x_1 x_3 - \theta_5 x_1 I_{fc} \quad (5)$$

$$\dot{x}_4 = 2\theta_6 x_4 x_1 + 2\theta_5 x_1 I_{fc} \quad (6)$$

$$\dot{x}_5 = -\theta_5 x_5 + \theta_6 I_{fc} \quad (7)$$

where $x_1 = T$ is a stack temperature; $x_2 = P_{H_2}$ is the partial pressure of hydrogen; $x_3 = P_{O_2}$ is the partial pressure of oxygen; $x_4 = P_{H_2O}$ is the partial pressure of water; $x_5 = V_{fc}$ is the output voltage of the PEM fuel cell; I_{fc} is the stack current; u_{P_A} is the channel pressure of hydrogen; u_{P_C} is the channel pressure of oxygen; u_{T_R} is room temperature; and the involved parameters and functions are given as follows:

$$L(x) = n_s\left[\left(\frac{2E_0^{Cell}}{M_{fc}C_{fc}}\right) + \left(\frac{Rx_1}{FM_{fc}C_{fc}}\right)\ln\left(\frac{x_2x_3^{0.5}}{x_4}\right) - V^{Act} - V^{Conc} - V^{O}\right]$$

$$\theta_1 = \frac{h_s n_s A_s}{M_{fc}C_{fc}}$$

$$\theta_2 = \left[\frac{\left(R(m_{H_2O})_{in}^{a}x_1\right)}{\left(V_a(P_{H_2O})_{in}^{a}\right)}\right]$$

$$\theta_3 = \left[\frac{Rx_1}{4V_cF}\right]$$

$$\theta_4 = \left[\frac{\left(R(m_{H_2O})_{in}^{c}x_1\right)}{\left(V_c(P_{H_2O})_{in}^{c}\right)}\right]$$

$$\theta_5 = \left[\frac{Rx_1}{4V_cF}\right]$$

$$\theta_6 = \left[\frac{\left(R(m_{H_2O})_{in}^{c}\left(P_{H_2O}^{in} - x_4\right)\right)}{\left(V_c(P_{H_2O})_{in}^{c}\right)}\right]$$

$$\theta_7 = \frac{1}{C(R_{ac} + R_{co})}$$

$$\theta_8 = \frac{1}{C} \tag{8}$$

2.2. Generic Model from MATLAB Toolbox (GMM)

A fuel cell stack block integrated into MATLAB/Simulink implements a generic model parameterized to represent the most popular types of fuel cell stacks fed with hydrogen and air. The block represents two versions of the stack model: a simplified model, and a detailed model. The user can switch between the two models by selecting the level in the mask under the model detail level in the block dialogue box. In this paper, we consider the detailed model represented by Figure 3. The notations used are the same as those from [29,30].

The fuel cell voltage is related to the fuel cell current as follows:

$$V_{fc} = E - R_i \times I_{fc} \tag{9}$$

where R_i is the internal resistance, and the controlled voltage source E is described by the following equation:

$$E = E_{oc} - NAln\left(\frac{I_{fc}}{i_0}\right) \times \frac{1}{s\frac{T_d}{3} + 1} \tag{10}$$

where s is the Laplace operator and E_{oc} is an open circuit voltage (V); N is the number of cells; A is a Tafel slope (V); i_0 is the exchange current (A); and T_d is the response time (at 95% of the final value) (s). In Equation (10), the parameters (E_{oc}, N, i_0) are updated online based on the input pressures and flow rates, stack temperature, and gas compositions [29].

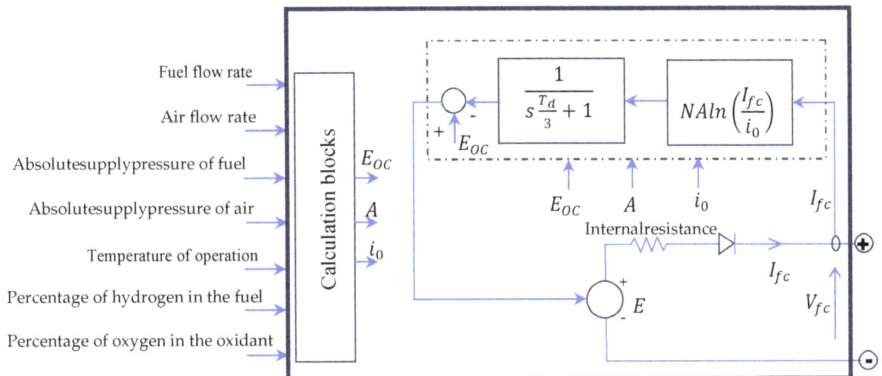

Figure 3. Circuit of the generic model of a fuel cell stack.

2.3. Equivalent Electrical RC Circuit (RCM)

Most dynamic models for PEMFCs are complex, and are not easy to use for control purposes. An equivalent electrical circuit could be used as a good alternative to model the fuel cell's dynamical behaviour as represented in Figure 4.

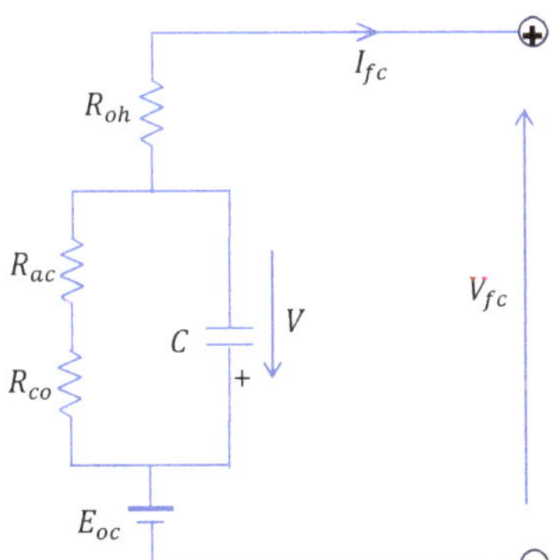

Figure 4. Equivalent electric RC circuit of a fuel cell stack.

From this figure, the fuel cell stack's static electrochemical behaviour can be represented by the following equations [15]:

$$V_{fc} = E_{oc} - V - R_{oh} I_{fc} \tag{11}$$

$$\frac{dV}{dt} = \frac{1}{C} I_{fc} - \frac{1}{\tau} V \tag{12}$$

where V represents the dynamical voltage across the equivalent capacitor; C is the equivalent electrical capacitance; R_{oh} is the ohmic resistance; and τ is the fuel cell electrical time constant, defined as follows:

$$\tau = (R_{ac} + R_{co})C \tag{13}$$

In Equation (11), E_{oc} is the open-circuit voltage, defined as follows:

$$E_{oc} = n_s(E_{Nernst} - V_{act}) \tag{14}$$

where n_s is the number of cells in series in the stack; E_{Nernst} is the thermodynamic potential of the cell, and represents its reversible voltage or Nernst potential; and V_{act} is the activation voltage drop. The quantities $E_{Nernst} V_{act}$ are given as follows [15,16,21,30]:

$$\begin{aligned} E_{Nernst} = \; & 1.229 - 8.5 \times 10^{-4} \times (T - 298.15) - 3.33 \times 10^{-3} I_{fc}(s) \tfrac{80s}{80s+1} \\ & +4.31 \times 10^{-5} \times T \times \left(\ln(P_{H2}) + \tfrac{1}{2}(P_{O2})\right) - 3.33 \times 10^{-3} I_{fc}(s) \tfrac{80s}{80s+1} \end{aligned} \tag{15}$$

$$V_{act} = -0.948 + T \times \left[2.86 \times 10^{-3} + 2 \times 10^{-4} \ln(A) + 4.3 \times 10^{-5} \ln(C_{H_2}) + 7.6 \times 10^{-5} \ln(C_{O_2})\right] \tag{16}$$

where P_{H_2} and P_{O_2} are the partial pressures (atm) of hydrogen and oxygen, respectively; T is the cell's absolute Kelvin temperature; and A is the cell's active area (cm^2). The terms C_{O_2} and C_{H_2} presented in Equation (16) are the oxygen concentration at the cathode membrane/gas interface (mol/cm^3), and the liquid phase concentration of hydrogen at the anode/gas interface (mol/cm^3), respectively. They can be obtained as follows [21]:

$$C_{O_2} = \frac{P_{O_2}}{5.08 \times 10^6 \exp\left(\frac{-498}{T}\right)} \tag{17}$$

$$C_{H_2} = \frac{P_{H_2}}{1.09 \times 10^6 \exp\left(\frac{77}{T}\right)} \tag{18}$$

It should be emphasized that the capacitor C in the RC model of Figure 4 affects the transient response of the PEMFC. Using the simulation RC model shown in Figure 4, the shape of the transient response of this model to the step load change is represented in Figure 5. It should be noted that when the load current steps up, the voltage drops simultaneously to some value due to the ohmic losses, and then it decays exponentially to its steady-state value due to the capacitor C. However, the experimental voltage of the PEMFC, as can be seen later (Section 5) and found in many works [12,31,32] has the form of Figure 5. This figure clearly illustrates a big difference between the experimental fuel cell voltage and the corresponding voltage given by the RC model. We conclude that the equivalent electrical RC circuit is not suitable for a PEM fuel cell. In the next section, we will present a new equivalent electrical model using an inductor instead of a capacitor, and we will show that the transient of the obtained model fits the experimental transient.

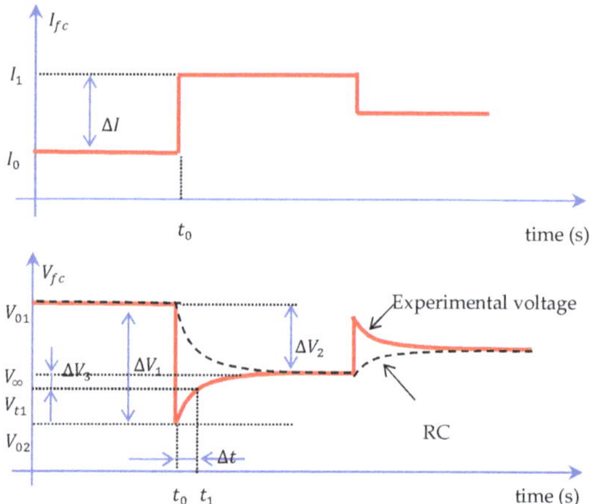

Figure 5. Comparison between the shape of the experimental FC voltage with the equivalent RC circuit.

2.4. Equivalent Electrical RL Circuit (RLM)

2.4.1. Proposed RL Circuit

As shown in the previous section, the equivalent electrical RC model is not appropriate for modelling the dynamics of a fuel cell, since its transient is different compared to the experimental data. In this paper, a new equivalent electrical circuit is proposed using an inductor and resistors, as shown in Figure 6. Note that the open-circuit voltage E_{oc} remains the same as in Equation (14).

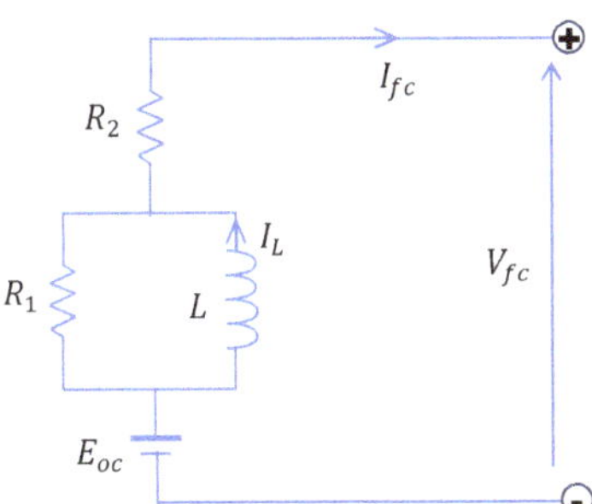

Figure 6. Equivalent electric RL circuit of a fuel cell stack.

2.4.2. Determination of Model Parameters

The electrical model presented in Figure 6 involves three parameters$L$, R_1, and R_2—which can be determined using the response of the fuel cell voltage to load current step change. Let us first introduce some useful electrical relationships based on the proposed circuit.

The fuel cell voltage is governed by the following equations:

$$V_{fc} = E_{oc} - R_2 I_{fc} - R_1 \left(I_{fc} - I_L \right) \tag{19}$$

$$L \frac{dI_L}{dt} = R_1 \left(I_{fc} - I_L \right) \tag{20}$$

$$L \frac{dI_L}{dt} + R_1 I_L = R_1 I_{fc} \tag{21}$$

For $t < t_0$ we suppose that $I_{fc} = I_0 = cte$, which corresponds to a constant fuel cell voltage $V_{fc} = V_0 = cte$ (see Figure 5). At the instant $t = t_0$, we apply a current step change from I_0 to I_1, and then the inductor current I_L will evolve, using Equation (21), according to the following equation:

$$I_L = I_1 + (I_0 - I_1) e^{-\frac{(t - t_0)}{\tau_L}} \tag{22}$$

where $\tau_L = \frac{L}{R_1}$ is a time constant of the RL circuit. Moreover, using Equations (19) and (22), the voltages V_{01}, V_{02}, and V_∞ represented in Figure 5 are given as follows:

$$\begin{gathered} V_{01} = E_{oc} - R_2 I_0 \\ V_{02} = E_{oc} - R_2 I_1 - R_1 (I_1 - I_0) \\ V_\infty = E_{oc} - R_2 I_1 \end{gathered} \tag{23}$$

It follows that the voltage variations ΔV_1 and ΔV_2, corresponding to the current variation $\Delta I = I_1 - I_0$, are given by:

$$\Delta V_1 = V_{01} - V_{02} = (R_1 + R_2)(I_1 - I_0) = (R_1 + R_2)\Delta I \tag{24}$$

$$\Delta V_2 = V_{01} - V_\infty = R_2 (I_1 - I_0) = R_2 \Delta I \tag{25}$$

Now, taking any instant $t_1 > t_0$ in the transient of the fuel cell voltage, using Equations (19), (22), and (23), one has:

$$V_{t1} = V_\infty - R_1(I_1 - I_0)e^{-\frac{(t_1 - t_0)}{\tau_L}} \tag{26}$$

which, in turn, taking into account Equations (24) and (25), gives:

$$\Delta V_3 = V_\infty - V_{t1} = R_1 \Delta I e^{-\frac{\Delta t}{\tau_L}} = (\Delta V_1 - \Delta V_2)e^{-\frac{\Delta t}{\tau_L}} \tag{27}$$

where $\Delta t = t_1 - t_0$.

Finally, the procedure for determining the RL model parameters can be summarized as follows:

1. From the plot of the fuel cell voltage corresponding to any current step change from I_0 to I_1, determine ΔI, V_{01}, V_{02}, and V_∞;
2. Take any instant $t_1 > t_0$ in the transient and determine its corresponding voltage V_{t1} and $\Delta t = t_1 - t_0$;
3. Calculate, using Equations (24), (25), and (27): ΔV_1, ΔV_2, and ΔV_3, respectively;
4. Calculate R_2 using Equation (25): $R_2 = \frac{\Delta V_2}{\Delta I}$;
5. Calculate R_1 using Equation (24): $R_1 = \frac{\Delta V_1}{\Delta I} - R_2$;
6. Calculate the inductance value L, using Equation (27), as follows: $L = \frac{\Delta t \times R_1}{\ln\left(\frac{\Delta V_1 - \Delta V_2}{\Delta V_3}\right)}$

3. Experimental Model (EXM)

In this section, we will determine the static and the dynamic behaviour of the Ballard Nexa 1200 fuel cell module, which has a rated power of 1.2 kW. To this end, an experimental bench was built, as shown in Figure 7; it consists—in addition to the fuel cell, with its monitoring software—of three metal hydride canisters from Heliocentris with storage capacities of 800 NL hydrogen, an H2 connection Kit 15 bar for connecting the metal canisters, a Nexa 1200 DC/DC converter, a power supply from BK Precision used for the fuel cell starter, Hall effect sensors to measure the voltage and current variables, a programmable DC electronic load from BK Precision and power resistors to make load changes, and a MicroLabBox-dSPACE with Control Desk software plugged into a personal computer for signal acquisition.

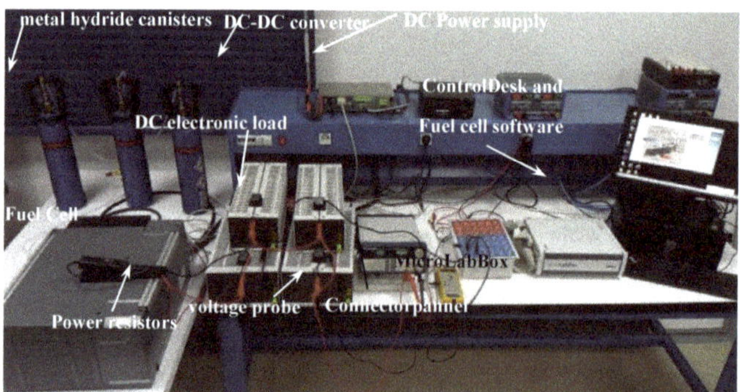

Figure 7. View of the experimental bench.

3.1. *Static Characteristics(i–v) of the Fuel Cell*

Several points representing the current and the voltage under static conditions of the fuel cell were determined, and are listed in Table 1. The obtained current–voltage characteristics are illustrated by Figure 8.

Table 1. Experimental data of the fuel cell current and voltage.

I_{fc}current (A)	0	1	2	3	4	5	6	7	8	9
V_{fc}voltage	33.58	32.7	31.93	31.26	30.68	30.18	29.75	29.38	29.06	28.78
I_{fc}current (A)	11	12	13	14	15	16	17	18	19	20
V_{fc}voltage	28.34	28.16	28	27.86	27.74	27.62	27.52	27.41	27.31	27.21
I_{fc}current (A)	22	23	24	25	26	27	28	29	30	31
V_{fc}voltage	27	26.9	26.79	26.68	26.56	26.44	26.32	26.19	26.07	25.94
I_{fc}current (A)	33	34	35	36	37	38	39	40	41	42
V_{fc}voltage	25.68	25.55	25.42	25.3	25.17	25.05	24.93	24.82	24.71	24.6
I_{fc}current (A)	44	45	46	47	48	49	50	51	52	53
V_{fc}voltage	24.39	24.29	24.19	24.09	23.98	23.88	23.75	23.62	23.47	23.31
I_{fc}current (A)	55	56	57	58	59	60				
V_{fc}voltage	22.89	22.64	22.34	22	21.6	21.13				

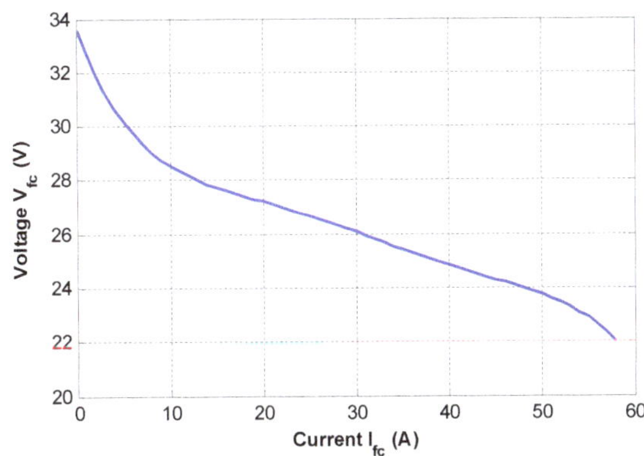

Figure 8. Obtained experimental (*i–v*) characteristics of the Nexa 1200 fuel cell module.

3.2. Dynamic Behaviour of the NEXA 1200Fuel Cell

Using a programmable DC electronic load and power resistors, a fuel cell current step change is operated from 13A to 33A after 65.1 s. The resulting fuel cell current and voltage are shown in Figures 8 and 9, respectively.

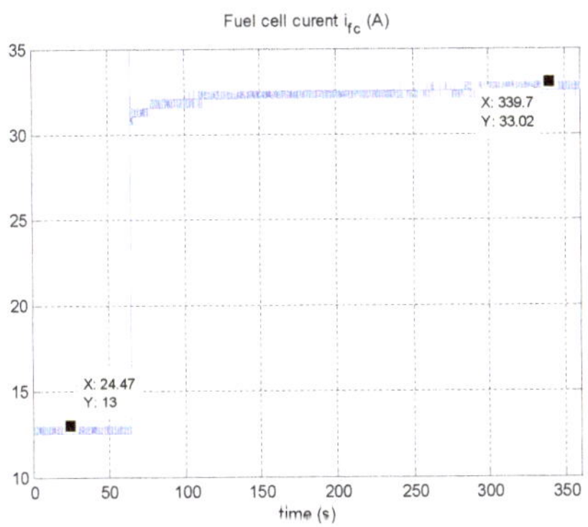

Figure 9. Fuel cell current step change.

3.3. RL Model Parameters

Using the procedure shown in Section 2.4, which describes the method for determining the parameters of the equivalent electric model RL, and using the experimental responses shown in Figures 8 and 9, we obtained the results listed in Table 2. Figure 10 shows a comparison between the experimental voltage of the fuel cell and that obtained from the RL model. One can show a good fit for the proposed RL model.

Table 2. RL circuit parameters.

ΔI (A)	ΔV_1(V)	ΔV_2(V)	ΔV_3(V)	Δt (s)	$R_1(m\Omega)$	R_2 $(m\Omega)$	L (H)
20	6.11	3.03	0.98	36.7	154	151.5	4.94

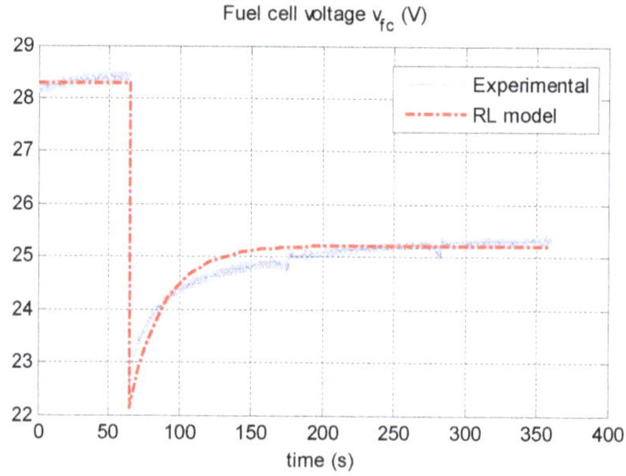

Figure 10. Comparison between the experimental and RL models.

4. Comparison between Different Models

In this section, we will evaluate the static and the dynamic behaviour of the studied models, and compare them to the experimental data of the used Nexa 1200 fuel cell module. All models are simulated using MATLAB/Simulink software. As a load of these models, we used a controlled current source whose variations were programmed similarly to those used for the experiment. Figure 11 illustrates the simulated models, while the experiments were carried out according to Figure 7. All of the parameters used for the simulation models are listed in Table 3.

Table 3. Simulation parameters.

Type of Model	Parameters
Nonlinear model (NLM)	Parameters used in [1]
Generic MATLAB mode (GMM)	Voltage at 0 A and 1 A [30] Nominal operating point (52 A, 24.23 V) Maximum operating point (100 A, 20 V)
RC model (RCM)	$E_{oc} = 28.32$ V; $R_{oh} = 2.89958$ $m\Omega$; $R_{ac} + R_{CO} = 155$ $m\Omega$; $C = 130$ F
Proposed RL model (RLM)	$E_{oc} = 28.32$ V; $R_1 = 157.70$ $m\Omega$; $R_2 = 156.17$ $m\Omega$; $L = 3.1078$ H

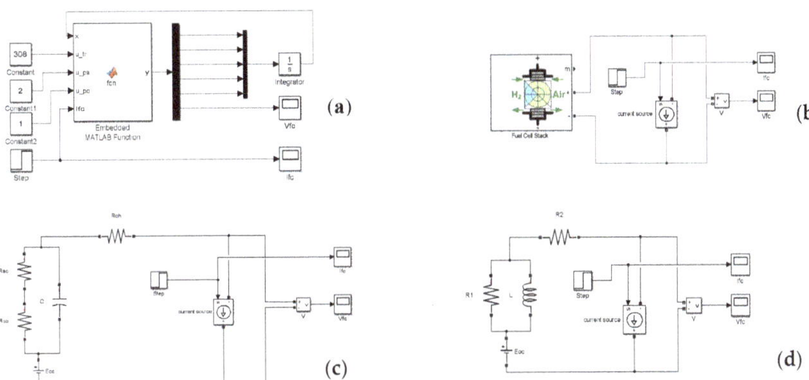

Figure 11. Simulated models in MATLAB/Simulink: (**a**) nonlinear model (NLM); (**b**) generic MATLAB model (GMM); (**c**) RC model (RCM); and (**d**) RL model (RLM).

4.1. Static Behaviour

The static behaviour of the simulated models is compared to the experimental results. Figure 12 illustrates the obtained current–voltage characteristics.

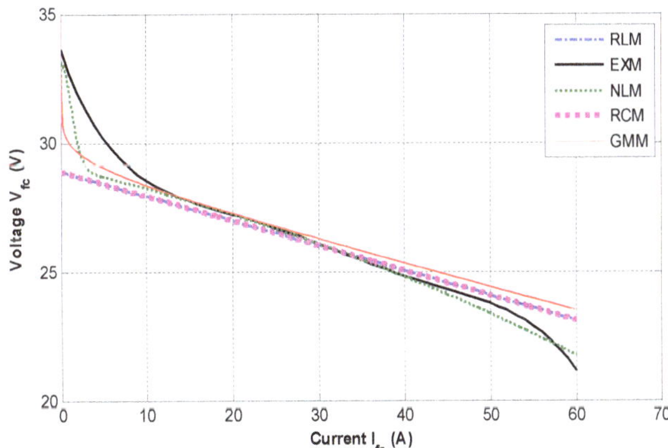

Figure 12. *i*–*v* characteristics of simulated models compared to the experiments.

To compare between different models, the following root-mean-square error (RMSE) criterion is selected, which is a frequently used measure of the differences between values predicted by a model and the values observed:

$$RMSE = \sqrt{\frac{\sum_{k=1}^{N}\left(V_{fcx}(k) - V_{fcm}(k)\right)^2}{N}} \tag{28}$$

where $(V_{fcx} - V_{fcm})$ is the error between the measured (experimental) and the modelled fuel cell voltage; and N is the total number of samples. The obtained results are summarized in Table 4. It is evident from the table that the nonlinear model is a better model; however, it requires a longer computational time, which is considered a drawback for real-time application and control purposes. The classic RC model and the proposed RL model have practically the same RMSE in static conditions. Nevertheless, we will later see the great supremacy of the proposed RL model in a dynamic regime.

Table 4. RMSE criteria for static (*i–v*) characteristic.

Type of Model	RMSE
Nonlinear model (NLM)	0.1852
Generic MATLAB model (GMM)	0.1961
RC model (RCM)	0.2382
Proposed RL model (RLM)	0.2319

4.2. Dynamic Behaviour

In this section, a comparison between the dynamic behaviour of four models and the experimental results in the presence of fuel cell current changes was studied. Figure 13 illustrates the dynamic behaviour of each model. As is clearly shown, the proposed RL equivalent electrical model presents the best dynamic behaviour compared to the conventional RC model used in the literature. The main advantages of this model lie in its simplicity and stability to produce the same behaviour as the fuel cell.

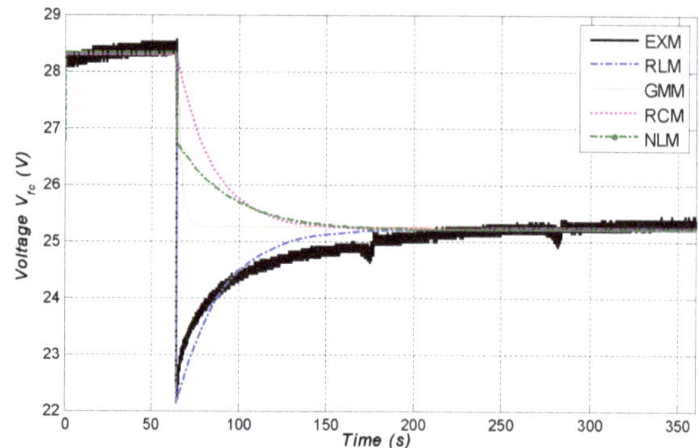

Figure 13. Dynamic behaviour of simulated models compared to the experiments.

Unlike the proposed RL equivalent electrical model, the mathematical model, the conventional generic MATLAB model and the RL model all provide the same static behaviour of the fuel cell, but do not produce a smooth transitory fuel cell behaviour. In addition, the disadvantages of the mathematical model are that it is complicated and requires a longer computational time for calculation. Consequently, this is a handicap for real-time control and implementation.

Therefore, the present modelling approach permits the researchers and students who do not own a real fuel cell to approximate its static and dynamic behaviour in a simple, fast, and affordable way.

5. Conclusions

The fuel cell is an electric component that is used more and more in distributed generation power fields, but also in hybrid electric vehicles. Its modelling is an important step not only to understand its dynamic behaviour, but also to develop advanced controllers for systems based on the fuel cell. Several models are proposed in the literature; this work re-investigates some of these most common models. The study shows that the equivalent electric RC circuit is not appropriate, since it presents a different behaviour from the experimental results. Therefore, the paper presents an analysis of a new equivalent electric RL circuit. A comparative study of different models with experimental data from a Nexa 1200 PEMFC reveals the advantages and disadvantages of each model. It turned out that the mathematical nonlinear model is better than the other investigated models, but requires a longer computational time, which is considered a drawback for real-time application and control purposes. The classic RC model and the proposed RL model have practically the same RMSE in static conditions. Nevertheless, the proposed equivalent electric RL model presents the advantages of providing the best transient behaviour compared to the classic RC model; indeed, contrary to the investigated models, the advantage of the proposed equivalent RL model lies essentially in its simplicity, and its ability to produce a transient behaviour similar to that of the commercial fuel cell used in this work.

Author Contributions: Conceptualization, H.E.F.; Formal analysis, F.Z.B.; and H.E.F.; Resources, M.K.; Software, Z.E.I.; Supervision, F.G.; Writing—review & editing, A.I. All authors have read and agreed to the published version of the manuscript.

Funding: This research was funded by the CNRST under grant number PPR/2015/36.

Acknowledgments: The authors gratefully acknowledge the support of the Moroccan Ministry of Higher Education (MESRSFC) and the CNRST under grant number PPR/2015/36.

Conflicts of Interest: The authors declare no conflict of interest.

Nomenclature

E_0^{Cell}	Reference potential at standard operating conditions (V).
h_s	Convective heat transfer coefficient (W/(m^{2}K)).
R	Universal gas constant (J/ (molK)).
T	Stack temperature (K).
F	Faraday's constant (C/mol).
P_{H_2}	The partial pressure of hydrogen (atm).
P_{O_2}	The partial pressure of oxygen (atm).
P_{H_20}	The partial pressure of water (atm).
$(P_{H_2O})_{in}^a$	The partial pressure of waterat the anode (atm).
$(P_{H_2O})_{in}^c$	The partial pressure of water at the cathode (atm).
$(m_{H_2O})_{in}^a$	A mole flow rate of water at the anode (mol/s).
$(m_{H_2O})_{in}^c$	A mole flow rate of water at the cathode (mol/s).
V^{Act}	Activation voltage loss (V).
V^{Conc}	Concentration voltage loss (V).
V^O	Ohmic voltage loss (V).
V_a	The volume of the anode (m^3).
C	Capacitance due to the charge double layer(F): equivalent electrical capacitance
R_{ac}	Equivalent resistance corresponding to activation voltage loss (Ω).
R_{co}	Equivalent resistance corresponding to concentration voltage loss (Ω).
R_{oh}	Ohmic resistance (Ω).
A_s	Area of a single cell (m^2).
M_{fc}	The total mass of the PEM fuel cell stack (kg).
C_{fc}	Specific heat capacity of the PEM fuel cell stack [J/(molK)].
I_{fc}	Stack current (A).
n_s	The number of PEM fuel cell stacks.
u_{T_R}	Room temperature (K).
u_{P_A}	Channel pressure of hydrogen (atm).
V_C	The voltage across the capacitor C (V).
V_{fc}	The output voltage of the PEM fuel cell (V)
NLM	State-space model (nonlinear model)
GMM	Generic MATLABmodel
RLM	RL model
RCM	RC model
EXM	Experimental model

References

1. Lai, J.; Ellis, M.W. Fuel Cell Power Systems and Applications. *Proc. IEEE* **2017**, *105*, 2166–2190. [CrossRef]
2. Magdalena, M.; Veziroglu, T.N. The properties of hydrogen as fuel tomorrow in sustainable energy system for a cleaner planet. *Int. J. Hydrog. Energy* **2005**, *30*, 795–802.
3. Samuelsen, S. Why the Automotive Future Will Be Dominated by Fuel Cells. Available online: https://spectrum.ieee.org/why-the-automotive-future-will-be-dominated-by-fuel-cells (accessed on 29 August 2021).
4. Cano, Z.P.; Banham, D.; Ye, S.; Hintennach, A.; Lu, J.; Fowler, M.; Chen, Z. Batteries and fuel cells for emerging electric vehicle markets. *Nat. Energy* **2018**, *3*, 279–289. [CrossRef]
5. Hbilate, Z.; Naimi, Y.; Takky, D. Modelling operation of proton exchange membrane fuel cells—A brief review of current status. *Mater. Today: Proc.* **2019**, *13*, 889–898. [CrossRef]
6. Springer, T.E.; Zawodzinski, T.A.; Gottesfeld, S. Polymer electrolytefuel cell model. *J. Electrochem. Soc.* **1991**, *138*, 2334–2342. [CrossRef]
7. Amphlett, J.C.; Baumert, R.M.; Mann, R.F.; Peppley, B.A.; Roberge, P.R.; Harris, T.J. Performancemodeling of the Ballard Mark IV solidpolymer electrolyte fuel cell: I. *Mech. Model Dev.* **1995**, *142*, 1–8.
8. Friede, W.; Rael, S.; Davat, B. Mathematical model and characterization of the transient behavior of a PEM fuel cell. *IEEE Trans. Power Electron.* **2004**, *19*, 1234–1241. [CrossRef]
9. Pasricha, S.; Shaw, S.R. A dynamic PEM fuel cell model. *IEEE Trans. Energy Convers.* **2006**, *21*, 484–490. [CrossRef]
10. Wang, C.; Nehrir, M.H.; Shaw, S.R. Dynamic models and model validation for PEM fuel cells using electrical circuits. *IEEE Trans. Energy Convers.* **2005**, *20*, 442–451. [CrossRef]
11. Tanrioven, M.; Alam, M.S. Modeling, Control, and Power Quality Evaluation of a PEM Fuel Cell-Based Power Supply System for Residential Use. *IEEE Trans. Ind. Appl.* **2006**, *42*, 1582–1589. [CrossRef]
12. Uzunoglu, M.; Alam, M.S. Dynamic modeling, design, and simulation of a combined PEM fuel cell and ultracapacitor system for stand-alone residential applications. *IEEE Trans. Energy Convers.* **2006**, *21*, 767–775. [CrossRef]
13. Na, W.K.; Gou, H.S.; Diong, B. Nonlinear Control of PEM Fuel Cells by Exact Linearization. *IEEE Trans. Ind. Appl.* **2007**, *43*, 1426–1433. [CrossRef]
14. Grasser, F.; Rufer, A. A Fully Analytical PEM Fuel Cell System Model for Control Applications. *IEEE Trans. Ind. Appl.* **2007**, *43*, 1499–1506. [CrossRef]
15. Puranik, S.V.; Keyhani, A.; Khorrami, F. State-Space Modeling of Proton Exchange Membrane Fuel Cell. *IEEE Trans. Energy Convers.* **2010**, *25*, 804–813. [CrossRef]
16. Wang, C.; Nehrir, M.H. A Physically Based Dynamic Model for Solid Oxide Fuel Cells. *IEEE Trans. Energy Convers.* **2007**, *22*, 887–897. [CrossRef]
17. Carnes, B.; Ned, D. Systematic parameter estimation for PEM fuel cell models. *J. Power Sources* **2005**, *144*, 83–93. [CrossRef]
18. Corrêa, J.M.; Farret, F.A.; Popov, V.A.; Simões, M.G. Sensitivity analysis of the modeling parameters used in simulation of proton exchange membrane fuel cells. *IEEE Trans. Energy Convers.* **2005**, *20*, 211–218. [CrossRef]
19. Dhirde, A.M.; Dale, N.V.; Salehfar, H.; Mann, M.D.; Han, T. Equivalent Electric Circuit Modeling and Performance Analysis of a PEM Fuel Cell Stack Using Impedance Spectroscopy. *IEEE Trans. Energy Convers.* **2010**, *25*, 778–786. [CrossRef]
20. Becherif, M.; Hissel, D.; Gaagat, S.; Wack, M. Electrical equivalent model of a proton exchange membrane fuel cell with experimental validation. *Renew. Energy* **2011**, *36*, 2582–2588. [CrossRef]
21. Restrepo, C.; Konjedic, T.; Garces, A.; Calvente, J.; Giral, R. Identification of a Proton-Exchange Membrane Fuel Cell's Model Parameters by Means of an Evolution Strategy. *IEEE Trans. Ind. Inform.* **2015**, *11*, 548–559. [CrossRef]
22. Saadi, A.; Becherif, M.; Hissel, D.; Ramadan, H.S. Dynamic modeling and experimental analysis of PEMFCs: A comparative study. *Int. J. Hydrog. Energy* **2017**, *42*, 1544–1557. [CrossRef]
23. Atlam, Ö.; Dûndar, G. A practical Equivalent Electrical circuit model for proton exchange membrane fuel cell (PEMFC) system. *Int. J. Hydrog. Energy* **2021**, *46*, 13230–13239. [CrossRef]
24. Wei, Y.; Zhao, Y.; Yun, H. Estimating PEMFC ohmic internal impedance based on indirect measurements. *Energy Sci. Eng.* **2021**, *9*, 1134–1147. [CrossRef]
25. Schumann, M.; Cosse, C.; Becker, D.; Vorwerk, D.; Schulz, D. Modeling and experimental parameterization of an electrically controllable PEM fuel cell. *Int. J. Hydrogen Energy* **2021**, *46*, 28734–28747. [CrossRef]
26. Qi, Z. *Proton Exchange Membrane Fuel Cells*; CRC Press: Boca Raton, FL, USA, 2021.
27. Larminie, J. *Fuel Cell Systems Explained*, 2nd ed.; Wiley: Hoboken, NJ, USA, 2003.
28. Njoya, S.M.; Tremblay, O.; Dessaint, L.-A. A generic fuel-cell model for the simulation of fuel cell vehicles. In Proceedings of the 2009 IEEE Vehicle Power and Propulsion Conference 2009, VPPC '09, Dearborn, MI, USA, 7–10 September 2009; IEEE: Piscataway, NJ, USA, 2009; pp. 1722–1729.
29. Motapon, S.N.; Tremblay, O.; Dessaint, L.A. Development of a generic fuel cell model: Application to a fuel cell vehicle simulation. *Int. J. Power Electron.* **2012**, *4*, 505–522. [CrossRef]
30. Mann, F.; Amphlett, J.C.; Hooper, M.A.; Jensen, H.M.; Peppley, B.A.; Roberge, P.R. Development and application of a generalized steady-state electrochemical model for a PEM fuel cell. *J. Power Sources* **2000**, *86*, 173–180. [CrossRef]

31. Yongping, H.; Mingxi, Z.; Gang, W. A transient semi-empirical voltage model of a fuel cell stack. *Int. J. Hydrog. Energy* **2007**, *32*, 857–862.
32. Edwards, R.L.; Demuren, A. Regression analysis of PEM fuel cell transient response. *Int. J. Energy Environ. Eng.* **2016**, *7*, 329–341. [CrossRef]

 micromachines

Article

Nonlinear Stabilization Controller for the Boost Converter with a Constant Power Load in Both Continuous and Discontinuous Conduction Modes

Juan Gerardo Parada Salado [1,†], Carlos Alonso Herrera Ramírez [2,†], Allan Giovanni Soriano Sánchez [3] and Martín Antonio Rodríguez Licea [3,*,†]

1 Department of Electronics, Celaya Institute of Technology, Celaya 38010, Mexico; d1903018@itcelaya.edu.mx
2 Robotic Engineering Department, Polytechnic University of Guanajuato, Cortazar 38496, Mexico; aherrera@upgto.edu.mx
3 CONACYT-Celaya Institute of Technology, Celaya 38010, Mexico; allan.soriano@itcelaya.edu.mx
* Correspondence: martin.rodriguez@itcelaya.edu.mx
† These authors contributed equally to this work.

Abstract: The operation of Boost converters in discontinuous conduction mode (DCM) is suitable for many applications due to the, among other advantages, inductor volume reduction, high efficiency, paralleling, and low cost. Uses in biomedicine, nano/microelectromechanical, and higher power systems, where wide ranges of input/output voltage and a constant power load (CPL) can coexist, are well-known examples. Under extremely wide operating ranges, it is not difficult to change to a continuous conduction mode (CCM) operation, and instability, chaos, or bifurcations phenomena can occur regardless of the conduction mode. Unfortunately, existing control strategies consider a single conduction mode or linearized models because only slight resistive/CPL power level or input/output voltage variations (and no conduction mode changes) were expected. In this paper, new mathematical models for the Boost converter (with resistive or CPL) that are conduction mode independent are presented and validated. Since the open-loop dynamics of the proposed CPL model is unstable, a nonlinear control law capable of stabilizing the boost converter regardless of the conduction mode is proposed. A stability analysis based on a common-Lyapunov function is provided, and numerical and experimental tests are presented to show the proposal's effectiveness.

Keywords: constant power load; boost converter; discontinuous conduction mode; nonlinear control; switched system

Citation: Parada Salado, J.G.; Herrera Ramírez, C.A.; Soriano Sánchez, A.G.; Rodríguez Licea, M.A. Nonlinear Stabilization Controller for the Boost Converter with a Constant Power Load in Both Continuous and Discontinuous Conduction Modes. *Micromachines* **2021**, *12*, 522. https://doi.org/10.3390/mi12050522

Academic Editor: Young-Ho Cho

Received: 25 March 2021
Accepted: 1 May 2021
Published: 6 May 2021

1. Introduction

It is well-known that the Boost converter can operate in three modes, named CCM, critical conduction mode (CRM), and DCM, related to the energy stored in the inductor. In CCM/DCM, the stored energy is strictly/non strictly greater than zero, and CRM stands for zero energy during an infinitesimal period. Many authors have considered the Boost converter's modeling and stabilization in CCM with a resistive load since no wide variations or constant power phenomena were expected at the load side. Some only mention that DCM operation should be avoided or treated with special care, and others have limited the DCM operation's study to its stationary behavior [1–6]. A brief survey of relevant DCM, CRM, and CCM models and controllers for the Boost converter is presented in the following.

The authors in [7] developed a low-output voltage Boost converter operating in DCM and designed a controller to stabilize the output by a steady-state analysis. Some authors analyzed the power quality in DCM operation for converters and rectifiers, for instance, in [8–13]. However, they based their analysis on steady-state behavior. Small signal linear models for DCM operation of the Boost converter, including power losses, were developed in [14]; however, the obtained transfer function includes many parameters that are not easy to acquire. In [15], a frequency control to operate the Boost converter in CRM was proposed. The authors based their study on a linear model of the Boost converter operating in DCM, obtained by a small-signal analysis. On the other hand, in [16], a numerical PSPICE implementation of an isolated full-bridge converter reproduced some CCM and DCM phenomena.

In [17], a nonlinear controller for DCM operation of the Boost converter was developed, whereby the authors obtained such controller from the discrete-time difference-approximation equations and the steady-state solution of the system. The authors in [18] proposed a controller for the output current of the Boost converter operating in DCM, using a steady-state estimation of the dynamics.

In [19], the authors developed a conduction mode independent (CMI) model for the non-inverting Buck-Boost converter. However, the proposed model depends on the inductor's discharge period, which is an implicit function of the duty cycle and parameters and cannot be extended to the Boost converter. Regardless of the previous, a back-stepping controller for the Buck-Boost converter with a resistive load (non-CPL) was developed.

Some authors have stated unwanted phenomena during the transition between conduction modes. In [20,21], proofs of open-loop bifurcation phenomena in the DCM-CCM boundary were provided. Chaos and bifurcation behaviors for the boost converter were also demonstrated in [22]. Controllers for other converters operating in DCM or CCM with no CPL were developed in [23–25].

Furthermore, it has been reported that a CPL can destabilize the dynamics of the Boost converter in any conduction mode. Recently developed control strategies considered a CPL with single conduction modes or linearized models [26–33]. Unfortunately, it is not difficult to achieve DCM to CCM changes in a Boost converter, with both a resistive load or a CPL and in addition, with a CPL, the conduction mode depends on their power level and output voltage/current. Moreover, linearized models are accurate only in a small region containing the operating point.

It is worth mentioning the work in [34], the authors reported that the Boost converter operating in DCM with a CPL is stable during steady-state. However, the transient stage and CCM to CCM changes cannot be neglected for applications in which wide ranges of input/output voltage and a constant power load (CPL) can coexist.

The authors in [35] presented a current control for the Boost converter feeding a CPL and included a passive compensation (paralleled RC network). They based their analysis on small-signal models valid only in a small region containing the operating point.

From the above state of the art, one can find that the complicated scenario that includes CPLs, and conduction mode changes in the Boost converter, has not been formally studied. The models presented until now are small-signal or steady-state. Table 1 summarizes relevant proposals for converters operating with a CPL or different conduction modes; the checkmark symbol means that the research considers the characteristic. There is a gap for the Boost converter's modeling and control with a CPL, operating in both DCM and CCM. Unfortunately, in biomedicine, microelectromechanical, and other applications, wide CPL and input/output voltage variation ranges could induce changes in the conduction mode and stability properties.

Hence, the contributions of this paper are as follows. Precise Boost converter CMI models (CPL-switched, CPL non-switched, resistive load switched, and resistive load non-switched) are presented. Through such models, it is shown that the Boost converter's operating point with a CPL is potentially unstable in open-loop; this has not been demonstrated from a CMI perspective to the authors' knowledge. Therefore, a nonlinear stabilizing controller is developed for the CMI model with a CPL. A stability analysis based on a common-Lyapunov function is provided, and numerical and experimental tests are presented to show the proposal's effectiveness. The controller does not depend on frequency control but the CPL power level and the output current, making it easy to implement.

This document is organized as follows. Section 2 presents the development of the mathematical models, and Section 3 its validation. Section 4 is devoted to instability demonstration for the Boost converter with a CPL. In Section 5, the controller design is presented, and in Section 6, numerical and experimental tests of the closed-loop system are presented. Finally, some conclusions are presented in Section 7.

Table 1. Summary of the literature review.

References	Boost	Model type	CPL	Controller
[1,2,5]	✓	CCM		✓
[3,4]	✓	Static		
[6]	✓	CCM w/loss		
[7–9,11–13,18]	✓	Static		✓
[19]		CMI (implicit)		✓
[14]	✓	DCM w/loss		
[15]	✓	CRM		✓
[16]		DCM/CCM		
[17]	✓	DCM		✓
[20,22]	✓	DCM/CCM		
[21]	✓	DCM/CCM		✓
[23]	✓	Static		
[24]	✓	Static		✓
[25]		Static		✓
[26]	✓	DCM	✓	
[27]		CCM	✓	✓
[28]	✓	CCM	✓	✓
[29]		CCM	✓	✓
[30]		CCM	✓	✓
[33]	✓	CCM	✓	✓
[34]	✓	DCM	✓	✓
[35]	✓	CCM	✓	✓
This proposal	✓	CMI	✓	✓

2. CMI Model of the Boost Converter

Consider the Boost converter schematic of Figure 1. Let us begin by introducing the well-known simplified mathematical model of the Boost converter with a CPL, obtained by an averaging technique in CCM, with infinite switching frequency as well as ideal components [1]:

$$L\frac{di}{dt} = -(1 - u_1)v + E \tag{1}$$

$$C\frac{dv}{dt} = (1 - u_1)i - \frac{P}{v} \tag{2}$$

where L, C, and P are the inductance, capacitance, and CPL power demand, respectively; v is the averaged output voltage, i is the averaged current flowing through the inductor, $R = v^2/P$ where P is the output power, and u_1 is the duty cycle of the PWM signal in Q. Note that $v = 0$ implies an indeterminate form in Equation (2), characteristic for the Boost converter with a CPL. Since protective circuits can avoid a high output current, it is reasonable to consider the output voltage and current, both greater than zero, for analysis purposes.

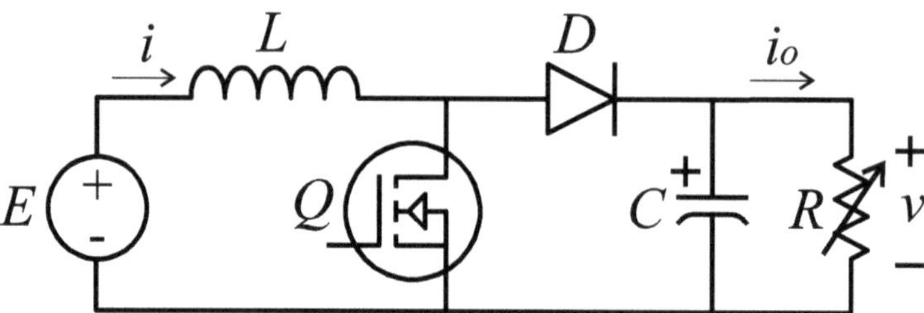

Figure 1. Basic Boost converter schematic with a variable resistive load whose value is modified to achieve a constant output power (CPL).

On the other hand, using $R = v^2/P$ in the well-known conduction-mode-inequality for the Boost converter [36], DCM occurs if:

$$k_{crit}(v) = \frac{2LfP}{v^2} < u_1(1-u_1)^2 \tag{3}$$

where f is the operating frequency of the PWM. In principle, the Boost converter cannot be readily designed to operate permanently in CCM or DCM with a CPL. If P varies to smaller values (low load), the converter could operate in DCM; conversely, a DCM to CCM change can occur if the CPL level has a high value. Recall that at this point, v and i both represent averaged values.

In an ideal resistive load scenario, a designer tries to ensure that the range for u_1 is as wide as possible to ensure either DCM or CCM (selecting the components to achieve $k_{crit,R} = 2fL/R < u(u-1)^2$ or $k_{crit,R} = 2fL/R > u(u-1)^2$, respectively). Figure 2 illustrates the conduction modes as a function of u_1 and $k_{crit,R}$. It is easy to notice that even in such a scenario, it is impossible to ensure a single conduction mode if the load is not a variable to be altered/bounded at will. Furthermore, the components' degradation or aging can alter $k_{crit,R}$, inducing a conduction mode change; hence, the dynamical model of Equations (1) and (2) obtained for CCM operation would not be appropriate for designing a controller. Indeed, many authors have demonstrated the bifurcation phenomena during the CCM to DCM change [20–22].

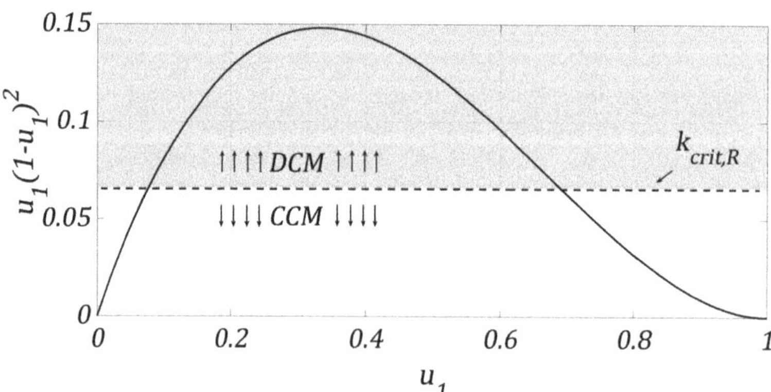

Figure 2. DCM/CCM dependence on u and $k_{crit,R}$.

Therefore, to stabilize/regulate the Boost converter with a CPL in any conduction mode, description (1) and (2) is not adequate. In the following, a complementary averaged model for DCM is used to obtain later CMI models for the Boost converter with/without a CPL. As far as the authors know, there are no CMI models such as the one described below.

For a DCM operation of the Boost converter, one can idealize Q as two switches $\bar{u}_1$, $\bar{u}_2$ as described in Figure 3; $\bar{u}_2 = 0$ stands for the zero current through the inductor (L). Reproducing the averaging methodology in [1] to obtain a DCM model, the three operating modes depicted in Figure 4 are possible. Mathematical expressions for each operating mode are presented in Table 2; note that X means "do not care" because $\bar{u}_2 = 0$ nullifies the current flow through L, and hence $\bar{u}_1$ has no effect on the equations. Note also that $\hat{v}$ and $\hat{i}$ are used to differentiate them from the averaged values v and i.

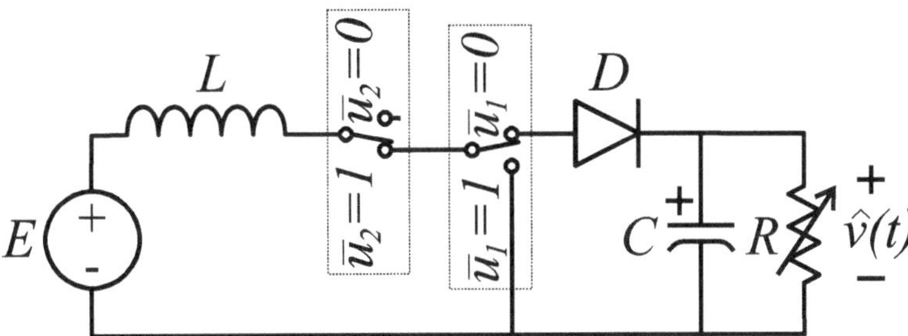

Figure 3. Idealization of the Boost converter operating in DCM through two ideal switches.

Table 2. Current flow modes of the inductor.

$\bar{u}_1$	$\bar{u}_2$	Mode	Equations
X	0	Holding	$L\frac{d\hat{i}}{dt} = 0; C\frac{d\hat{v}}{dt} = -\frac{P}{\hat{v}}$
0	1	Discharging	$L\frac{d\hat{i}}{dt} = -\hat{v} + E; C\frac{d\hat{v}}{dt} = \hat{i} - \frac{P}{\hat{v}}$
1	1	Charging	$L\frac{d\hat{i}}{dt} = E; C\frac{d\hat{v}}{dt} = \hat{i} - \frac{P}{\hat{v}}$

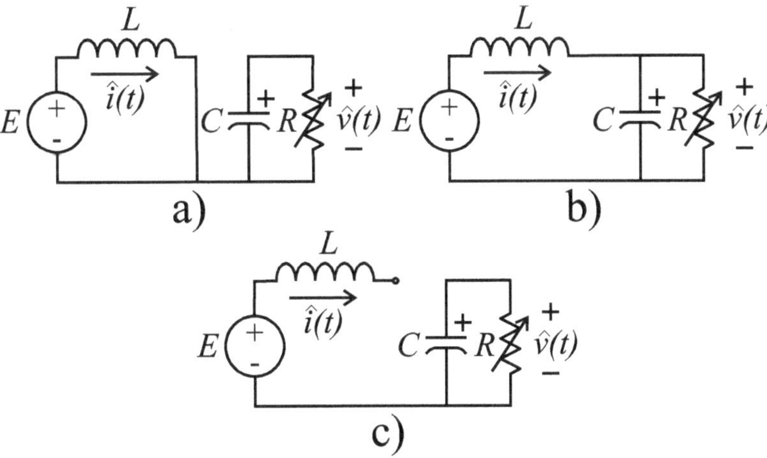

Figure 4. Illustration of the inductor current flow modes: (**a**) Charging, (**b**) discharging, and (**c**) holding.

An elementary analysis allows obtaining an averaged model of the charging, discharging, and holding operating modes as:

$$L\frac{di}{dt} = -(1-u_1)v + E(1+2u_1u_2) \tag{4}$$

$$C\frac{dv}{dt} = (1-u_1)i - \frac{P}{v} \tag{5}$$

where $u_1 \in [0,1]$ is the duty cycle (percentage of T in charging mode with $T = 1/f$ as the PWM period), and u_2 is the percentage of T in holding mode (zero inductor current); these periods should not be confused with $\bar{u}_1$, $\bar{u}_2$ since these last only represent the switches in Figure 3. See Table 2 and Figure 5 for an illustration; u_d is used to describe the period's percentage in discharge mode. Note that $u_2 = 0$ for CCM and $u_2 > 0$ for DCM; hence, there is no ambiguity on using u_1, v, and i from the CCM model of Equations (1) and (2), for this model also.

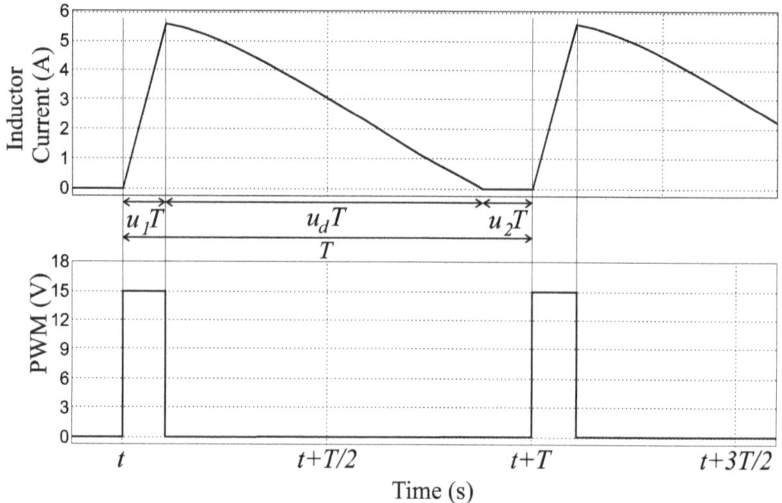

Figure 5. Exemplification of the inductor current behavior in DCM concerning the triggering on Q. PWM stands for the voltage signal in the gate pin.

From inductor and diode current waveforms in steady-state, it is well known that the load current is [36]:

$$\frac{v}{R} = \frac{P}{v} = \frac{Eu_1u_d}{2fL}. \tag{6}$$

Furthermore, a time balance over the inductor current can be represented as (see Figure 5):

$$T = u_1T + u_dT + u_2T \tag{7}$$

such that:

$$u_2 = 1 - u_1 - u_d = 1 - u_1 - \frac{2PLf}{Evu_1}. \tag{8}$$

Then, it is easy to see that:

$$u_2 = 1 - u_1 - \frac{2PLf}{Evu_1} \tag{9}$$

is a C^1-diffeomorphism (for the inverse, it is enough to take only the positive root to have a bijection; this inverse is differentiable) for Equations (4) and (5) with $v > 0$ and $1 \geq u_1 > 0$ ($u_1 = 0$ is a trivial case), allowing to eliminate the dependency on u_2 [37]:

$$L\frac{di}{dt} = -(1-u_1)v - \frac{4PLf}{v} + (1+2u_1-2u_1^2)E \tag{10}$$

$$C\frac{dv}{dt} = (1-u_1)i - \frac{P}{v}. \tag{11}$$

The following section provides validations showing that the result of this transformation is quantitatively favorable in terms of the mean squared error (MSE).

By comparison of Equations (10) and (11) with Equations (1) and (2) and from inequality (3), one can design/approximate (recall that the CCM and DCM models were obtained using an infinite frequency consideration, while the change between conduction modes (3) is obtained from a steady-state analysis. Since a common Lyapunov function is used to design a controller and demonstrate stability even with arbitrary switching, it is not relevant to obtain a precise design of the switching signal; these following switching signals are presented for completeness purposes) a switching signal as $\gamma(u_1, v) \in \{0,1\}$:

$$\gamma(u_1, v) \triangleq 0.5\left(1 + sign\left(\left(u_1(1-u_1)^2 - \frac{2LfP}{v^2}\right)\right)\right) \tag{12}$$

where:

$$sign(x) = \begin{cases} -1, & x < 0 \\ 0, & x = 0 \\ 1, & x > 0 \end{cases}. \tag{13}$$

That is, $\gamma(u_1, v) = 1$ for DCM, and $\gamma(u_1, v) = 0$ for CCM; the switched model is then:

$$L\frac{di}{dt} = -(1-u_1)v - \frac{4PLf}{v}\gamma + (1+2\gamma u_1-2\gamma u_1^2)E \tag{14}$$

$$C\frac{dv}{dt} = (1-u_1)i - \frac{P}{v}. \tag{15}$$

Alternatively, one can approximate the switching signal with a C^∞ function $0 \leq \rho \leq 1$ to obtain a continuous (non-switched) CMI model of the Boost converter:

$$\rho(u_1, v) = 0.5\left(1 + tanh\left(a\left(u_1(1-u_1)^2 - \frac{2LfP}{v^2}\right)\right)\right) \tag{16}$$

with $a \gg 1$, and the switched model is:

$$L\frac{di}{dt} = -(1-u_1)v - \frac{4PLf}{v}\rho + (1+2\rho u_1-2\rho u_1^2)E \tag{17}$$

$$C\frac{dv}{dt} = (1-u_1)i - \frac{P}{v}. \tag{18}$$

Note that for $\gamma = 0$ and $\rho = 0$, the operating point congruently coincides with that of the classic CCM model.

For the non-CPL case (resistive load), the switched/CMI model can be obtained by a similar procedure:

$$L\frac{di}{dt} = -\left(1 - u_1 + \frac{4Lf}{R}\sigma\right)v + (1+2\sigma u_1-2\sigma u_1^2)E \tag{19}$$

$$C\frac{dv}{dt} = (1-u_1)i - \frac{v}{R} \tag{20}$$

where:

$$\sigma(u_1) \triangleq 0.5\left(1 + sign\left(\left(u_1(1-u_1)^2 - \frac{2Lf}{R}\right)\right)\right), \tag{21}$$

or alternatively:

$$L\frac{di}{dt} = -\left(1 - u_1 + \frac{4Lf}{R}\varphi\right)v + (1 + 2\varphi u_1 - 2\varphi u_1^2)E \tag{22}$$

$$C\frac{dv}{dt} = (1 - u_1)i - \frac{v}{R} \tag{23}$$

with,

$$\varphi(u_1) = 0.5\left(1 + tanh\left(a\left(u_1(1 - u_1)^2 - \frac{2Lf}{R}\right)\right)\right). \tag{24}$$

3. Validation of the CMI Model for the Boost Converter

In this section, the validations for the non-CPL model developed in Equations (19) and (20) are presented (validation of the model of Equations (22) and (23) is not presented here because clearly, it provides a very smooth CCM-DCM switching in comparison with the model of Equations (19) and (20); the dynamic behavior is almost the same for other conditions except during this switching); this is done by comparing the classic CCM model in [1] (Equations (1) and (2) in this paper), and the s-domain linear DCM model presented in [14]. PSIM 2020a is used to generate the control group data.

Consider $E = 100$ V, $L = 15$ µH, $C = 100$ µF, $R = 10$ Ω, and $f = 20$ kHz; $0.0693 < u_1 < 0.7091$ values allow DCM while other values allow CCM operation. Zero initial conditions and steps from $u_1 = 0.10$ to $u_1 = 0.90$ are introduced. It can be easily noted from Figure 6 that the CMI model shows minimum voltage error compared to the other models. The DCM model shows minimal error only around the linearization point ($u_1 = 0.35$, which is in the middle of the DCM range). The CCM classic model is inaccurate if the converter operates in DCM (shadowed area in Figure 2). To show the error quantitatively, Figure 7 shows a comparative plot of the MSE against u_1 for the three models. MSE for the inductor current, although not presented here, shows very similar results.

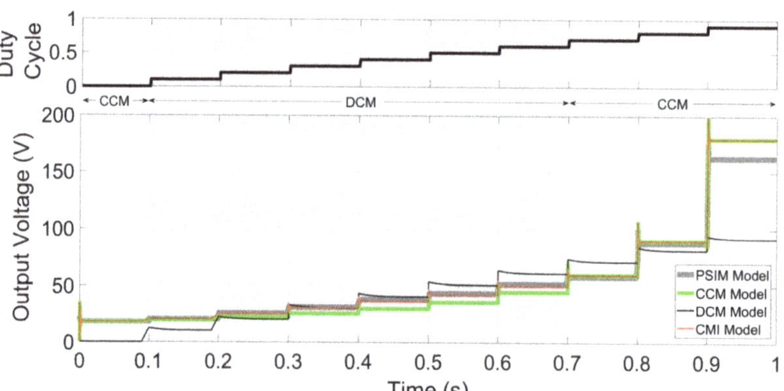

Figure 6. Comparative of the output voltage for PSIM (averaged), CCM, DCM, and CMI models, using steps from $u_1 = 0.0$ to $u_1 = 0.90$.

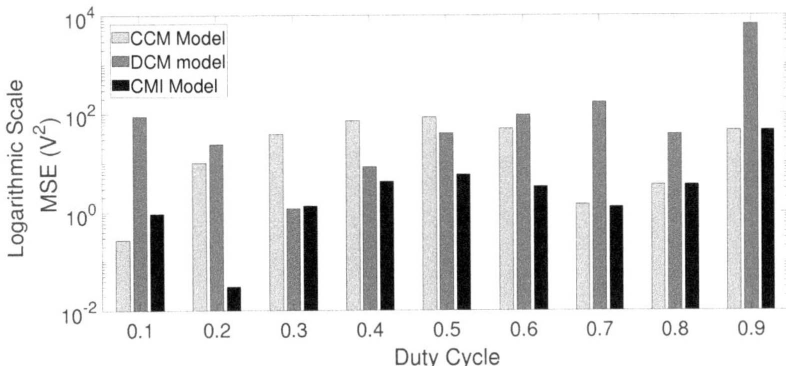

Figure 7. Voltage mean-squared error for CCM, DCM, and CMI models, in logarithmic scale. The CMI model presented in this paper provides the lowest MSE in almost all the u_1 range.

Figure 8 shows the comparatives of v and i during the switch/change between conduction modes as a function of u_1. A slow ramp with a positive slope is introduced to force the DCM to CCM change about 0.1 s, showing no impulsive effects in both PSIM and CMI model dynamics. Note that the model presented in this paper can accurately follow the slope change during the DCM to CCM switch; this is not the case for the classic CCM model or the DCM model.

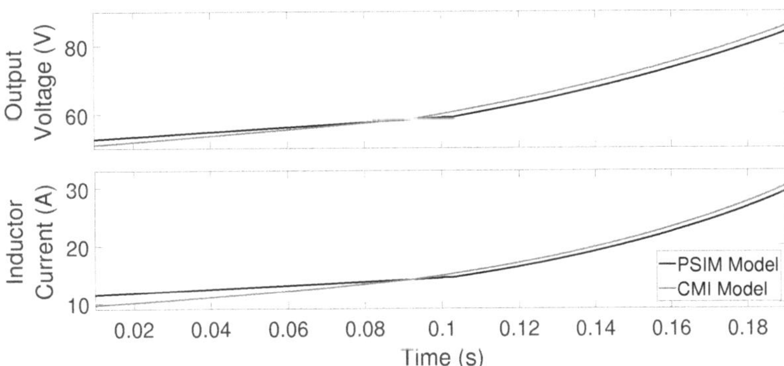

Figure 8. A comparison of the output voltages and inductor's currents obtained with the PSIM (averaged) and CMI models for a ramp input with unitary slope. The model presented in this paper can accurately follow the response-curves change during the DCM to CCM switch.

As an additional validation, an experimental 200-W Boost converter is built with the parameters mentioned above (see Figure 9) to validate the non-CPL CMI model. The inductor's current is measured with a current transducer from LEM USA Inc., and the load (resistive for this test) is emulated with a BK PRECISION-8510 electronic load. Figure 10 shows representative output voltages and inductor currents for several values of u_1; from experimental and CMI model simulation data, the current error is on average 0.223 A, and the RMS voltage error is approximately 3.12 V for $u_1 = 0.1$ up to $u_1 = 0.8$.

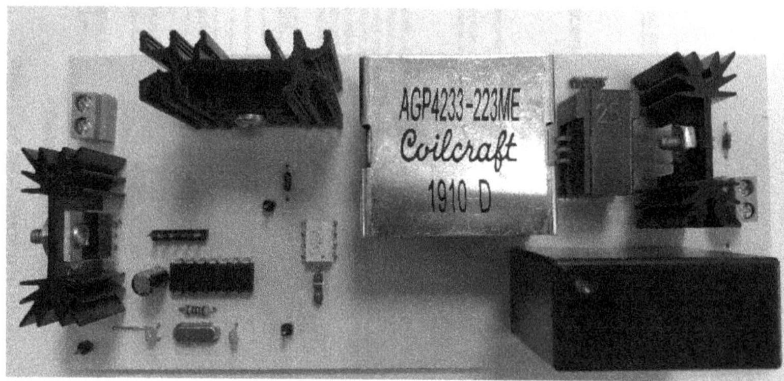

Figure 9. PCB used for the experimental tests presented in this paper.

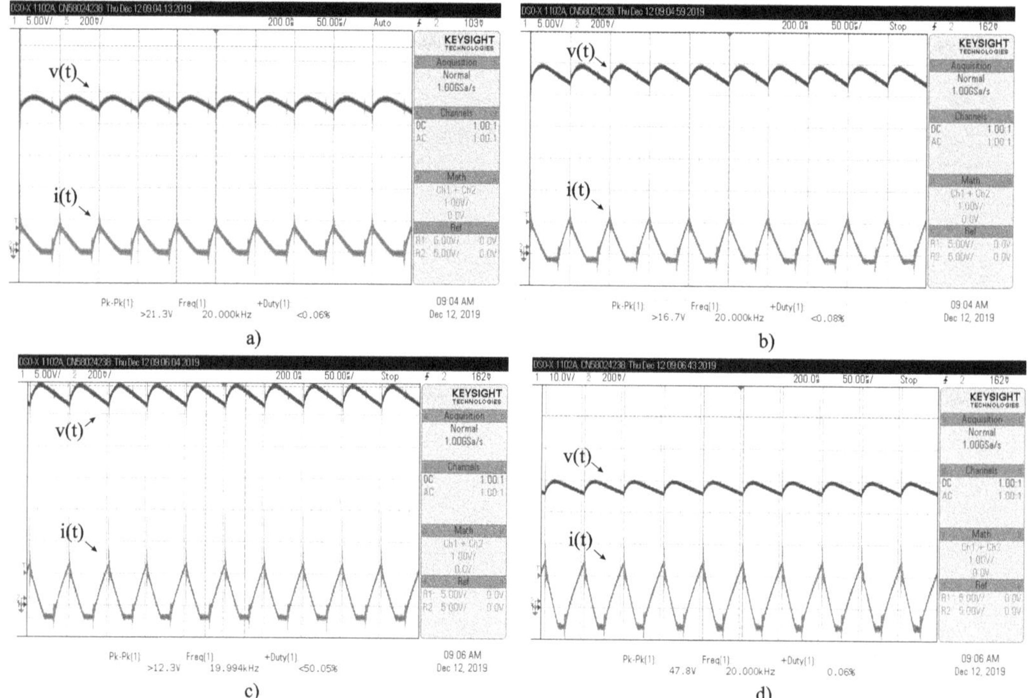

Figure 10. Output voltage and inductor current for the experimental validation, with (**a**) $u = 0.1$, (**b**) $u = 0.2$, (**c**) $u = 0.3$, and (**d**) $u = 0.4$. The current error is at an average of 0.223 A, and the RMS voltage error is approximately 3.12 V.

It is important to mention at this point that the CMI model with a CPL can hardly be validated experimentally in an open-loop scenario (essentially, for high power and low duty-cycle values, the converter's operating point becomes unstable, as illustrated in Figure 11). As shown below, a CPL induces instability, and it is preferable and less risky to validate a stable dynamic instead.

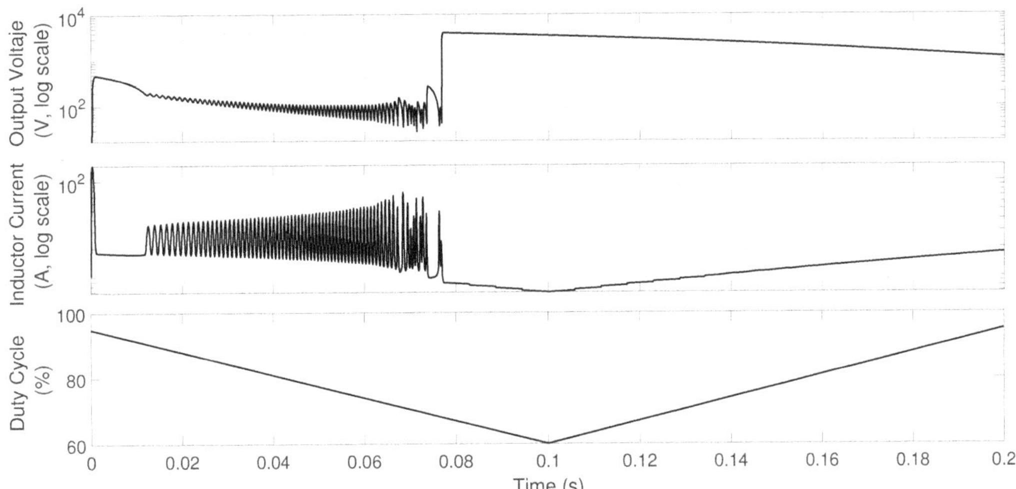

Figure 11. Illustration of open-loop, unstable behavior dependence on u_1. For this example, the same previous parameters are used with $P = 500$ W. Average output voltage and inductor current are obtained by PSIM (averaged values) and shown in logarithmic scales. u_1 is introduced as a (slow-dynamics) triangular wave to show how values lower than approximately 72% destabilizes the converter with a CPL; even worse, even if the active cycle increases again, stability cannot be recovered. Note that for high values of u_1, the voltage and current do not show large amplitude oscillations, but as the cycle decreases, the oscillations increase considerably.

4. Open-Loop Instability of the Boost Converter Dynamics with a CPL

Consider the dynamic system represented by Equations (14) and (15). The following shows that an operating point of such a system can be unstable either in DCM or CCM operation.

Proposition 1. *At least, one operating point of the dynamic system represented by the Equations* (14) *and* (15), *is Lyapunov-unstable for* $P, L > 0$.

Proof. See Appendix A.1. □

It is worth mentioning that different operating points may be open-loop stable; however, in general, open-loop stability cannot be ensured by a constant control input. The interested reader will surely obtain parameters and constant values of u_1 for which the operating point is stable. However, the question remains whether there is a control signal u_1 capable of avoiding unstable system dynamics.

5. Nonlinear Stabilization Controller for the Boost Converter with a CPL for CMI Operation

This section shows that a nonlinear control law can stabilize the system represented by Equations (14) and (15). At present, stability for general nonlinear switched systems under arbitrary switching can be ensured only by a common Lyapunov function approach; see, for instance [38,39] and the references therein.

Theorem 1. *Let* $P > 0$, $v > 0$, $i > 0$, *and* $v \geq E$. *The gain scheduling control law:*

$$1 - u_1 = \frac{k_1 P}{v} - k_2 P - \frac{k_3 \dot{v}}{v^2} - k_4 \dot{v}, \tag{25}$$

$$\text{using for } \gamma = 1 \begin{cases} k_1 = \frac{3k_3 - 12CELf}{8CPLf} \\ k_2 > 0 \\ k_3 > 0 \text{ (small enough)} \\ k_4 = \frac{3k_3}{8LPf} \end{cases}, \tag{26}$$

$$\text{and for } \gamma = 0 \begin{cases} k_1 > 0 \text{ (small enough)} \\ k_2 > \frac{1}{P} \\ k_3 = 0 \\ k_4 = 0 \end{cases}, \tag{27}$$

stabilizes the switched system (14) *and* (15) *for an arbitrary switching law.*

Proof. See the Appendix A.2. □

It is worth mentioning that the small enough constants can be selected by establishing an upper bound for i; this is feasible because the inductor has a saturation current.

6. Closed-Loop Numerical and Experimental Validations

Firstly, numerical validations for the system operating in a closed-loop with the proposed controller are presented; this is done by using two CPL sweeps and the parameters established in Section 3. The first sweep consists of introducing 40 W steps in the CPL level, from 0 up to 200 W (PSIM), as shown in the upper plot of Figure 12. The proposed controller in Equation (25) is used to calculate the Boost converter input control u_1, and the dynamic behavior is shown in the lower plots of such a figure. The second and third plots show the output voltage and output current dynamics, respectively; the controller adequately adapts voltage and current levels to achieve the CPL demand. Since power levels are achievable, the controller, besides stabilizing the converter, provides stable operating points for voltage and currents (inductor and output). In this test, the boost converter operates almost in DCM at all times because the duty cycle is calculated within zero and 0.85. For completeness purposes in the two last plots of Figure 12, the inductor current and the calculated duty cycle are shown. It is worth highlighting that using gains that do not accomplish conditions (26) and (27) or using regular proportional-integral controllers, unstable voltage, and current levels are obtained (not shown here). Regularly in such situations, the voltage suddenly increases, and the current falls, and the probability of devices' damage is very high.

A second CPL sweep of larger amplitudes is introduced to corroborate the controller's operation during the conduction mode changes. The associated dynamic behavior is presented in Figure 13. Note that in this case, up to 1 kW is demanded, and inductor-current peaks up to 200 A are reached. There are clear commutations from DCM to CCM and vice versa, in addition to significant changes in the inductor current. Despite the above, the CPL can be powered with acceptable voltage and current variations at the output. Although in the previous sweep, it was confirmed that fast changes to CCM were obtained, this test clearly proves that CCM is reached for long periods. In this severe scenario, the voltage gain limit of the converter is intentionally reached; oscillations are expected because the controller looks to increase the current to feed extremely high CPL levels.

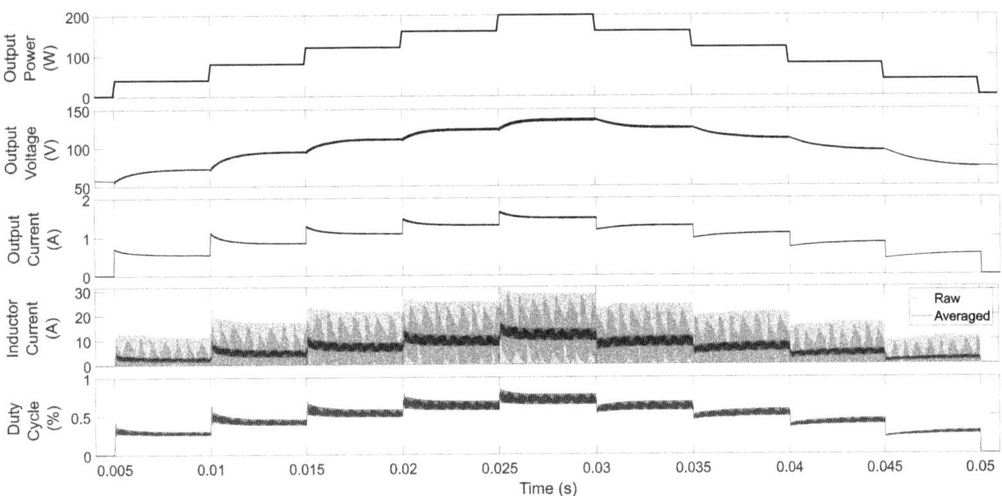

Figure 12. CPL sweep in closed-loop operation. The duty cycle never reaches the upper limit (set to 0.95), and a DCM operation is achieved at almost all times.

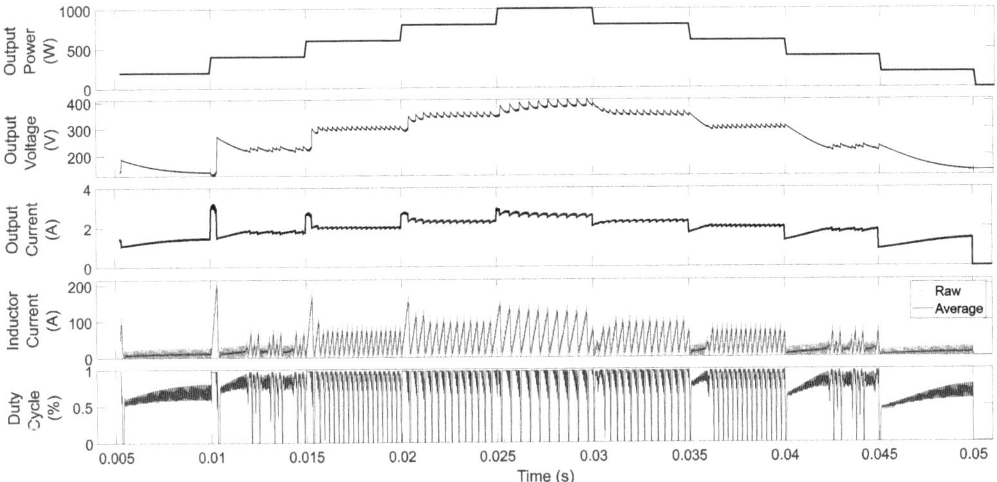

Figure 13. CPL sweep in the closed-loop operation of the Boost converter. The duty cycle reaches the upper limit (set to 0.95), and it is not possible to compensate for the small oscillations in output voltage and current. This is an inherent characteristic of the Boost converter.

In the following, the results of the experimental tests in closed-loop are presented. The same Boost converter used for model validation in Section 3 is employed to perform the following experiments; however, the measurement of the output voltage and current must be performed to calculate the controller action (Equation (25)). The BK PRECISION-8510 electronic load is now programmed to consume constant power levels of 25, 50, 100, and 150 W. Figure 14 shows the four scenarios described; note that these signals are not averaged and are shown in different scales. At all the levels of CPL's power demand, it is possible to stabilize both the output voltage and the inductor current, and the CPL's power demand level is satisfied accordingly. It is worth mentioning that hard switching without any filters or snubbers is used; therefore, noise and a small transient staring at each MOSFET switch is expected. The used semiconductors are also generic (regular Si and not SiC/GaN chemistry). Table 3 shows the efficiency at each level of power demand. The design of a high-efficiency production-level prototype is left for future research.

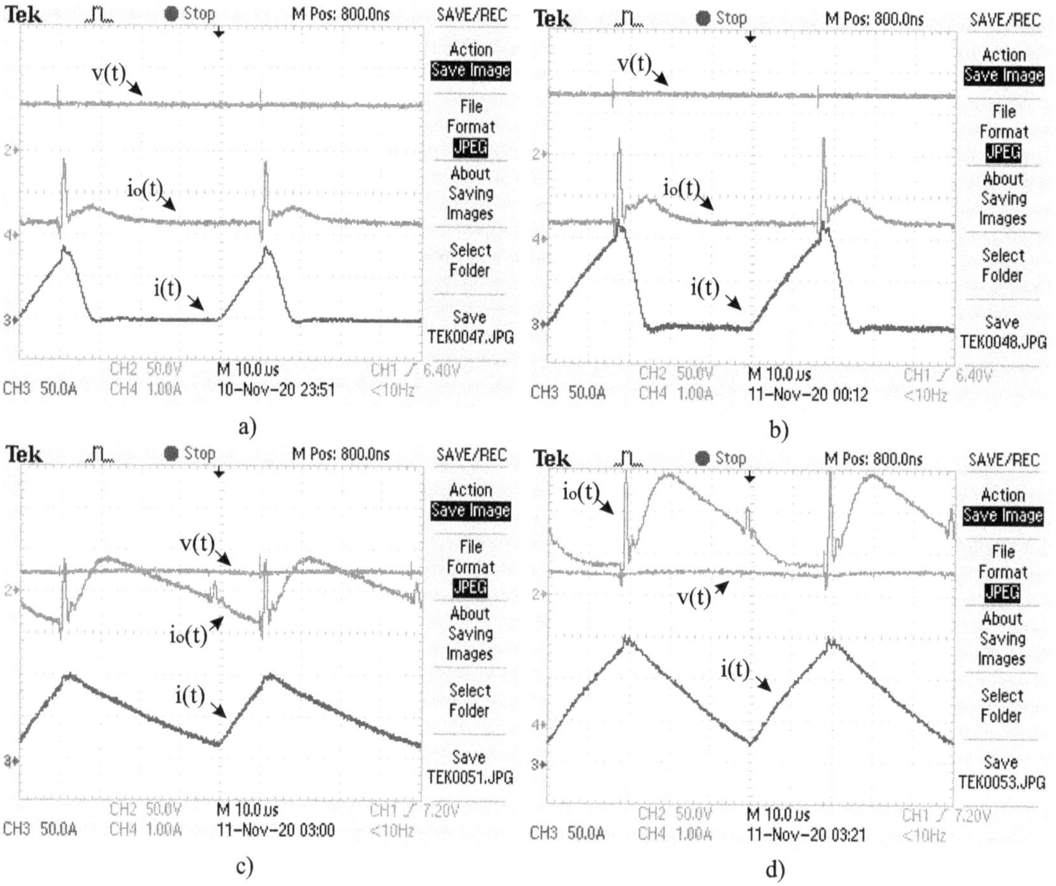

Figure 14. Output voltage (v), output current (io), and inductor current (i) for the Boost converter operating in closed-loop, with a CPL power demand of (**a**) 20, (**b**) 40, (**c**) 100, and (**d**) 135 W. 50 V/div is used for v, 1 A/div for io, and 5 A/div for i (the probe is configured for x/10). The output voltage and both currents are stabilized in the four scenarios despite the conduction mode type.

Table 3. Efficiency of the Boost converter operating in closed-loop.

CPL Power-Level (W)	Efficiency (%)	Conduction Mode
20	86.14	DCM
40	77.17	DCM
100	91.03	CCM
135	89.87	CCM

7. Conclusions

From the results provided in this paper, one can conclude that care must be taken during the design of controllers for the Boost converter, especially if the load can have several distant operating points or behave suddenly as a CPL or could occur conduction mode changes.

It was possible to corroborate both analytically and experimentally that the mathematical large-signal models presented here are accurate and could also be used to develop new predictive and robust strategies or for different control objectives. Regardless of the above, the nonlinear control law proposed here can stabilize the Boost converter with a CPL regardless of the conduction mode; this controller is easy and cheap to implement. Future research includes the analogous study of the stability for parallel boost stages, the analysis of robust stability, and the use of higher frequency with Sic or GaN devices to improve the converter's efficiency.

Author Contributions: Conceptualization, M.A.R.L. and J.G.P.S.; methodology, M.A.R.L.; software, J.G.P.S.; validation, M.A.R.L. and J.G.P.S.; formal analysis, M.A.R.L. and J.G.P.S.; investigation, M.A.R.L. and J.G.P.S.; resources, M.A.R.L., C.A.H.R. and A.G.S.S.; writing–original draft preparation, M.A.R.L., C.A.H.R. and A.G.S.S.; writing–review and editing, M.A.R.L.; visualization, M.A.R.L. and J.G.P.S.; supervision, M.A.R.L.; project administration, M.A.R.L.; funding acquisition,M.A.R.L., C.A.H.R. and A.G.S.S. All authors have read and agreed to the published version of the manuscript.

Funding: This research received no external funding.

Acknowledgments: The authors would like to thank CONACYT México for the Cátedras ID 4155 and 6782, and the scholarship of J.G.P.S.; also to the Robotic Engineering Department, affiliated with the Polytechnic University of Guanajuato.

Conflicts of Interest: The authors declare no conflict of interest.

Appendix A

Appendix A.1. Proof of Proposition 1

Proof. From Theorem 4.7 in [40], if any eigenvalue of the Jacobian matrix has a positive real part in some operating point, then such an operating point is Lyapunov-unstable. Consider $u_1 \to 0$ such that the equilibrium point is $i_e \to P/E, v_e \to E$. The Jacobian matrix at this operating point is:

$$J \approx \begin{bmatrix} 0 & -\frac{1}{L} \\ \frac{1}{C} & \frac{P}{CE^2} \end{bmatrix} \tag{A1}$$

where $d(sign(x))/dt = \delta(x)$ is used, and δ is the Dirac's delta function. The eigenvalues of J are:

$$\lambda_{1,2} \approx \frac{PL \pm \sqrt{L^2P^2 - 4LCE^4)}}{2LCE^2} \tag{A2}$$

of which, at least one has a positive real part with $P, L > 0$, demonstrating instability of the operating point. □

Appendix A.2. Proof of Theorem 1

Proof. From Theorem 8 in [38], consider the common Lyapunov candidate function $V = \frac{Cv^2}{L} + \frac{i^2}{2}$; hence, $V(0) = 0$ and $V \in \mathcal{C}^\infty$. For $\gamma = 1$ and replacing Equation (25) on Equations (14) and (15), the time derivative along the system's trajectory is sought to comply with:

$$\begin{aligned} \dot{V} &= -\frac{2CPv^3 - C\Delta Eiv^3 - \Delta Ek_3i^2v + Pk_4iv^3 - \Delta Ek_4i^2v^3}{Lv(ik_3 + Cv^2 + ik_4v^2 + Pk_3iv)} \\ &\quad -\frac{4LPfk_3i^2 - CPk_1iv^3 + CPk_2iv^4 + 4LPfk_4i^2v^2}{Lv(ik_3 + Cv^2 + ik_4v^2)} \qquad \text{(A3)} \\ &\quad -\frac{4CLPfiv^2}{Lv(ik_3 + Cv^2 + ik_4v^2)} < 0 \qquad \text{(A4)} \end{aligned}$$

where $E \le \Delta E = (1 + 2u_1 - 2u_1^2)E \le \frac{3E}{2}$. Since $v, i > 0$ the denominator is greater than zero, and using k_1, k_4 from (26) and (27) one can look equivalently for:

$$\begin{aligned} - &\quad CPk_2iv^4 + \frac{9Ek_3}{16LPf}i^2v^3 - 2CPv^3 - 4CLPfiv^2 - k_3Piv \\ - &\quad 4LPfk_3i^2 - \frac{3k_3}{2}i^2v^2 + \frac{3k_3E}{2}i^2v < 0. \end{aligned} \quad \text{(A5)}$$

From the last two terms of the previous inequality, with $k_3 > 0$, and from expressions (26) and (27):

$$\begin{aligned} -\frac{3k_3}{2}i^2v^2 + \frac{3k_3E}{2}i^2v &< 0 \\ \frac{3k_3E}{2}i^2v &< \frac{3k_3}{2}i^2v^2 \\ Ei^2v &< i^2v^2 \\ E &< v \end{aligned} \quad \text{(A6)}$$

which is true for $P > 0$; that is, in the Boost converter this condition is feasible under regular parameterization, and one looks for stability under this condition. Hence, inequality (A5) turns equivalently in:

$$-CPk_2iv^4 + \frac{9Ek_3}{16LPf}i^2v^3 - 2CPv^3 - 4CLPfiv^2 - k_3Piv - 4LPfk_3i^2 < 0. \quad \text{(A7)}$$

Taking the v^3 terms, one can find that their sum is negative if:

$$\begin{aligned} \frac{9Ek_3}{16LPf}i^2v^3 - 2CPv^3 &< 0 \\ i^2 &< \frac{32CLfP^2}{9Ek_3}. \end{aligned} \quad \text{(A8)}$$

Note that $k_3 > 0$ can be arbitrarily small to allow any finite i, therefore inequality (A7) turns in:

$$-CPk_2iv^4 - 4CLPfiv^2 - k_3Piv - 4LPfk_3i^2 < 0 \quad \text{(A9)}$$

which is always negative.

On the other hand, for $\gamma = 0$ and replacing Equation (25) on Equations (14) and (15), the time derivative along the system's trajectory is sought to comply with:

$$\dot{V} \;=\; Ei - 2P + Pk_1i - Pk_2iv < 0. \quad \text{(A10)}$$

Separating into two inequalities, one has:

$$\begin{aligned} Ei - Pk_2iv &< 0, \\ -2P + Pk_1i &< 0. \end{aligned} \quad \text{(A11)}$$

For the first one:

$$\begin{aligned} Ei < Pk_2iE &< Pk_2iv, \\ \frac{1}{P} &< k_2, \end{aligned} \quad \text{(A12)}$$

and for the second one:

$$\begin{aligned} Pk_1 i &< 2P, \\ i &< \frac{2}{k_1}. \end{aligned} \quad \text{(A13)}$$

Since V is a common Lyapunov function, the proof is complete. □

References

1. Sira-Ramirez, H.J.; Silva-Ortigoza, R. *Control Design Techniques in Power Electronics Devices*; Springer Science & Business Media: Berlin/Heidelberg, Germany, 2006.
2. Khaligh, A.; Nie, Z.; Lee, Y.J.; Emadi, A. *Integrated Power Electronic Converters and Digital Control*; CRC Press: Boca Raton, FL, USA, 2009.
3. Mazda, F.F. *Power Electronics Handbook: Components, Circuits and Applications*; Elsevier: Amsterdam, The Netherlands, 2016.
4. Sum, K.K. *Switch Mode Power Conversion: Basic Theory and Design*; Routledge: London, UK, 2017.
5. Luo, F.L.; Ye, H. *Power Electronics: Advanced Conversion Technologies*; CRC Press: Boca Raton, FL, USA, 2018.
6. Ayachit, A.; Kazimierczuk, M.K. Averaged small-signal model of PWM DC-DC converters in CCM including switching power loss. *IEEE Trans. Circuits Syst.* **2018**, *66*, 262–266. [CrossRef]
7. Man, T.Y.; Mok, P.K.; Chan, M.J. A 0.9-V input discontinuous-conduction-mode boost converter with CMOS-control rectifier. *IEEE J. Solid-State Circuits* **2008**, *43*, 2036–2046. [CrossRef]
8. Liu, K.H.; Lin, Y.L. Current waveform distortion in power factor correction circuits employing discontinuous-mode boost converters. In Proceedings of the 20th Annual IEEE Power Electronics Specialists Conference, Milwaukee, MI, USA, 26–29 June 1989; pp. 825–829.
9. Kolar, J.W.; Ertl, H.; Zach, F.C. Space vector-based analytical analysis of the input current distortion of a three-phase discontinuous-mode boost rectifier system. *IEEE Trans. Power Electron.* **1995**, *10*, 733–745. [CrossRef]
10. Lazar, J.; Cuk, S. Open loop control of a unity power factor, discontinuous conduction mode boost rectifier. In Proceedings of the 17th International Telecommunications Energy Conference, The Hague, The Netherlands, 29 October–1 November 1995; pp. 671–677.
11. Chan, C.; Pong, M. Input current analysis of interleaved boost converters operating in discontinuous-inductor-curent mode. In Proceedings of the 28th Annual IEEE Power Electronics Specialists Conference. Formerly Power Conditioning Specialists Conference 1970. Power Processing and Electronic Specialists Conference 1972, St. Louis, MO, USA , 27 June 1997; pp. 392–398.
12. Kolar, J.W.; Ertl, H.; Zach, F.C. A comprehensive design approach for a three-phase high-frequency single-switch discontinuous-mode boost power factor corrector based on analytically derived normalized converter component ratings. *IEEE Trans. Ind. Appl.* **1995**, *31*, 569–582. [CrossRef]
13. Ye, Z.Z.; Jovanovic, M.M. Implementation and performance evaluation of DSP-based control for constant-frequency discontinuous-conduction-mode boost PFC front end. *IEEE Trans. Ind. Electron.* **2005**, *52*, 98–107. [CrossRef]
14. Reatti, A.; Kazimierczuk, M.K. Small-signal model of PWM converters for discontinuous conduction mode and its application for boost converter. *IEEE Trans. Circuits Syst. Fundam. Theory Appl.* **2003**, *50*, 65–73. [CrossRef]
15. Lo, Y.K.; Lin, J.Y.; Ou, S.Y. Switching-frequency control for regulated discontinuous-conduction-mode boost rectifiers. *IEEE Trans. Ind. Electron.* **2007**, *54*, 760–768. [CrossRef]
16. Tsai, F.S. Small-signal and transient analysis of a zero-voltage-switched, phase-controlled PWM converter using averaged switch model. *IEEE Trans. Ind. Appl.* **1993**, *29*, 493–499. [CrossRef]
17. Tse, C.; Adams, K. Qualitative analysis and control of a DC-to-DC boost converter operating in discontinuous mode. *IEEE Trans. Power Electron.* **1990**, *5*, 323–330. [CrossRef]
18. Ferdowsi, M.; Emadi, A. Estimative current mode control technique for DC-DC converters operating in discontinuous conduction mode. *IEEE Power Electron. Lett.* **2004**, *2*, 20–23. [CrossRef]
19. Salimi, M.; Soltani, J.; Markadeh, G.A.; Abjadi, N.R. Indirect output voltage regulation of DC–DC buck/boost converter operating in continuous and discontinuous conduction modes using adaptive backstepping approach. *IET Power Electron.* **2013**, *6*, 732–741. [CrossRef]
20. Parui, S.; Banerjee, S. Bifurcations due to transition from continuous conduction mode to discontinuous conduction mode in the boost converter. *IEEE Trans. Circuits Syst. Fundam. Theory Appl.* **2003**, *50*, 1464–1469. [CrossRef]
21. Hai-Peng, R.; Ding, L. Bifurcation behaviours of peak current controlled PFC boost converter. *Chin. Phys.* **2005**, *14*, 1352. [CrossRef]
22. Ying, Z.B.Q. The Precise Discrete Mapping of Voltage-Fed DCM Boost Converter and Its Bifurcation and Chaos. *Trans. China Electrotech. Soc.* **2002**, *3*, 43–47.
23. Lai, J.S.; Chen, D. Design consideration for power factor correction boost converter operating at the boundary of continuous conduction mode and discontinuous conduction mode. In Proceedings of the Eighth Annual Applied Power Electronics Conference and Exposition, San Diego, CA, USA, 7–11 March 1993; pp. 267–273.

24. De Gusseme, K.; Van de Sype, D.M.; Van den Bossche, A.P.; Melkebeek, J.A. Digitally controlled boost power-factor-correction converters operating in both continuous and discontinuous conduction mode. *IEEE Trans. Ind. Electron.* **2005**, *52*, 88–97. [CrossRef]
25. Kancherla, S.; Tripathi, R. Nonlinear average current mode control for a DC-DC buck converter in continuous and discontinuous conduction modes. In Proceedings of the 2008 IEEE Region 10 Conference, Hyderabad, India, 19–21 November 2008; pp. 1–6.
26. Benadero, L.; El Aroudi, A.; Martínez-Salamero, L.; Tse, C.K. Period Doubling Route to Chaos in Open Loop Boost Converters under Constant Power Loading and Discontinuous Conduction Mode Conditions. In Proceedings of the 2020 IEEE International Symposium on Circuits and Systems (ISCAS), Seville, Spain, 12–21 October 2020; pp. 1–5.
27. Zheng, C.; Dragičević, T.; Zhang, J.; Chen, R.; Blaabjerg, F. Composite Robust Quasi-Sliding Mode Control of DC-DC Buck Converter With Constant Power Loads. *IEEE J. Emerg. Sel. Top. Power Electron.* **2020**, *9*, 1455–1464. [CrossRef]
28. Martinez-Trevino, B.A.; El Aroudi, A.; Valderrama-Blavi, H.; Cid-Pastor, A.; Vidal-Idiarte, E.; Martinez-Salamero, L. PWM Nonlinear Control with Load Power Estimation for Output Voltage Regulation of a Boost Converter with Constant Power Load. *IEEE Trans. Power Electron.* **2020**, *36*, 2143–2153. [CrossRef]
29. Gheisarnejad, M.; Farsizadeh, H.; Tavana, M.R.; Khooban, M.H. A Novel Deep Learning Controller for DC/DC Buck-Boost Converters in Wireless Power Transfer Feeding CPLs. *IEEE Trans. Ind. Electron.* **2020**, *68*, 6379–6384. [CrossRef]
30. Hassan, M.A.; He, Y. Constant Power Load Stabilization in DC Microgrid Systems Using Passivity-Based Control With Nonlinear Disturbance Observer. *IEEE Access* **2020**, *8*, 92393–92406. [CrossRef]
31. Tang, G.; Zhang, T.; Zhang, X. An Active Stabilization Control Strategy for DC-DC Converter Feeding Constant Power Load in Electric Vehicle. In Proceedings of the 2020 IEEE 4th Information Technology, Networking, Electronic and Automation Control Conference (ITNEC), Chongqing, China, 12–14 June 2020; pp. 375–379.
32. Li, X.; Zhang, X.; Jiang, W.; Wang, J.; Wang, P.; Wu, X. A Novel Assorted Nonlinear Stabilizer for DC-DC Multilevel Boost Converter with Constant Power Load in DC Microgrid. *IEEE Trans. Power Electron.* **2020**, *35*, 11181–11192. [CrossRef]
33. Martinez-Treviño, B.A.; El Aroudi, A.; Cid-Pastor, A.; Martinez-Salamero, L. Nonlinear control for output voltage regulation of a boost converter with a constant power load. *IEEE Trans. Power Electron.* **2019**, *34*, 10381–10385. [CrossRef]
34. Rahimi, A.M.; Emadi, A. Discontinuous-conduction mode DC/DC converters feeding constant-power loads. *IEEE Trans. Ind. Electron.* **2009**, *57*, 1318–1329. [CrossRef]
35. Li, Y.; Vannorsdel, K.R.; Zirger, A.J.; Norris, M.; Maksimovic, D. Current mode control for boost converters with constant power loads. *IEEE Trans. Circuits Syst. Regul. Pap.* **2011**, *59*, 198–206. [CrossRef]
36. Erickson, R.W.; Maksimovic, D. *Fundamentals of Power Electronics*; Springer Science & Business Media: Berlin/Heidelberg, Germany, 2007.
37. Neumann, K.; Steil, J.J. Learning robot motions with stable dynamical systems under diffeomorphic transformations. *Robot. Auton. Syst.* **2015**, *70*, 1–15. [CrossRef]
38. Moulay, E.; Bourdais, R.; Perruquetti, W. Stabilization of nonlinear switched systems using control lyapunov functions. *Nonlinear Anal. Hybrid Syst.* **2007**, *1*, 482–490. [CrossRef]
39. Zhao, X.; Kao, Y.; Niu, B.; Wu, T. *Control Synthesis of Switched Systems*; Springer: New York, NY, USA, 2017.
40. Khalil, H.K. *Nonlinear Systems*; Prentice Hall: Upper Saddle River, NJ, USA, 2002.

 micromachines

Article

Hybrid PWM Techniques for a DCM-232 Three-Phase Transformerless Inverter with Reduced Leakage Ground Current

Gerardo Vazquez-Guzman [1], Panfilo R. Martinez-Rodriguez [2,*], Jose M. Sosa-Zuñiga [1], Dalyndha Aztatzi-Pluma [3], Diego Langarica-Cordoba [2], Belem Saldivar [4,5], Rigoberto Martínez-Méndez [4]

1 Laboratory of Electrical and Power Electronics, Tecnologico Nacional de Mexico/ITS de Irapuato, Irapuato 36821, GTO, Mexico; gerardo.vazquez@ieee.org (G.V.-G.); jmsosa@ieee.org (J.M.S.-Z.)
2 School of Sciences, Universidad Autonoma de San Luis Potosi (UASLP), San Luis Potosi 78295, SLP, Mexico; diego.langarica@uaslp.mx
3 Department of Mechatronics Engineering, Tecnologico Nacional de Mexico/ITS de Abasolo, Abasolo 36976, GTO, Mexico; dalyndha.ap@abasolo.tecnm.mx
4 Faculty of Engineering, Autonomous University of the State of Mexico, Toluca 50130, MEX, Mexico; mbsaldivarma@conacyt.mx (B.S.); rmartinezme@uaemex.mx (R.M.-M.)
5 Cátedras CONACYT, Ciudad de Mexico 03940, CDMX, Mexico
* Correspondence: pamartinez@ieee.org

Abstract: Pulse Width Modulation (PWM) strategies are crucial for controlling DC–AC power converters. In particular, transformerless inverters require specific PWM techniques to improve efficiency and to deal with leakage ground current issues. In this paper, three hybrid PWM methods are proposed for a DCM-232 three-phase topology. These methods are based on the concepts of carrier-based PWM and space vector modulation. Calculations of time intervals for active and null vectors are performed in a conventional way, and the resulting waveforms are compared with a carrier signal. The digital signals obtained are processed using Boolean functions, generating ten signals to control the DCM-232 three-phase inverter. The performance of the three proposed PWM methods is evaluated considering the reduction in leakage ground current and efficiency. The proposed modulation techniques have relevant performances complying with international standards, which make them suitable for transformerless three-phase photovoltaic (PV) inverter markets. To validate the proposed hybrid PWM strategies, numerical simulations and experimental tests were performed.

Keywords: PV generation; space vector modulation; transformerless inverters; grid connection; leakage ground current

Citation: Vazquez-Guzman, G.; Martinez-Rodriguez, P.R.; Sosa-Zũniga, J.M.; Aztatzi-Pluma, D.; Langarica-Cordoba, D.; Saldivar, B.; Martínez Méndez, R. Hybrid PWM Techniques for a DCM-232 Three-Phase Transformerless Inverter with Reduced Leakage Ground Current. *Micromachines* **2022**, *13*, 36. https://doi.org/10.3390/mi13010036

Academic Editor: Francisco J. Perez-Pinal

Received: 7 December 2021
Accepted: 23 December 2021
Published: 28 December 2021

Publisher's Note: MDPI stays neutral with regard to jurisdictional claims in published maps and institutional affiliations.

1. Introduction

Transformerless photovoltaic (PV) systems have increased their popularity due to their high performance in terms of efficiency, size, and price. Nevertheless, the loss of galvanic isolation involves other challenges, for instance, reducing or eliminating leakage currents (LKC) that appear in the ground path.

In three-phase transformerless PV systems, conventional topologies such as the six-switch three-phase inverter or the three-phase cascade multilevel inverter (3P-CMI) generate high-frequency common-mode voltage (CMV) components, which due to the structure of the system, cause the common-mode current (CMC), also known as LKC. The CMC is the major issue in transformerless grid connected PV systems, as it can lead to additional power losses, protection tripping, important safety problems, and high total harmonic distortion (THD) of the current injected into the grid. Due to these issues, international norms have been developed to limit the CMC circulation through the PV system to guarantee the security of the system and humans in contact with it, for instance, the international standard DIN VDE 0126-1-1 establishes the maximum limit of the *RMS* value of the CMC in 300 mA. The scientific community has reported several techniques based on different approaches to mitigate the CMC. Among these, pulse width modulation (PWM) solutions have been popular because it is not necessary to increase the number of semiconductors in the topology or to implement complex control systems. Furthermore, additional functions such as voltage balancing of capacitors can be performed in split DC bus topologies as in neutral point clamped (NPC) inverters, and total system efficiency can be improved through a proper modulation design.

The six-switch three-phase inverter (3P-FB) has been widely studied under space vector modulation techniques for transformerless grid connection and motor drive applications. In [1], a near-state PWM (NSPWM) method was proposed; here, a comparison with similar PWM methods is performed. The NSPWM method makes use of a set of three voltage vectors to match the output and volt-second references. The three voltage vectors are two adjacent vectors together with a near-neighbor vector; then, nonzero voltage vectors are utilized. The sectors are displaced from each other by 30°; therefore, new regions are defined with respect to the conventional distribution. As no zero voltage vectors are used, the CMV does not take values produced by those vectors; therefore, CMV variations are reduced. In the case of PV systems, in [2], an evaluation of three-phase converters without galvanic isolation is reported. The analysis considers the conventional 3P-FB, the 3P-FB with split capacitor (3P-FBSC), and the three-phase NPC inverter (3P-NPC). The results demonstrate that the 3P-FBSC and 3P-NPC inverters produce low CMC, which make them two suitable solutions in three-phase transformerless inverters.

Derived from the 3P-FB inverter topology, some modified topologies have been also proposed to solve the CMC issue. For example, by adding passive components, as in the Z-source inverter topology presented in [3], it is possible to avoid the use of a boost stage at the input of the system. Additionally, the CMC magnitude can be reduced by modifying the PWM strategy as in [4], where the authors proposed a modified Z-source inverter and a space vector based modulation (SVM) technique that reduces the CMC magnitude. Moreover, in [5–7], a family of topologies called Quasi Z-source inverters is presented, the main idea is to reduce the component rating, the source stress, and component count and to make some contributions to simplify the control strategies. Another approach consists in the addition of active components in order to deal with the CMC, for instance, diodes, IGBTs, or MOSFETs as in [8], where additional diodes connected as a three-phase rectifier plus an IGBT are used to implement the null vectors in an SVM technique. The main idea is to connect the output of the inverter to $Vdc/3$ or to $2Vdc/3$, thus avoiding some CMV transitions and reducing the magnitude of the CMC. Another alternative based on the traditional 3P-FB inverter is to use the idea of DC decoupling developed in the H5 and H6 topologies for single-phase systems [9,10]. In this case, [11,12] proposed the H7 topology and a study of several SVM techniques to reduce the CMC, and [13] proposed that the H8 topology and its SVM technique are presented. In both cases, the main idea is to disconnect the DC bus during the null vectors, which in combination with an adequate sequence of active vectors allows us to mitigate the CMC. Another topology that is a combination of a NPC topology and the 3P-FB inverters is presented in [14] where a type of NPC circuit is added at the output of the inverter following the idea of the HERIC single-phase inverter [15]. The null vectors are now implemented in a freewheeling circuit, which reduces the transitions of the CMV in the circuit, obtaining a reduction in the CMC magnitude.

The DCM-232 topology and its space vector PWM strategy have been designed to deal with the CMC issue. The main objective as in the aforementioned cases is to reduce the fast changes in the CMV by as much as possible and consequently to decrease the magnitude of the CMC. This inverter is based also on the 3P-FB topology at the AC output side, while on the DC input side, there are two DC sources that can be completely decoupled from the 3P-FB circuit by means of two semiconductors switched at the same time [16]. The PWM strategy is based on the space vector PWM technique, where the main difference is that the active vectors are implemented in the 3P-FB circuit and that the zero vectors are implemented by decoupling the DC sources using power semiconductor switches. In the literature, some PWM strategies have also been designed for the DCM-232 inverter; see for instance [17], where a carrier-based PWM is proposed to solve the CMC issue using the principle explained above.

In this paper, three PWM methods based on the Space Vector PWM (SVPWM) technique are studied to reduce the CMC components. The proposed modulation strategies are used to control the three-phase DCM-232 topology. The time intervals to control on and off conditions for each switch are defined using the waveforms obtained by means of the SVPWM concept and then a comparison with a triangular carrier signal is performed. Finally, the resulting signals are processed by a Boolean function implemented in a Complex Programable Logic Device (CPLD) to determine the final sequence for each switch. The DCM-232 topology consists in a 3P-FB inverter plus four switches that decouples the signal to generate two independent DC sources, i.e., two PV generators. The main idea is to control the decoupling switches in order to keep the CMV constant, thereby achieving a reduction in the CMC. The CMV evaluation is performed by driving the Common Mode Model (CMM) of the DCM-232 inverter. In addition, the paper considers an efficiency analysis based on numerical results obtained by means of the implementation of the real models of the semiconductors. Finally, a comparative analysis between the proposed PWM techniques and some solutions available in the literature is performed.

2. Topology Description and Proposed Space Vector PWM Techniques

A simplified circuit of the DCM-232 topology considered for the design of the SVPWM techniques is shown in Figure 1. One of the main considerations for the simplified circuit is that the DC sources are assumed constant. However, in a real PV transformerless system, the voltage magnitude is slightly variable and, in that case, it is necessary to implement a solution, for example, modifying the modulation strategy or implementing a balance control loop; nevertheless, this topic is out of the scope of this paper and is left for a future research. Three additional important considerations of the simplified diagram are that only the stray capacitance generated by the PV panel are considered; the drive circuits for the semiconductors and control system are not included because these elements do not affect the common mode behavior. Finally, in the case of the ground impedance, the capacitive and inductive effects are disregarded in the system. Therefore, the impedance is considered mainly resistive [18]. The different states that can be proposed to control the DCM-232 inverter in which the structure consists of ten switches are summarized in Table 1. As it can be observed, there are eight possible states. These states produce the following voltage levels between phases and the neutral connection: $V_{DC1} = V_{DC2}$, $2V_{DC}/3$, $V_{DC}/3$, $0V$, $-V_{DC}/3$, and $-2V_{DC}/3$. It is important to note that all of the states considered here are exactly the same as in the conventional 3P-FB inverter. However, the main difference is that the switches on the DC side, S_{7a},S_{7b}, S_{8a}, and S_{8b}, are used to make a decoupling action when certain active or null vectors supply the load.

2.1. DCM-232 Common Mode Model

In order to determine the CMV behavior, a CMM for the DCM-232 topology is derived. Considering the directions given in [19], the simplified CMM shown in Figure 2 can be obtained. As it can be observed, there are two separated circuits, the circuit in Figure 2a corresponds to the DC source V_{DC1}, while the simplified circuit shown in Figure 2b corresponds to the DC source V_{DC2}. Since the power sources are isolated from each other with a common load, the obtained model is also separated and is essentially the model obtained in [2] for the 3P-FB inverter. Based on that, the common mode voltage in this topology can be calculated in each DC source using (1) and (2). Considering (1) and (2), the CMV can be calculated for each state defined in Table 1, and the results are shown in Table 2. It can be noted that the CMV maintains the same magnitude throughout the switching period when the odd vectors (V_1, V_3 and V_5) or the even vectors (V_2, V_4 and V_6) are connected to the load using V_{DC1} and V_{DC2}, respectively. On the other hand, when zero vectors are generated (V_0 and V_7), the CMV keeps the previous value because both DC sources are decoupled from the load.

$$V_{CMV1} = \frac{V_{aZ1} + V_{bZ1} + V_{cZ1}}{3}, \quad (1)$$

$$V_{CMV2} = \frac{V_{aZ2} + V_{bZ2} + V_{cZ2}}{3}. \tag{2}$$

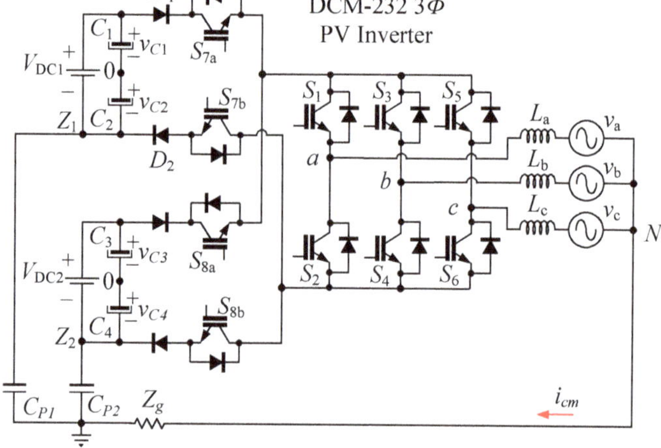

Figure 1. DCM-232 three-phase transformerless PV inverter topology.

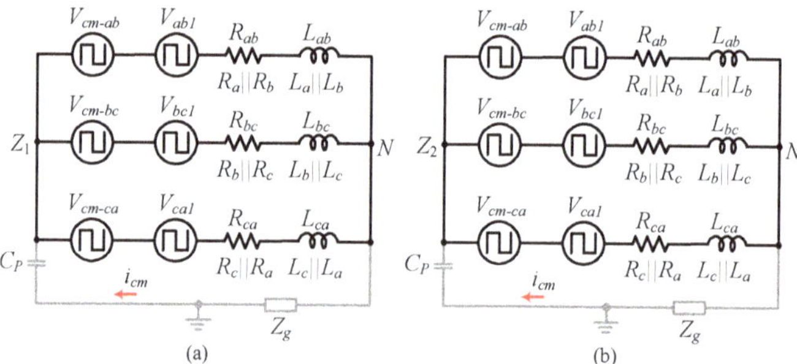

Figure 2. DCM-232 common mode model: (**a**) V_{DC1} and (**b**) V_{DC2}.

Table 1. Switching configurations of the three-phase DCM-232 transformerless inverter.

State (Vector)	S_1	S_3	S_5	S_{7a}/S_{7b}	S_{8a}/S_{8b}	v_{aN} (V)	v_{bN} (V)	v_{cN} (V)
V_0	0	0	0	0	0	0	0	0
V_1	1	0	0	1	0	$\frac{2V_{DC1}}{3}$	$\frac{-V_{DC1}}{3}$	$\frac{-V_{DC1}}{3}$
V_2	1	1	0	0	1	$\frac{V_{DC2}}{3}$	$\frac{V_{DC2}}{3}$	$\frac{-2V_{DC2}}{3}$
V_3	0	1	0	1	0	$\frac{-V_{DC1}}{3}$	$\frac{2V_{DC1}}{3}$	$\frac{-V_{DC1}}{3}$
V_4	0	1	1	0	1	$\frac{-2V_{DC2}}{3}$	$\frac{V_{DC2}}{3}$	$\frac{V_{DC2}}{3}$
V_5	0	0	1	1	0	$\frac{-V_{DC1}}{3}$	$\frac{-V_{DC1}}{3}$	$\frac{2V_{DC1}}{3}$
V_6	1	0	1	0	1	$\frac{V_{DC2}}{3}$	$\frac{-2V_{DC2}}{3}$	$\frac{V_{DC2}}{3}$
V_7	1	1	1	0	0	0	0	0

Table 2. Switching configurations of the three-phase DCM-232 transformerless inverter.

State (Vector)	V_{CMV1}	V_{CMV2}
V_0 to V_7	$\frac{V_{DC}}{3}$	$\frac{2V_{DC}}{3}$

2.2. Proposed Space Vector PWM Techniques

The circuit theory states that a three-phase system can be represented on a α-β plane by means of the Clarke transformation, as shown in (3)–(5). Note that a simplification using cosine functions for the three-phase voltage components (v_a, v_b, and v_c) is considered.

$$\begin{bmatrix} v_\alpha \\ v_\beta \end{bmatrix} = \frac{2}{3}\begin{bmatrix} 1 & -\frac{1}{2} & -\frac{1}{2} \\ 0 & \frac{\sqrt{3}}{2} & -\frac{\sqrt{3}}{2} \end{bmatrix}\begin{bmatrix} v_a \\ v_b \\ v_c \end{bmatrix}, \tag{3}$$

$$v_\alpha = V_m \cos(\omega t), \tag{4}$$

$$v_\beta = V_m \sin(\omega t). \tag{5}$$

Based on (4) and (5), the module and the angle of the reference vector, V_{ref}, can be calculated as

$$|V_{ref}| = \sqrt{{v_\alpha}^2 + {v_\beta}^2}, \tag{6}$$

$$\tan(\theta) = \frac{v_\beta}{v_\alpha}. \tag{7}$$

Then, substituting (4) and (5) in (7), the module of the reference vector can be redefined as

$$|V_{ref}| = \sqrt{(V_m \cos(\omega t))^2 + (V_m \sin(\omega t))^2},$$

$$|V_{ref}| = V_m, \tag{8}$$

and finally, from (7), the angle θ can be calculated as follows:

$$\theta = \tan^{-1}\left(\frac{v_\beta}{v_\alpha}\right). \tag{9}$$

From the above analysis, the eight states of the DCM-232 inverter can be represented on the α-β plane, as shown in Figure 3. Considering this representation, it should be noted that this is similar to that of the space vector representation for a 3P-FB inverter; however, in this particular case, the zero vectors imply the decoupling of the DC sources performed by switches S_{7a}, S_{7b}, S_{8a}, and S_{8b}.

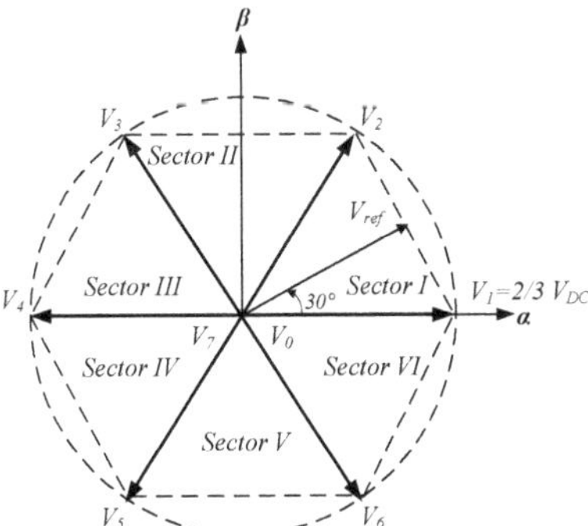

Figure 3. Space vector representation of the DCM-232 states.

Considering the polar representation, the eight vectors in the complex plane can be written as

$$V_n = \begin{cases} \frac{2}{3}V_{DC}e^{j\frac{\pi}{3}(n-1)} & n = 1,\ldots,6 \\ 0 & n = 0,7 \end{cases} \tag{10}$$

For a balanced three-phase system, V_{ref} can be expressed as

$$V_{ref} = Ve^{j\omega t}. \tag{11}$$

In order to synthesize V_{ref}, three successive space vectors can be applied along a switching period ($T_s = \frac{1}{f_s}$). Therefore, the addition of the applied vectors (active and/or null) must satisfy

$$V_a t_a + V_b t_b + V_N t_0 = V_{ref} T_s, \tag{12}$$

Notice that the switching period is the sum of the times of each applied vector:

$$t_a + t_b + t_0 = T_s. \tag{13}$$

To determine the duty cycles for each applied vector, the complex components (V_A and V_B) in the α-β plane for V_{ref} can be defined as

$$V_{ref} = \begin{cases} V_A = \frac{1}{T_s} V_a t_a \\ V_B = \frac{1}{T_s} V_b t_b. \end{cases} \tag{14}$$

The complex components V_A and V_B defined in (14) can be represented in Sector 1, as shown in Figure 4. As can be observed, $V_a = V_1$ and $V_b = V_2$. Therefore, analyzing and performing the projections of V_{ref} over the α and β axis yields

$$|V_A|\angle 0^\circ,$$

$$|V_B|\angle 60^\circ,$$

$$|V_{ref}| = V_m,$$

$$|V_{ref}|\cos(\theta) = |V_A| + |V_B|\sin(60^\circ), \tag{15}$$

$$|V_{ref}|\sin(\theta) = |V_B|\sin(60^\circ). \tag{16}$$

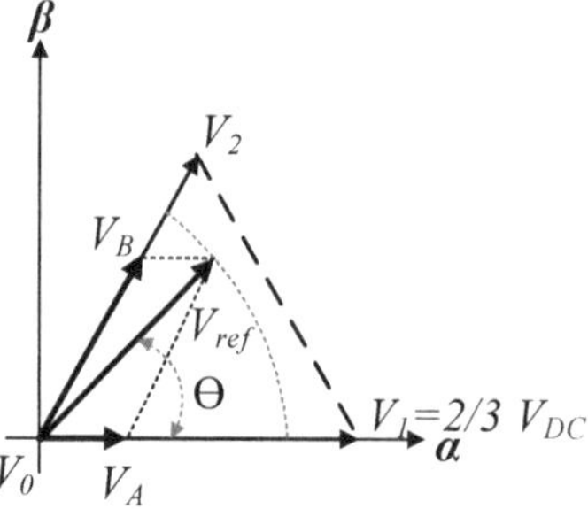

Figure 4. Analysis of V_{ref} at the Sector 1.

Solving for $|V_B|$ from (16),

$$|V_B| = \frac{|V_{ref}|\sin(\theta)}{\sin(60^\circ)}; \tag{17}$$

solving for $|V_A|$ from (15); and substituting (17) yields

$$|V_A| = |V_{ref}|\cos(\theta) - |V_B|\sin(60^\circ),$$

$$|V_A| = |V_{ref}|\left(\frac{\sin(60^\circ)\cos(\theta) - \cos(60^\circ)\sin(\theta)}{\sin(60^\circ)}\right).$$

Then, using the following trigonometric identity,

$$\sin(a-b) = \sin(a)\cos(b) - \cos(a)\sin(b),$$

yields

$$|V_A| = \frac{|V_{ref}|\sin(60^\circ - \theta)}{\sin(60^\circ)}. \tag{18}$$

Substituting the components (14) in (17) and (18),

$$\frac{1}{T_s}V_b t_b = \frac{|V_{ref}|\sin(\theta)}{\sin(60^\circ)},$$

$$\frac{1}{T_s}V_a t_a = \frac{|V_{ref}|\sin(60^\circ - \theta)}{\sin(60^\circ)}.$$

Now, solving for t_a,

$$t_a = \frac{|V_{ref}|T_s\sin(60^\circ - \theta)}{V_a\sin(60^\circ)},$$

and then solving for t_b,

$$t_b = \frac{|V_{ref}|T_s\sin(\theta)}{V_b\sin(60^\circ)},$$

and since $V_a = V_b = \frac{2}{3}V_{DC}$ and $\sin(60^\circ) = \frac{\sqrt{3}}{2}$,

$$t_a = \frac{\sqrt{3}|V_{ref}|T_s\sin(60^\circ - \theta)}{V_{DC}} \tag{19}$$

and

$$t_b = \frac{\sqrt{3}|V_{ref}|T_s\sin(\theta)}{V_{DC}}. \tag{20}$$

Finally, solving for t_0 from (13) yields

$$t_0 = T_s - t_a - t_b. \tag{21}$$

Equations (19)–(21) are a general solution for t_a, t_b, and t_0 since the times used for each active or null vector along the grid period are the same. The evolution of the calculated times and PWM switching signals along a grid period is depicted in Figure 5, where A is the waveform for the evolution of t_a, B is the waveform for the evolution of t_b, and C is the waveform for the evolution of t_0.

By using these time calculations and by considering that the main objective of the DCM-232 topology is to reduce the CMC by decoupling the DC sources, a simple way to obtain a constant common mode voltage for different vector sequences is proposed. Considering the active vectors given in Table 1 and the operation states of the DCM-232 inverter as logic states, the following Boolean expressions are obtained:

$$S_7 = S_1\overline{S_3 S_5} + \overline{S_1}S_3\overline{S_5} + \overline{S_1 S_3}S_5, \tag{22}$$

$$S_8 = S_1 S_3\overline{S_5} + \overline{S_1}S_3 S_5 + S_1\overline{S_3}S_5. \tag{23}$$

According to (22) and (23), the switches S_7 and S_8 are in the active state when the corresponding logic states involved in each equation comply with the logic conditions. Therefore, only when the active vectors appear in the modulation sequences are S_7 and S_8 turned on.

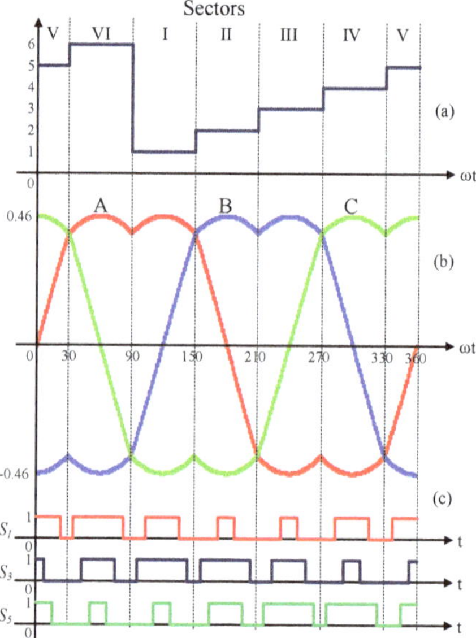

Figure 5. From top to bottom, (**a**) sectors; (**b**) evolution of the switching times t_a, t_b, and t_0; and (**c**) switching sequences for S_1, S_3, and S_5 in a conventional three-phase full-bridge inverter.

Based on the above analysis, three different modulation strategies are proposed in this paper to control the DCM-232 topology using the proposed technique. Note that any vector sequence can be adopted to control the inverter. In this paper, the proposed SVM strategies are based in the conventional SVM for a three-phase full-bridge inverter, named Conventional Symmetric Space Vector Modulation (CSSVM), Conventional Asymmetric Space Vector Modulation (CASVM), and Discontinuous Space Vector Modulation Maximum (DSVMMAX). The switching patterns for these three proposed SVM strategies are depicted in Figure 6. Note that the switching pattern for CSSVM and CASVM strategies is the same, and the main difference is the way in which the times t_a, t_b, and t_0 are computed, as shown below.

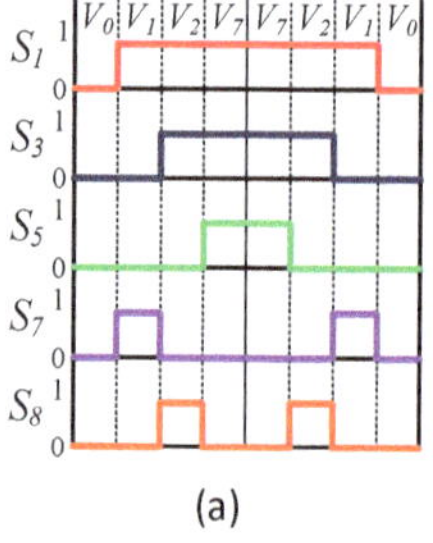

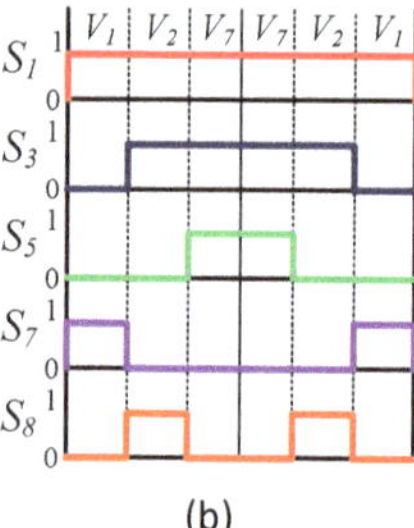

Figure 6. Switching patterns for (**a**) CSSVM and CASVM, and (**b**) DSVMMAX.

3. Numerical Results

To validate the proposed SVM technique for the three-phase DC-232 inverter, the numerical results are reported using the parameters shown in Table 3. It is important to highlight that the numerical results are obtained in an open loop configuration, since the main objective is to validate the proposed SVM techniques operating with this topology. The general scheme implemented in the PSIM software is shown in Figure 7, where (a) the three-phase signals, (b) the Clark transformation, (c) the module and angle of the reference vector, (d) time vector calculation, (e) the reference signals for the space vectors, and (f and g) PWM signal generation are presented. In the particular case of the block (e), the reference signals are defined in different ways for the three-proposed SVM strategies. In Figure 8, the reference signals for CSSVM, CASVM, and DSVMMAX are depicted. Note that the reference signals for CSSVM have only positive values, CASVM is centered at zero and has positive and negative values, and DSVMMAX is centered at zero but has an unsymmetrical waveform. Note that these reference signals are generated by the addition of the time intervals calculated for each vector along each sector and their magnitude is related to the switching period. Moreover, the block depicted in (f) is dedicated to generate the PWM signals, in particular, the signals for the switches on the DC side are generated using the digital circuit depicted in Figure 9 according to (22) and (23).

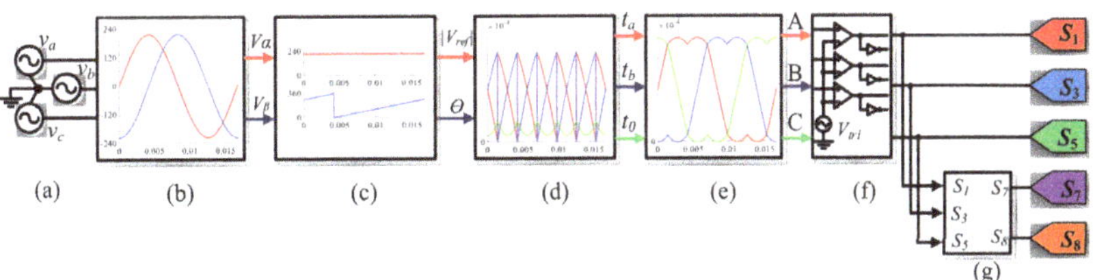

Figure 7. Block diagram of the implemented system in the PSIM software: the block (**a**) represents the three-phase signals; (**b**) is the Clarke transformation; (**c**) is the module and angle of the reference vector; (**d**) is the time vectors calculation; (**e**) reference signals for space vectors; and (**f**,**g**) represent the PWM signal generation.

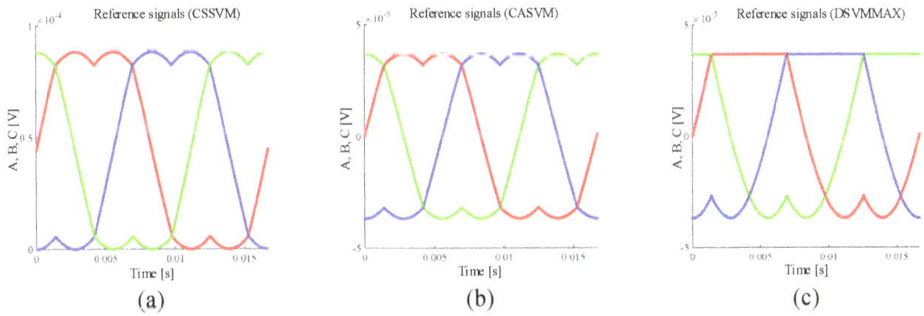

Figure 8. Reference signals for (**a**) CSSVM, (**b**) CASVM, and (**c**) DSVMMAX.

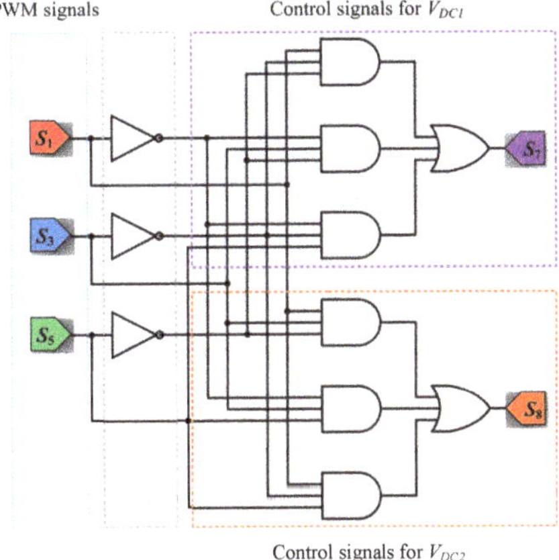

Figure 9. Digital circuit to generate the PWM signals for S7 and S8 switches.

Table 3. Simulation and experimental parameters.

Parameter	Value
V_{DC1}, V_{DC2}	400 V
f_s	10 and 12 kHz
t_d	1 µs
L_a, L_b, L_c	2 mH
R_a, R_b, R_c	71.43 Ω
Z_g	22 Ω
C_1, C_2, C_3, C_4	2200 µF
$C_{p1}, C_{p2}, C_{p3}, C_{p4}$	160 nF
M_i	0.8

The numerical results were obtained for the three SVM techniques; however, the waveforms for the output currents and voltages are very similar in the three cases. Therefore, for brevity, only the waveforms for the CSSVM technique are included. Figure 10 shows the simulation results for the three-phase output currents, line-to-neutral voltages, and line-to-line voltages. As observed, these waveforms are similar to those typical waveforms of a three-phase conventional inverter. It can be observed also that the switching ripple appears at the sinusoidal current waveforms in which peak current is around 2.5 A. It is important to note that, under these conditions, the ripple magnitude is large and the measured THDi is around 16%, which is not allowed by the international standard, for instance, IEEE 519-2014 (considering a low grid voltage). It should be also noted that, in this case, a first-order low-pass filter is used at the output of the inverter (see Figure 11), so this can be improved by implementing a third-order filter. It is possible to increase the switching frequency or the rated power to improve the THDi as well; however, in this case, these parameters are limited by the experimental setup. On the other hand, the common mode voltage and current are also obtained by simulations; in this case, the three sets of results are presented in Figure 12. As it can be observed, the results show that these two parameters are similar for the three proposed cases. In the case of the CMV, the magnitude is predominantly constant, and in the case of the CMC, the maximum value is around 40 mA, which is well below the limits imposed by the DIN VDE 0126-1-1 international standard, which is up to 300 mA (RMS). It should be noted that the numerical simulations were performed considering a balanced DC-bus, that is $V_{DC1} = V_{DC2}$. However, in a transformerless PV application, when the irradiance changes along a day, the voltage at the maximum power point also changes. This variation, which is typically around 10% to 20%, produces a DC component at the output of the DCM-232 inverter. Under these conditions, the inverter is still capable of operating but with a DC component which is not allowed. To solve this problem, a balance technique should be implemented, and this can be solved either by implementing a balance control loop or by modifying the modulation strategy; however, this issue is out of the scope of this paper.

To better compare the three SVM strategies under study, an efficiency analysis was performed. For this purpose, the IGBT model was loaded into the Thermal Module of the PSIM software and the total losses of the DCM-232 converter were calculated. The model loaded in the software considers the parameters provided by the manufacturer; then, the behavior of the power losses is expected to be close to the real behavior. The results of the switching and conduction losses are shown in Figure 13. As can be observed, the sum of the switching and conduction power losses is greater for the DSVMMAX with respect to the other techniques, while the CASVM presents the lowest total power losses. Therefore, the CASVM should be expected to present the highest efficiency. To validate this parameter, the efficiency of the system was also measured and the results are presented in Table 4, where, as expected, the CASVM technique presents the best efficiency.

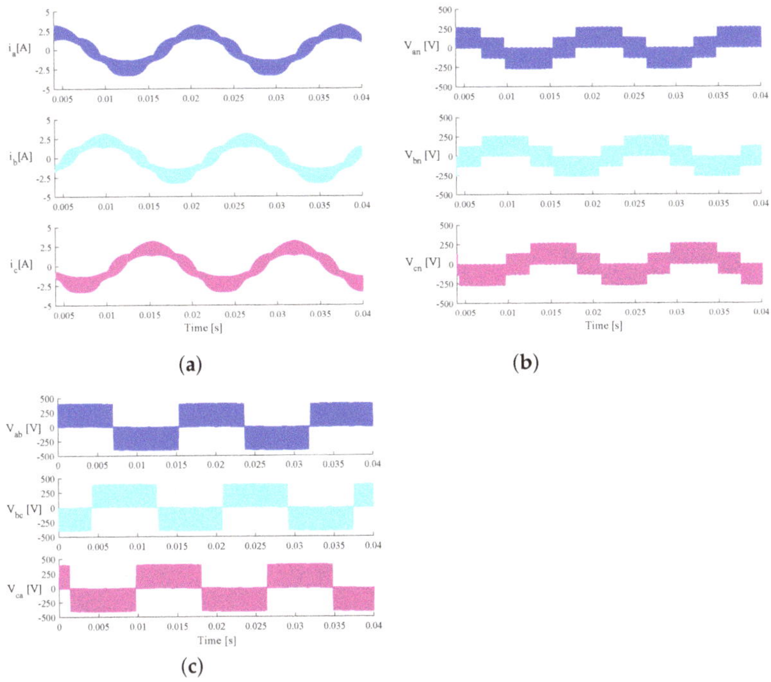

Figure 10. Output currents and voltages of the three-phase DCM-232 inverter under SVM-CSSVM technique. (**a**) Three-phase output currents, (**b**) line-to-neutral voltages, and (**c**) line-to-line voltages.

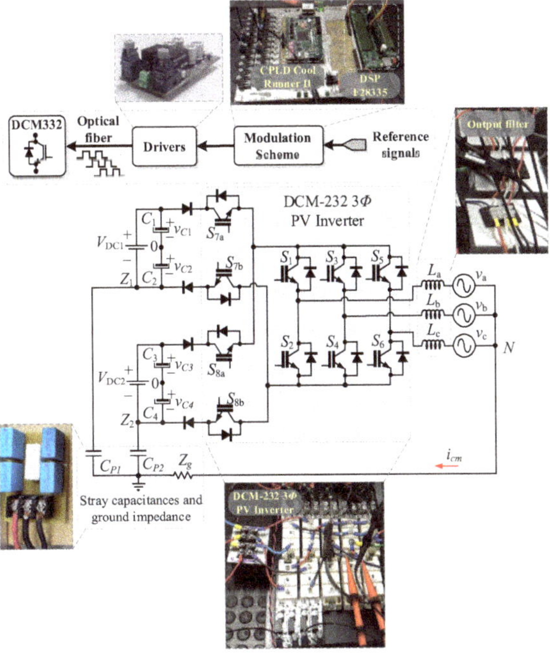

Figure 11. Experimental setup of the DCM-232 inverter.

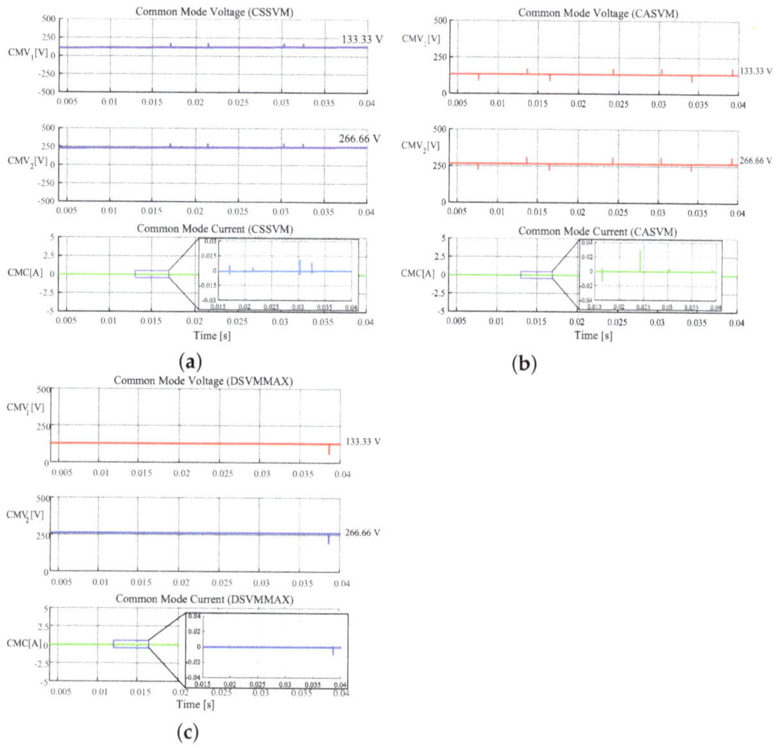

Figure 12. Simulation results of common mode voltage (CMV) and common mode current (CMC) of the three-phase DCM-232 inverter under the (**a**) CSSVM, (**b**) CASVM, and (**c**) DSVMMAX techniques.

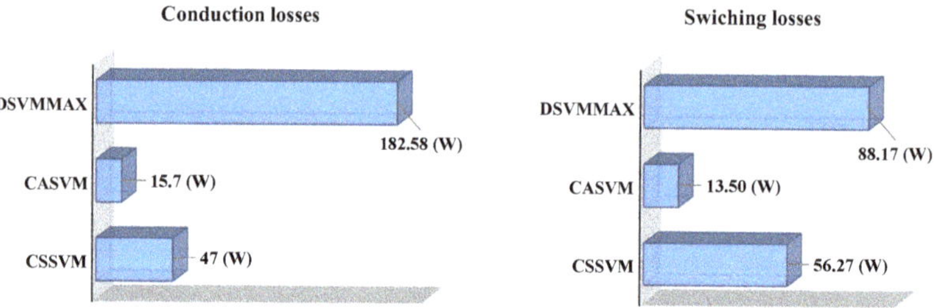

Figure 13. Switching and conduction losses of the DCM-232 inverter under the three SVM strategies.

Table 4. Efficiency of the DCM-232.

SVM Strategy	Efficiency (%)
CSSVM	85.87
CASVM	95.85
DSVMMAX	85.50

Considerations for a Practical Implementation

The DCM-232 three-phase inverter was implemented as a laboratory prototype to validate the proposed SVM strategies. A flow chart of the implementation process is shown in Figure 14. The algorithms for the SVM strategies were implemented using a Digital Signal Processor (DSP) TMS320F28335 device together with the PSIM software.

Additionally, the digital functions for the PWM signals were implemented in a complex programmable logic device (CPLD) CoolRunner-II according with Figure 14. The power module SKM50GB12T4 was used to implement the DCM-232 three-phase inverter, and the diodes D_1 to D_4 were implemented using the power module 200RD4TVL. The electrical parameters are in accordance with the parameters used for the simulation test, as listed in Table 3. A simplified block diagram of the experimental setup is depicted in Figure 11. The ground path was implemented by connecting the neutral point of the RL load to the available terminal of the parasitic capacitors C_{p1}, C_{p2}, C_{p3}, and C_{p4} through a resistance with a value equal to 22 Ω.

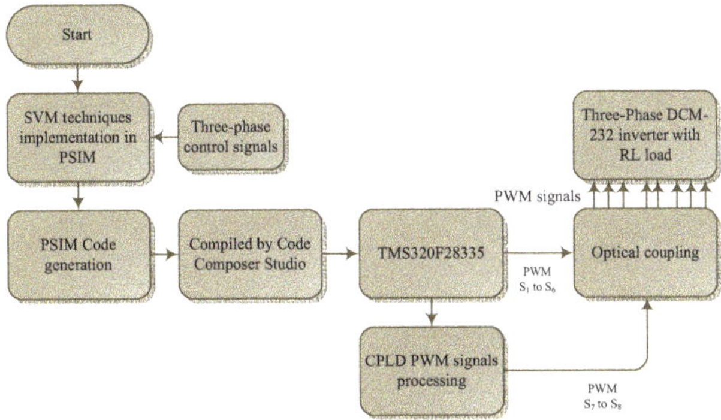

Figure 14. Flowchart for the experimental implementation.

4. Experimental Validation

The experimental implementation was performed considering the parameters and conditions described in Section 3 and in Table 3. In Figure 15, the output currents in (a), the line-to-neutral voltages in (b), and the line-to-line voltages in (c) for the CSSVM strategy are presented. As can be observed, the waveforms are similar to those obtained for the simulation results. Namely, the output current for the three phases is sinusoidal plus the switching ripple, the voltage between each phase and the neutral connection has the typical five levels, and the voltages between lines are also the typical of a full bridge three-phase system. Since these waveforms are close to the waveforms of the other two proposed SVPWM strategies, only the results for the CSSVM are included. In Figure 16, the results obtained for the common mode behavior for (a) CSSVM, (b) CASVM, and (c) DSVMMAX at the DC bus 1 and the common mode current are shown. Notice that, in all cases, the CMC has a value below 300 mA, which is established by the international norm DIN VDE 0126-1-1 as the maximum allowable limit. Note that the CMV_1 waveform contains a noise component, which is due to the oscilloscope internal calculations and the effect of parasitic components during the switching process. Moreover, the CMV regarding the second DC bus CMV_2 has also been obtained. The average value of this parameter is close to $266.66V$, which is the value obtained by means of simulations; however, and considering that the waveform is similar to the signal presented for the CMV_1, the last is not included in the paper.

In order to better compare the results obtained by the implementation of the proposed SVM algorithms regarding CMC, the measures are summarized in Table 5. As can be noted, the RMS value for the CMC for the three proposed cases is similar and complies with the international standard mentioned before. Moreover, the results regarding CMC were compared with the conventional full-bridge three-phase inverter (3PFB-VSI) under SVM, and it can be noted that the CMC has a larger magnitude regarding the proposed modulation topology and SVPWM algorithms.

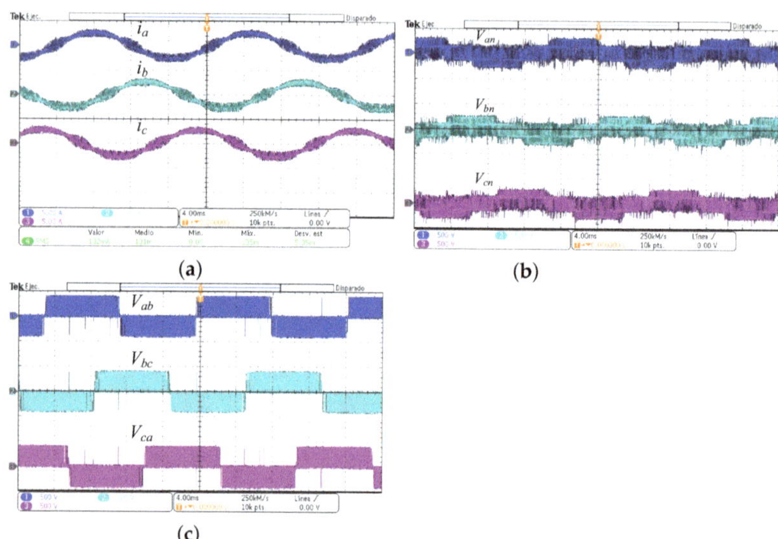

Figure 15. Experimental output currents and voltages of the three-phase DCM-232 inverter under the SVM-CSSVM technique. (**a**) Output currents, (**b**) line-to-neutral voltages, and (**c**) line-to-line voltages.

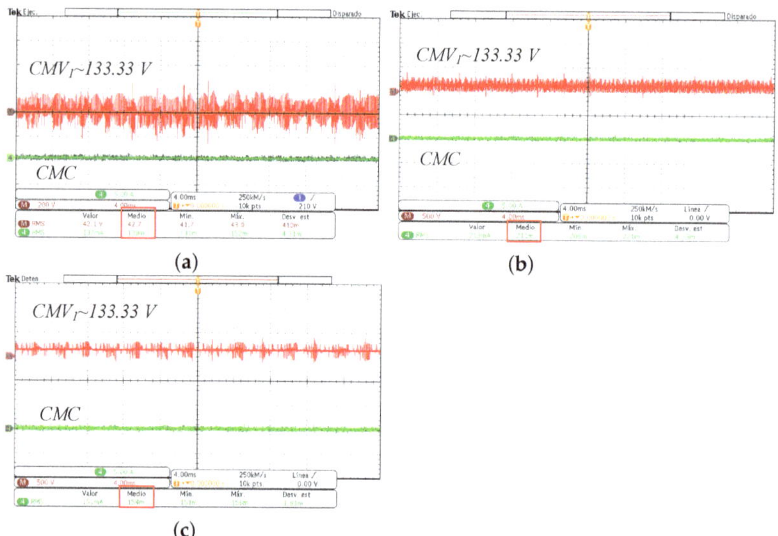

Figure 16. Experimental common mode voltage (CMV_1) and common mode current (CMC) of the three-phase DCM-232 inverter under the (**a**) CSSVM, (**b**) CASVM, and (**c**) DSVMMAX techniques.

Table 5. CMC magnitudes of the DCM-232 under the proposed SVPWM strategies.

SVM Strategy	CMC (mA_{RMS})
CSSVM	140
CASVM	145
DSVMMAX	156
3PFB-VSI	1630

5. Conclusions

In this paper, three hybrid modulation strategies were proposed. The proposed modulation strategies combine the time calculations of the space vector modulation technique and the comparison of the reference signals with a carrier triangular waveform, as does the sinusoidal pulse width modulation strategies. The results show that the SVM algorithms for the DCM-232 can be implemented using this combined method, which is convenient for practical implementation in a digital platform. Moreover, an evaluation of the efficiency was performed, and the results show that the modulation sequence has an important effect on this parameter. This analysis demonstrates that the CASVM technique has the highest efficiency (95.85%) among the proposed modulations algorithms, although all of the proposed methods present an efficiency above 85.5%. Finally, it can be also concluded that any of the proposed techniques is capable of reducing the magnitude of the common mode current (i.e., up to 140 mA_{RMS} for the CSSVM technique), which is a very interesting feature for transformerless photovoltaic applications since the power inverter operation complies with the international standards.

Author Contributions: All authors contributed to the development of the overall document, space vector modulation design, analysis, simulations, and experimental results. All authors have read and agreed to the published version of the manuscript.

Funding: This research was partially funded by Tecnologico Nacional de Mexico under the project 10138.21-PD.

Conflicts of Interest: The authors declare no conflict of interest.

Abbreviations

The following abbreviations are used in this paper:

PWM	Pulse Width Modulation
DC	Direct Current
DC-AC	Direct Current–Alternating Current
PV	Photovoltaic
LKC	Leakage Currents
3P-CMI	Three-Phase Cascade Multilevel Inverter
CMV	Common-Mode Voltage
CMC	Common-Mode Current
THD	Total Harmonic Distortion
RMS	Root Mean Square
NPC	Neutral Point Clampled
3P-FB	Six-Switch Three-Phase Inverter
NSPWM	Near-State Pulse Width Modulation
3P-FBSC	Six-Switch Three-Phase Inverter with Split Capacitor
3P-NPC	Three-Phase Neutral Point Clamped Inverter
SVM	Space Vector Modulation
IGBT	Isolated Gate Bipolar Transistor
MOSFET	Metal-Oxide Semiconductor Field-Effect Transistor
HERIC	Highly Efficient and Reliable Inverter Concept
SVPWM	Space Vector Pulse Width Modulation
CPLD	Complex Programmable Logic Device
CMM	Common-Mode Model
CSSVM	Conventional Symmetric Space Vector Modulation
CASVM	Conventional Asymmetric Space Vector Modulation
DSVMMAX	Discontinuous Space Vector Modulation Maximum
DSP	Digital Signal Processor

References

1. Un, E.; Hava, A.M. A Near-State PWM Method With Reduced Switching Losses and Reduced Common-Mode Voltage for Three-Phase Voltage Source Inverters. *IEEE Trans. Ind. Appl.* **2009**, *45*, 782–793. [CrossRef]
2. Kerekes, T.; Teodorescu, R.; Liserre, M.; Klumpner, C.; Sumner, M. Evaluation of Three-Phase Transformerless Photovoltaic Inverter Topologies. *IEEE Trans. Power Electron.* **2009**, *24*, 2202–2211. [CrossRef]
3. Huang, Y.; Shen, M.; Peng, F.Z.; Wang, J. Z-Source Inverter for Residential Photovoltaic Systems. *IEEE Trans. Power Electron.* **2006**, *21*, 1776–1782. [CrossRef]
4. Bradaschia, F.; Cavalcanti, M.C.; Ferraz, P.E.P.; Neves, F.A.S.; dos Santos, E.C.; da Silva, J.H.G.M. Modulation for Three-Phase Transformerless Z-Source Inverter to Reduce Leakage Currents in Photovoltaic Systems. *IEEE Trans. Ind. Electron.* **2011**, *58*, 5385–5395. [CrossRef]
5. Park, J.; Kim, H.; Nho, E.; Chun, T.; Choi, J. Grid-connected PV System Using a Quasi-Z-source Inverter. In Proceedings of the 2009 Twenty-Fourth Annual IEEE Applied Power Electronics Conference and Exposition, Washington, DC, USA, 15–19 February 2009; pp. 925–929.
6. Li, Y.; Anderson, J.; Peng, F.Z.; Liu, D. Quasi-Z-Source Inverter for Photovoltaic Power Generation Systems. In Proceedings of the 2009 Twenty-Fourth Annual IEEE Applied Power Electronics Conference and Exposition, Washington, DC, USA, 15–19 February 2009; pp. 918–924.
7. Anderson, J.; Peng, F.Z. Four quasi-Z-Source inverters. In Proceedings of the 2008 IEEE Power Electronics Specialists Conference, Rhodes, Greece, 15–19 June 2008; pp. 2743–2749.
8. Vazquez, G.; Kerekes, T.; Rocabert, J.; Rodriguez, P.; Teodorescu, R.; Aguilar, D. A photovoltaic three-phase topology to reduce Common Mode Voltage. In Proceedings of the 2010 IEEE International Symposium on Industrial Electronics, Bari, Italy, 4–7 July 2010; pp. 2885–2890.
9. Li, W.; Gu, Y.; Luo, H.; Cui, W.; He, X.; Xia, C. Topology Review and Derivation Methodology of Single-Phase Transformerless Photovoltaic Inverters for Leakage Current Suppression. *IEEE Trans. Ind. Electron.* **2015**, *62*, 4537–4551. [CrossRef]
10. Xiao, H.; Xie, S.; Chen, Y.; Huang, R. An optimized transformerless photovoltaic grid-connected inverter. *IEEE Trans. Ind. Electron.* **2011**, *58*, 1887–1895. [CrossRef]
11. Kheng Suan, F.T.; Rahim, N.A.; Ping, H.W. Three-phase transformerless grid-connected photovoltaic inverter to reduce leakage currents. In Proceedings of the IEEE Conference on Clean Energy and Technology (CEAT), Lankgkawi, Malaysia, 18–20 November 2013; pp. 277–280.
12. Freddy, T.K.S.; Rahim, N.A.; Hew, W.; Che, H.S. Modulation Techniques to Reduce Leakage Current in Three-Phase Transformerless H7 Photovoltaic Inverter. *IEEE Trans. Ind. Electron.* **2015**, *62*, 322–331. [CrossRef]
13. Rahimi, R.; Farhangi, S.; Farhangi, B.; Moradi, G.R.; Afshari, E.; Blaabjerg, F. H8 Inverter to Reduce Leakage Current in Transformerless Three-Phase Grid-Connected Photovoltaic systems. *IEEE J. Emerg. Sel. Top. Power Electron.* **2018**, *6*, 910–918. [CrossRef]
14. Zhou, L.; Gao, F.; Shen, G.; Xu, T.; Wang, W. Low leakage current transformerless three-phase photovoltaic inverter. In Proceedings of the 2016 IEEE Energy Conversion Congress and Exposition (ECCE), Milwaukee, WI, USA, 18–22 September 2016; pp. 1–5.
15. Kerekes, T.; Teodorescu, R.; Rodriguez, P.; Vazquez, G.; Aldabas, E. A New High-Efficiency Single-Phase Transformerless PV Inverter Topology. *IEEE Trans. Ind. Appl.* **2011**, *58*, 184–191. [CrossRef]
16. Rodriguez, P.; Munoz-Aguilar, R.S.; Vazquez, G.; Candela, I.; Aldabas, E.; Etxeberria-Otadui, I. DCM-232 converter: A PV transformerless three-phase inverter. In Proceedings of the IECON 2011-37th Annual Conference of the IEEE Industrial Electronics Society, Melbourne, Australia, 7–10 November 2011; pp. 3876–3881.
17. Guo, X.; Xu, D.; Wu, B. New control strategy for DCM-232 three-phase PV inverter with constant common mode voltage and anti-islanding capability. In Proceedings of the 2014 IEEE Energy Conversion Congress and Exposition (ECCE), Pittsburgh, PA, USA, 14–18 September 2014; pp. 5613–5617.
18. Noroozi, N.; Zolghadri, M.R.; Yaghoubi, M. A novel modulation method for reducing common mode voltage in three-phase inverters. In Proceedings of the Environment and Electrical Engineering and 2017 IEEE Industrial and Commercial Power Systems Europe (EEEIC/I&CPS Europe), Italy, Milan, 6–9 June 2017; pp. 1–5.
19. Gubia, E.; Sanchis, P.; Ursua, A.; Lopez, J.; Marroyo, L. Ground currents in single-phase transformerless photovoltaic systems. *Prog. Photovolt. Res. Appl.* **2007**, *15*, 629–650. [CrossRef]

 *micromachines*

Article

Fractional-Order Approximation of PID Controller for Buck–Boost Converters

Allan G. S. Sánchez [1,*], Josué Soto-Vega [2], Esteban Tlelo-Cuautle [3] and Martín Antonio Rodríguez-Licea [1]

1 CONACYT—Instituto Tecnológico de Celaya, Guanajuato 38010, Mexico; martin.rodriguez@itcelaya.edu.mx
2 Instituto Tecnológico de Celaya, Guanajuato 38010, Mexico; m2003021@itcelaya.edu.mx
3 Department of Electronics, INAOE, Puebla 72840, Mexico; etlelo@inaoep.mx
* Correspondence: allan.soriano@itcelaya.edu.mx

Abstract: Viability of a fractional-order proportional–integral–derivative (PID) approximation to regulate voltage in buck–boost converters is investigated. The converter applications range not only to high-power ones but also in micro/nano-scale systems from biomedicine for energy management/harvesting. Using a classic closed-loop control diagram the controller effectiveness is determined. Fractional calculus is considered due to its ability at modeling different types of systems accurately. The non-integer approach is integrated into the control strategy through a Laplacian operator biquadratic approximation to generate a flat phase curve in the system closed-loop frequency response. The controller synthesis considers both robustness and closed-loop performance to ensure a fast and stable regulation characteristic. A simple tuning method provides the appropriate gains to meet design requirements. The superiority of proposed approach, determined by comparing the obtained time constants with those from typical PID controllers, confirms it as alternative to controller non-minimum phases systems. Experimental realization of the resulting controller, implemented through resistor–capacitor (RC) circuits and operational amplifiers (OPAMPs) in adder configuration, confirms its effectiveness and viability.

Keywords: fractional-order PID controller; DC–DC converters; Non-minimum phase systems; experimental validation

Citation: S. Sánchez, A.G.; Soto-Vega, J.; Tlelo-Cuautle, E.; Rodríguez-Licea, M.A. Fractional-Order Approximation of PID Controller for Buck–Boost Converters. *Micromachines* **2021**, *12*, 591. https://doi.org/10.3390/mi12060591

Academic Editor: Young-Ho Cho

Received: 22 April 2021
Accepted: 18 May 2021
Published: 21 May 2021

1. Introduction

DC–DC (direct current) conversion is one of the most studied and applied functionalities from power electronics. DC–DC elementary conversion modes step down and up the converter input voltage using power semiconductor devices, operated with high-speed control, to produce the well-known buck and boost topologies, respectively. By cascading both elementary conversion modes the buck–boost topology is obtained. With the same quantity of elements, the resulting converter produces a smaller or greater output voltage than its input power source, with inverse polarity.

The versatility of buck–boost converter to transform its supply voltage into a higher/lower output makes it an alternative for applications requiring DC voltage regulation, ranging from light-emitting diode (LED) lighting [1] to renewable energy sources [2,3], microgrids [4], and battery charging [5,6], to mention the most relevant. Biomedicine applications [7–9] take special relevance due to their impact in human's health, since the need for appropriate energy management/harvesting/storage strategies to be applied in micro/nano-scale is one of the main drawbacks of emerging cardiac technologies, for instance. Some of the most recent and significant results on strategies to regulate voltage in a buck–boost converter are the following: in [10], a deep learning based approach was used to stabilize voltage in the converter. A sliding mode-based observer combined with an optimization algorithm, a deep reinforcement learning technique, and a neural network were suggested to estimate converter unknown dynamics, while controller gains were adjusted online. Good transient behavior and output-robust stabilization against reference changes were the major improvements. In [11], an intelligent control with metaheuristic optimization was proposed. A fish-swarm algorithm is used to optimally tune a PI controller to regulate voltage in the converter. The simplicity of the approach is the strongest advantage of this contribution. The effectiveness of the controller was determined experimentally, where a fast-tracking characteristic was the main improvement. A fuzzy logic controller to stabilize voltage through a buck–boost converter in a turbine generation unit was proposed in [12]. Takagi–Sugeno-type rules were employed due to their wider range for gain variations and versatility. Even though authors compare open-loop vs. closed-loop performance to determine effectiveness, the latter exhibited acceptable time constants, and good tracking performance.

A passivity-based control was suggested in [13] to stabilize voltage fed to a microgrid. The proposed strategy achieved Lyapunov asymptotic stability through a state-feedback control law, which required an online invariance and immersion power estimator. In spite of authors compared their results with a pure PI to determine controller effectiveness, acceptable tracking characteristic, good transient performance and negligible steady-state error for both conversion modes were the improvements. With a similar approach but targeting a different application, a passivity-based controller with active disturbance rejection was proposed in [14]. A generalized proportional integral observer was used to provide accurate estimations to the controller. Good tracking characteristic and disturbances rejected effectively are the main contribution.

In [15], a modified sliding mode controller was proposed. The control strategy was divided into two portions, a linear approach based on a PI controller for voltage control loop and a nonlinear one for current loop based on hysteresis. The resulting regulated output voltage described a smooth response with overshoot absence, acceptable robustness against load variations, and good tracking performance. In [16], a predictive control approach was proposed. The authors combined a quadratic programming optimization algorithm, to predict the control signal at every sampling time, with a predictive controller, to consider load variations in the model, thus ensuring robustness and stability. The appropriate control law was generated by predicting the future behavior of the plant, resulting in a fast response, minimum overshoot and good tracking characteristic. On the other hand, due to the inherent closed-loop instability of non-minimum phase systems, PI/PID controllers are used in combination with some of the above-described techniques or some optimization algorithms. The above derived in control strategies, although efficient, with high computational/implementation complexity [17–19].

In this paper, a fractional-order PID controller approximation to regulate voltage in a buck–boost converter is proposed. In addition to the accuracy modeling real systems, robustness against parameter variations, and noise-level reduction through lower-order derivatives from fractional calculus, exploring its effectiveness controlling non-minimum phases systems is the main reason to consider a non-integer approach in the control strategy. The controller synthesis is achieved through a biquadratic module that exhibits a flat phase response. Its design considers both robustness and closed-loop performance, while a simple tuning method allows us to determine appropriate gains to achieve design requirements. The controller structure to operate in both conversion modes is generalized to generate its electrical representation directly. Numerical and experimental results are provided to corroborate effectiveness of the proposed approach.

The paper is organized as follows: in Section 2 preliminaries on buck–boost converter and the methodology to approximate fractional-order differentiator are provided. In Section 3 the algorithm to synthesize the controller is described step by step. Numerical simulations, including a comparison with a typical PID controller, a generalization of the synthesized controller, and experimental results of the obtained electrical arrangement are presented in this section as well. A discussion on the most significant results, including future work and some conclusions/remarks, are given in Sections 4 and 5, respectively.

2. Materials and Methods

In this section, the necessary preliminaries on buck–boost converter, the small-signal linearization of its model, and fractional-order approximation of Laplacian operator are briefly described.

2.1. Buck–Boost Converter Model

The DC–DC buck–boost converter was derived from the combination of elementary converters buck and boost. The resulting configuration can provide an output voltage of inverse polarity, either greater or smaller than the input voltage, with the same amount of elements.

Figure 1 shows the electrical diagram of the buck–boost converter. Assuming ideal components and continuous conduction mode (CCM), the averaged model of the buck–boost converter is obtained as follows [20],

$$\begin{aligned} L\frac{di_L}{dt} &= V_g\bar{D} + (1-\bar{D})v_C, \\ C\frac{dv_C}{dt} &= -(1-\bar{D})i_L - \frac{v_C}{R}, \end{aligned} \tag{1}$$

where $\bar{D} \in [0,1]$, V_g is the DC power supply, L, C and R are the inductance, capacitance and the resistance, respectively.

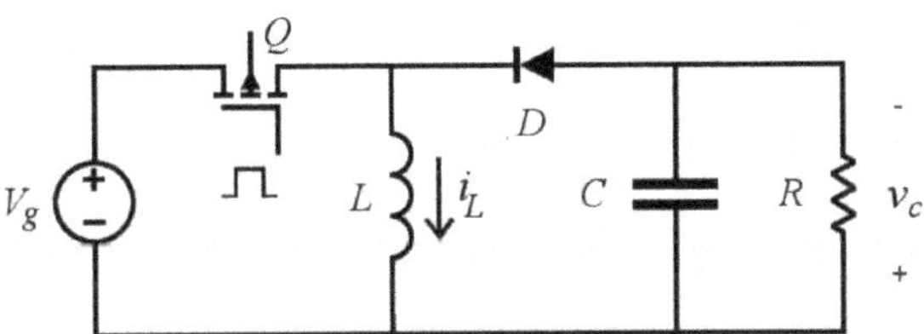

Figure 1. Electrical diagram of the DC–DC (direct current) buck–boost converter.

Linearization of (1) is performed through the small-signal technique, which consists of perturbing the original model signals to generate its DC and alternating current (AC) components [21]. The resulting AC component will be the linearized small-signal average model of the buck–boost converter, whose state space representation $\dot{x} = Ax + Bu$ and $y = Cx$ around the equilibrium point $[i_L, v_C] = [V_g\bar{D}/(R(1-\bar{D})^2), V_g\bar{D}/(1-\bar{D})]$ is given by [21],

$$\begin{bmatrix} \dot{\hat{i}}_L \\ \dot{\hat{v}}_C \end{bmatrix} = \begin{bmatrix} 0 & \frac{(1-\bar{D})}{L} \\ \frac{-(1-\bar{D})}{C} & \frac{-1}{RC} \end{bmatrix} \begin{bmatrix} \hat{i}_L \\ \hat{v}_C \end{bmatrix} + \begin{bmatrix} \frac{V_g}{L(1-\bar{D})} & \frac{\bar{D}}{L} \\ \frac{V_g\bar{D}}{RC(1-\bar{D})^2} & 0 \end{bmatrix} \begin{bmatrix} \hat{d} \\ \hat{v}_g \end{bmatrix}, \tag{2}$$

and

$$y = \begin{bmatrix} 0 & 1 \end{bmatrix} \begin{bmatrix} \hat{i}_L \\ \hat{v}_C \end{bmatrix}, \tag{3}$$

where $\hat{i}_L$, $\hat{v}_C$, $\hat{d}$ and $\hat{v}_g$ are the perturbation terms of i_L, v_C, $\bar{D}$ and V_g, respectively.

The transfer function of buck–boost converter $G_p(s)$ will be given by the relation $\hat{d}$-to-$\hat{v}_c$ as follows,

$$G_p(s) = C(sI - A)^{-1}B_1, \tag{4}$$

where $B_1 = \begin{bmatrix} \frac{V_g}{L(1-\bar{D})} & \frac{V_g\bar{D}}{RC(1-\bar{D})^2} \end{bmatrix}^T$. Thus, the system transfer function will be given by,

$$G_p(s) = \frac{\left(\frac{V_g\bar{D}}{RC(1-\bar{D})^2}\right)s - \left(\frac{V_g}{CL}\right)}{s^2 + \left(\frac{1}{RC}\right)s + \frac{(1-\bar{D})^2}{CL}}, \tag{5}$$

which has a right-half plane (RHP) zero, thus it is a non-minimum phase transfer function. The buck–boost converter transfer function (5) can be divided into factors as follows,

$$G_p(s) = G_{pm}(s)G_{nm}(s), \tag{6}$$

where

$$G_{pm}(s) = \frac{\left(\frac{V_g\bar{D}}{RC(1-\bar{D})^2}\right)\left(s + \frac{R(1-\bar{D})^2}{L\bar{D}}\right)}{s^2 + \left(\frac{1}{RC}\right)s + \frac{(1-\bar{D})^2}{CL}}, \tag{7}$$

$$G_{nm}(s) = \frac{s - \frac{R(1-\bar{D})^2}{L\bar{D}}}{s + \frac{R(1-\bar{D})^2}{L\bar{D}}}, \tag{8}$$

are the minimum phase and normalized non-minimum phase parts of $G_p(s)$, respectively. Please note that $G_{nm}(s)$ is an all-pass system, i.e., $|G_{nm}(j\omega)| = 1$, thus, the converter dynamic is given by the minimum phase part $G_{pm}(s)$, which will be considered to be the uncontrolled plant. The non-minimum phase part commonly introduces a delay, but it is also responsible for the output polarity inversion.

In the following, the methodology to approximate the non-integer PID controller is described.

2.2. Fractional-Order Approximation of Laplacian Operator

In this section, the approximation of the fractional-order Laplacian operator is described.

The integro-differential operator $s^{\pm\alpha}$ can be approximated as follows [22,23],

$$s^{\alpha} \approx T\left(\frac{s}{\omega_c}\right) = \frac{a_0\left(\frac{s}{\omega_c}\right)^2 + a_1\left(\frac{s}{\omega_c}\right) + a_2}{a_2\left(\frac{s}{\omega_c}\right)^2 + a_1\left(\frac{s}{\omega_c}\right) + a_0}, \quad 0 < \alpha < 1, \tag{9}$$

which is a biquadratic module that exhibits a flat phase response, where ω_c stands for the center frequency and a_0, a_1, a_2 are alpha-dependent real constants defined as follows,

$$\begin{aligned} a_0 &= \alpha^{\alpha} + 3\alpha + 2, \\ a_2 &= \alpha^{\alpha} - 3\alpha + 2, \\ a_1 &= 6\alpha \tan\frac{(2-\alpha)\pi}{4}. \end{aligned} \tag{10}$$

By assuming $\omega = \omega_c$ and considering constants (10), the integro-differential operator (9) will be described as

$$s^{\alpha} \approx T\left(\frac{j\omega_c}{\omega_c}, \alpha\right) = \frac{(a_2 - a_0) + ja_1}{-(a_2 - a_0) + ja_1} = \frac{-6\alpha + j6\alpha \tan\frac{(2-\alpha)\pi}{4}}{6\alpha + j6\alpha \tan\frac{(2-\alpha)\pi}{4}}, \tag{11}$$

whose phase contribution will be given by $\arg\{T(j1,\alpha)\} = -2\tan^{-1}\left(\tan\frac{(2-\alpha)\pi}{4}\right)$. Thus, the phase contribution of approximation (9) will be given by [22,23],

$$\arg\{T(s/\omega_c)|_{j\omega_c}\} = \pm\alpha\pi/2, \tag{12}$$

which means that depending on the value of α, the biquadratic approximation of $s^{\pm\alpha}$ contributes with $0 < \arg\{T(s/\omega_c)|_{j\omega_c}\} < \pm 90^\circ$.

Equation (9) will behave as a fractional-order differentiator around ω_c as long as $a_0 > a_2 > 0$, i.e., $\arg\{T(s/\omega_c)\} > 0$. Conversely, the effect of fractional-order integrator can be produced by ensuring that $0 < a_0 < a_2$, which produces $\arg\{T(s/\omega_c)|_{j\omega_c}\} < 0$. The latter is consistent with the location of zeros/poles of (9) in the complex plane, where zeros lead poles for $a_0 > a_2$ and poles lead zeros for $a_0 < a_2$, which confirms derivative/integral effects.

In Figure 2 the frequency response of $s^{\pm 0.5}$ is shown, where $a_0 = 4.2071$, $a_1 = 7.2427$ and $a_2 = 1.2071$ to ensure derivative (Figure 2a) and integral (Figure 2b) effects.

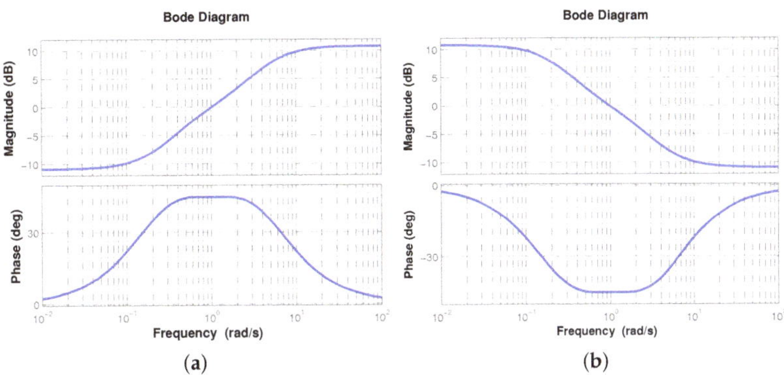

Figure 2. Frequency response of $s^{\pm 0.5}$, where $\arg\{T(j1)\} = \pm\alpha\pi/2 = \pm 45°$ for both (**a**) derivative and (**b**) integral effect of (9).

The synthesis of controller structure, using the fractional-order approximation of Laplacian operator, is described in the following section.

2.3. Synthesis of Fractional-Order PID Approximation

The standard representation of a PID controller $G_c(s) = k_p(1 + 1/(T_i s) + T_d s)$ was modified to consider the non-integer approach in its integral/derivative modes. The fractional-order PI$^{\alpha}$D$^{\mu}$ structure is a special case of PID controller with additional degrees of freedom which is described as follows [24]

$$G_c(s) = k_p\left(1 + \frac{1}{T_i s^{\alpha}} + T_d s^{\mu}\right), \tag{13}$$

where $0 < \alpha, \mu < 1$, k_p is the proportional gain, T_i and T_d are the integral and derivative time constants, respectively.

Different optimization strategies have suggested that T_i and T_d are related through $T_i = \eta T_d$, where η is a constant. The first suggestion derived in $\eta = 4$ aiming to achieve a compromise between controller performance and its viability to be implemented [25]. Motivated by the need to ensure unique solutions of (13), some other results showed that smaller values of η produced significant improvements [25,26]. By slightly modifying (13) and by setting $\alpha = \mu$, $\eta = 1$, (13) can be expressed as follows,

$$G_c(s) = \frac{k_c(T_i s^{\alpha} + 1)^2}{s^{\alpha}}, \tag{14}$$

which directly simplifies the PID structure through a perfect square trinomial [27], where $k_c = k_p/T_i$.

The phase of the plant to be controlled ϕ_{pm}, the controller phase ϕ_c and phase margin ϕ_m are related through $\phi_c + \phi_{pm} = -\pi + \phi_m$ at the phase crossover frequency ω_{pc}, which implies that $\phi_c = \phi_m - \pi - \phi_{pm}$, thus,

$$\alpha = \frac{(-\pi - \phi_{pm} + \phi_m)}{(\pi/2)}. \tag{15}$$

In the next section, the fractional-order PID controller approximation is validated numerically and experimentally. A generalization of its structure for buck and boost modes of the converter is derived as well.

3. Results

In this section, the methodology to synthesize a fractional-order approximation of a PID controller is described step by step. The generalization of its structure for buck and boost conversion modes is derived. Numerical and implementation results are provided to show its effectiveness.

3.1. Control Design and Numerical Results

The suggested algorithm to synthesize the approximation of the fractional-order PID (FOPID) controller for the buck–boost converter comprises the following steps:

1. Consider buck–boost converter transfer function divided into minimum and non-minimum phase parts.
2. Think on the minimum phase transfer function $G_{pm}(s)$ as the plant to be controlled.
3. Determine uncontrolled plant phase ϕ_{pm} and the phase margin ϕ_m.
4. Compute the required controller fractional-order α through (15).
5. Compute fractional-order approximation s^α through (9).
6. Generate controller structure $G_c(s)$ as a function of integral time constant T_i and gain k_c.
7. Determine T_i and k_c values that produce the required effect.
8. Determine regulation/tracking performance of the closed-loop response.

Since buck–boost converter operates in both elementary conversion modes and considering that its transfer function is one of varying parameters, controller design is developed under the following conditions: buck–boost converter transfer function (5) and (6), whose parameter values are shown in Table 1, desired stability margins of $g_m \geq 10$ dB, $30^\circ \leq \phi_{dm} \leq 60^\circ$ and equilibrium points generated with the average duty cycle $\bar{D} = 0.375$ and $\bar{D} = 0.583$ for buck and boost conversion modes to produce an output voltage of $V_o = 15$ V and $V_o = 35$ V, respectively.

Table 1. Parameter values of buck–boost converter in Figure 1.

Element	Notation	Value
DC voltage source	V_g	25 V
Capacitor	C	30 µF
Inductor	L	10 mH
Resistance	R	10 Ω
Switching frequency	f_s	20 kHz

By considering the described conditions, the minimum phase transfer function $G_{pm}(s)$ of buck–boost converter is linearized around equilibrium points $[i_L,\ v_C] = [2.4,\ -15]$ for buck conversion mode and $[i_L,\ v_C] = [8.4,\ -35]$ for boost conversion mode. In Table 2, computation of phase margin ϕ_m, uncontrolled plant phase ϕ_{pm}, and fractional-order α for buck and boost conversion modes are provided.

Table 2. Values of ϕ_m, ϕ_{pm} and α for buck/boost conversion modes.

Parameter	Notation	Buck	Boost
Phase margin	ϕ_m	90.7°	90.5°
Uncontrolled plant phase	ϕ_{pm}	−89.3°	−89.5°
Fractional-order	α	0.6745	0.6727

Therefore, the controller phase contributions are −60.7° and −60.5° for buck and boost conversion modes, respectively. Please note that the algorithm for controller design provides a very similar result for both conversion modes up to this point. However, it should be kept in mind that the transfer function is of varying parameters; thus, its frequency response changes depending on the equilibrium point considered in the linearization. In Figure 3, the fractional-order approximations that are generated with the computed values of α are shown. One can see their similarity in shape and phase contribution, but they differ in their frequency band.

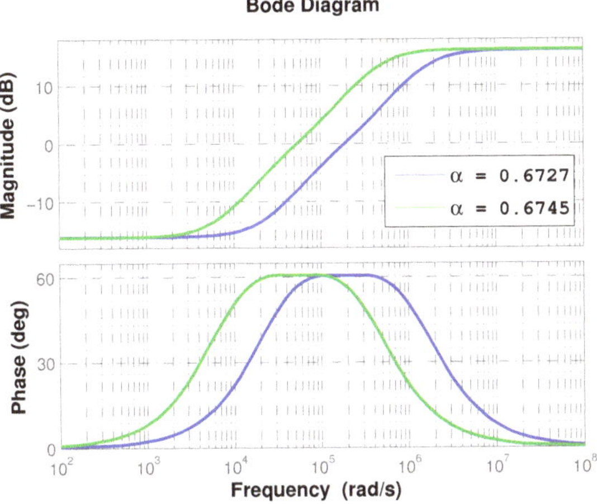

Figure 3. Frequency response of fractional-order approximation for buck/boost conversion modes.

The controller $G_c(s)$ can be obtained by manipulating (9) as $s^{\alpha} \approx T(s) \equiv N(s)/D(s)$ and then substituting in (14), thus the controller structure will be given by,

$$G_c(s) = k_c \frac{(T_i N(s) + D(s))^2}{N(s)D(s)}, \tag{16}$$

from which one concludes that controller effect can be varied between integral and derivative depending on the value of T_i. In Figure 4 the transition between both effects are shown. As can be seen, integral (derivative) effect is achieved as $T_i \to 0$ ($T_i \to \infty$).

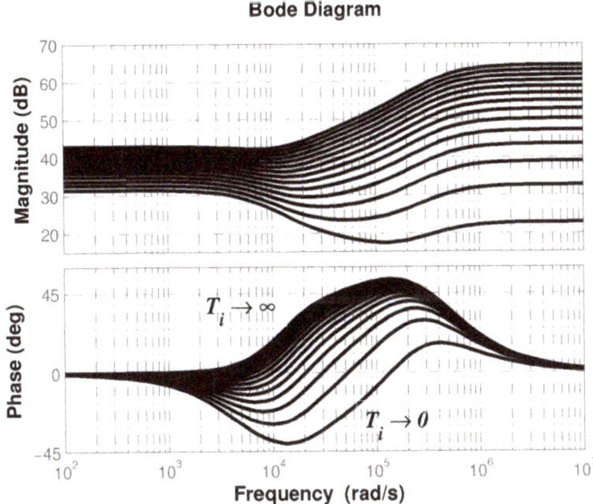

Figure 4. Transition between integral and derivative effect as function of T_i.

By setting $T_i = 0.001$ and $k_c = 3$, the controller $G_c(s)$ produces the necessary effect to fulfill the required stability margins in both conversion modes. The resulting controller structure for converter in Figure 1 is given by,

$$G_c(s) = k\frac{s^4 + \rho_1 s^3 + \rho_2 s^2 + \rho_3 s + \rho_4}{s^4 + \psi_1 s^3 + \psi_2 s^2 + \psi_3 s + \psi_4}, \tag{17}$$

whose parameters are listed in Table 3.

Table 3. Coefficients for approximation of the fractional-order PID controller (17).

Coefficient	**Buck**	**Boost**
ρ_1/ψ_1	$9.866 \times 10^5/5.729 \times 10^5$	$3.434 \times 10^6/1.996 \times 10^6$
ρ_2/ψ_2	$2.798 \times 10^{11}/5.694 \times 10^{10}$	$3.391 \times 10^{12}/6.941 \times 10^{11}$
ρ_3/ψ_3	$1.798 \times 10^{16}/1.629 \times 10^{15}$	$7.607 \times 10^{17}/6.954 \times 10^{16}$
ρ_4/ψ_4	$3.321 \times 10^{20}/8.092 \times 10^{18}$	$4.907 \times 10^{22}/1.214 \times 10^{21}$
k	0.4714	0.4749

The step response of the closed-loop system will allow us to determine the effectiveness of the synthesized controller. In Figure 5, the regulation capacity of the proposed controller can be confirmed. One can see that the response exhibits a fast and stable tracking characteristic for both conversion modes, which can be corroborated quantitatively through performance parameters in Table 4, column 2.

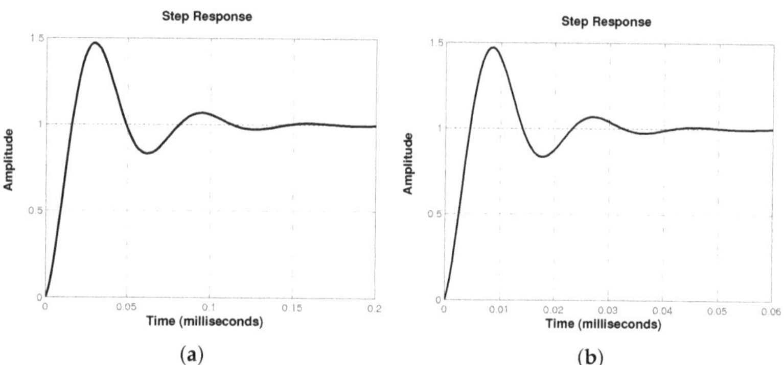

Figure 5. Closed-loop step response for both (**a**) buck and (**b**) boost conversion modes of converter in Figure 1.

Table 4. Closed-loop response performance parameters for buck/boost conversion modes, where e_{ss}, τ, t_r, t_p, t_s and $\%M$ stand for steady-state error, time constant, rising time, peak time, settling time and overshoot, respectively.

Notation	**FOPID**	**PID 1st Option**	**PID 2nd Option**
e_{ss}	0	0	0
τ	10.6/3.03 µs	99.8/84.2 µs	98.4/8.27 µs
t_r	11.6/3.31 µs	113/93.8 µs	99.5/8.29 µs
t_p	30/8.55 µs	299/245 µs	286/21.9 µs
t_s	134/38.3 µs	1.71/1.36 ms	2.94/0.32 ms
$\%M$	47%	52/50%	69/74%
ϕ_{dm}	60°	60°	30°

Although regulation velocity of the response is acceptable, the proposed controller performs different depending on the selected conversion mode. These differences are attributed to the operating frequency band of the approximation, being the one for boost conversion mode of a higher frequency range, as shown in Figure 3.

A comparison of the fractional-order PID approximation with a typical PID controller allows us to determine that the former outperforms the latter. By using MatLab algorithm to tune PID controller, targeting 60° phase margin and a compromise between robustness and performance, typical PID controllers were tuned, resulting $[k_p,\ T_i,\ T_d]_{\text{buck}} = [0.068,\ 3.85 \times 10^{-5},\ 9.92 \times 10^{-9}]$ and $[k_p,\ T_i,\ T_d]_{\text{boost}} = [0.021,\ 2.59 \times 10^{-5},\ 8.04 \times 10^{-9}]$ for both conversion modes. Performance of inter-order PID controller is quantified in Table 4 column 3. By comparing columns 2 and 3, the proposed approach superiority can be determined through time constants since the ones obtained with the approximation of fractional-order PID controller are much smaller. Although typical PID achieves the output regulation with zero steady-state error, it took longer in both conversion modes to reach the reference value.

To determine if a typical PID structure can equalize performance of the proposed approach and considering that an increase in performance reduces robustness and vice versa, a second option of PID controller was tuned with a less restrictive phase margin aiming to obtain performance improvement, resulting $[k_p,\ T_i,\ T_d]_{\text{buck}} = [0.034,\ 1.51 \times 10^{-5},\ 3.78 \times 10^{-6}]$ and $[k_p,\ T_i,\ T_d]_{\text{boost}} = [0.123,\ 1.37 \times 10^{-6},\ 3.43 \times 10^{-7}]$ for both conversion modes. In Table 4, column 4 performance parameters produced by the second PID controller are provided. Please note that despite the increase in performance, the second option of integer-order PID controller produces time constants that are not competitive with those of the proposed approach.

The comparison is moved to the frequency domain to corroborate the stability margins and determine the effect of controller on the magnitude/phase curves of the closed-loop response. In Figure 6, frequency response of closed-loop system with fractional-order PID and typical PID controllers, operating in buck and boost conversion modes, are shown. One can see that both controllers were able to achieve the desired phase margin. Note the shape similarity of the magnitude curves. They both have their peaks around the same value, which corroborates that overshoot is similar for both controllers. The operating frequency band is wider for the system controlled with the fractional-order PID approximation. The latter is consistent with the response velocity measured for system controlled with the proposed controller, since the wider the bandwidth the shorter the rising time, due to the higher-frequency signals pass through the system more easily.

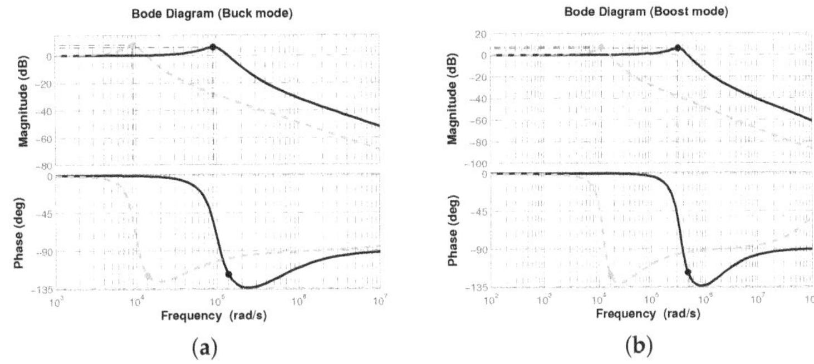

Figure 6. Frequency response of closed-loop system for both (**a**) buck and (**b**) boost conversion modes.

Lastly, frequency response of the sensitivity and complementary sensitivity functions allows us to determine the robustness of the controller through its disturbance and noise rejection characteristics. Recalling that sensitivity function $S = 1/(1 + G_pG_c)$, and complementary sensitivity function $T = G_pG_c/(1 + G_pG_c)$ determine how disturbances/perturbations and noise affect respectively the output, it is expected the controller G_c to produce a curve with attenuation in low frequencies for S and a curve with attenuation in high ones for T ([28], Chap. 4). In Figure 7, frequency response of sensitivity S and complementary sensitivity T functions is shown for both conversion modes, where $L = G_p G_c$ is the loop gain, when using fractional-order PID and typical PID controllers.

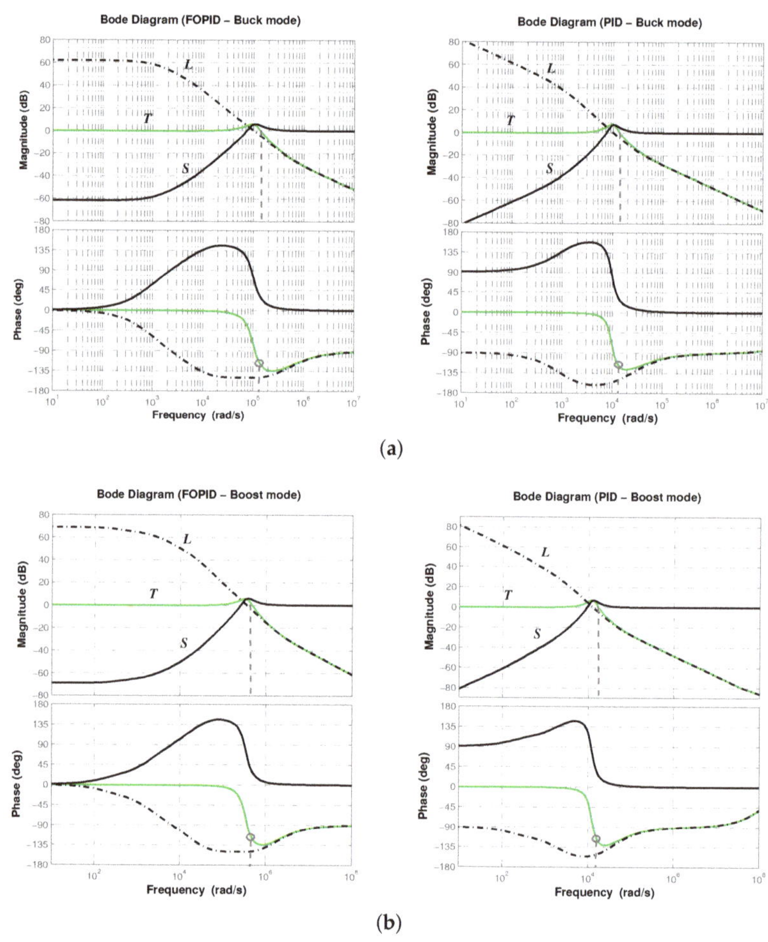

Figure 7. Frequency response of sensitivity S, complementary sensitivity T functions and loop gain L for both (**a**) buck and (**b**) boost conversion modes.

Note the magnitude flatness of sensitivity function S produced by the fractional-order PID controller, which implies that it attenuates better disturbances/perturbations for a wider frequency band. On the other side, complementary sensitivity function T successfully attenuates high-frequency noise. Thus, a better disturbance/perturbation and noise rejection characteristic is obtained with the fractional-order PID controller approximation and therefore, the closed-loop system will exhibit a robust performance.

In the following section, a generation of the fractional-order PID approximation to facilitate controller implementation is derived. Experimental results are also provided and described in that section.

3.2. Generalization of Controller for the Implementation

Since the converter of Figure 1 operates in buck and boost conversion modes depending on the value of duty cycle D, a general structure for the controller must be determined. The objective is to investigate and determine if the proposed controller appropriately regulates voltage in either conversion mode.

Electrical implementation of the controller (16) requires a simpler mathematical representation. By using the partial fraction expansion of (16), one obtains mathematical expressions whose electrical equivalence is standard and well known. Before synthesizing the electrical arrangement that describes the controller (16), it is necessary to determine the type of roots that will be obtained in a general way. Therefore, by considering the effect of T_i previously described in Figure 4, the controller will be given by (16), $G_c(s) = k_c N(s)/D(s)$ $(T_i \to \infty)$ or $G_c(s) = k_c D(s)/N(s)$ $(T_i \to 0)$.

Please note that in all cases, the controller $G_c(s)$ depends on $N(s)$ and $D(s)$, which are numerator and denominator of approximation s^α. Since both $N(s)$ and $D(s)$ of approximation (9) are quadratic polynomials, as long as $a_1^2 > 4a_2a_0$, the roots of controller $G_c(s)$ will be real. Knowing that a_0, a_1 and a_2 depend on $0 < \alpha < 1$, Figure 8 proofs that condition $a_1^2 > 4a_2a_0$ holds for every value of α, therefore, the partial fraction expansion of $G_c(s)$ will be given in terms of real poles only as follows,

$$G_c(s) = \left(\frac{A_1}{\gamma_1 s + 1}\right) + \left(\frac{A_2}{\gamma_2 s + 1}\right) + \left(\frac{A_3}{\gamma_3 s + 1}\right) + \left(\frac{A_4}{\gamma_4 s + 1}\right) + A_5. \tag{18}$$

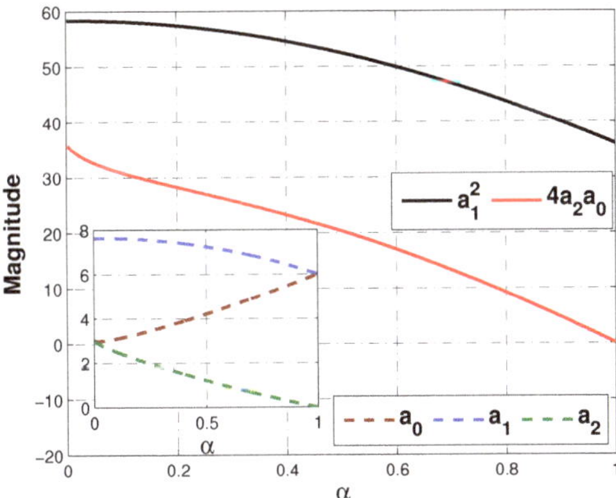

Figure 8. Approximation parameters $a_0(\alpha)$, $a_1(\alpha)$, $a_2(\alpha)$ and values a_1^2, $4a_2a_0$ that ensure $a_1^2 > 4a_2a_0$.

Since the first four terms resemble an RC circuit transfer function, the partial fraction expansion of the controller (18) can be directly generated through RC circuits and OPAMPs as inverting amplifiers and in adder configuration. In Figure 9a, the electrical arrangement to implement the controller (18) is shown. Gamma coefficients are the equivalence of multiplying R_1C_1 to R_4C_4, and constants A's are the gains of inverting amplifiers obtained by dividing R_5 to R_9 over R.

Constant values to represent fractional-order PID controller approximation (17), whose coefficients are given in Table 3 for buck and boost conversion modes, in its partial fraction expansion (18) are shown in Table 5 columns 1 to 3. Due to resulting gain values for the controller (18), the electrical circuit of Figure 9a is rearranged to consider the sign of A_1 and A_3, thus resulting in the electrical circuit of Figure 9b, whose parameter values are provided in Table 5 columns 4 to 6.

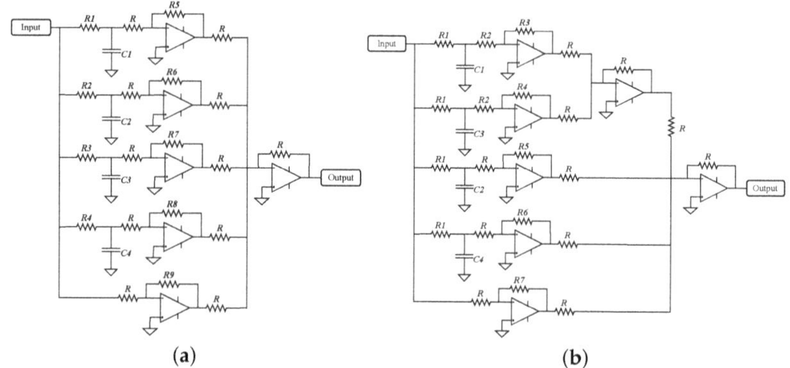

Figure 9. (**a**) Electrical representation of partial fraction expansion of controller (18). (**b**) Electrical representation of controller (18) for parameter values of Table 5.

Table 5. Constants A's and γ's of controller (18) and parameter values for electrical arrangement of Figure 9b for buck and boost conversion modes.

Constant	Buck	Boost	Element	Buck	Boost
A_1	-1.7675×10^{-5}	-1.7523×10^{-5}	R_1	100 Ω	10 Ω
A_2	1.1981	1.2018	R_2	100 kΩ	100 kΩ
A_3	-1.1981×10^{-6}	1.2018×10^{-6}	R_3	1.76 Ω	1.75 Ω
A_4	17.6752	17.5233	R_4	0.12 Ω	0.12 Ω
A_5	0.4714	0.4749	R_5	1.2 kΩ	1.2 kΩ
γ_1	2.19×10^{-6}	0.631×10^{-6}	R_6	17.7 kΩ	17.5 Ω
γ_2	14.14×10^{-6}	4.04×10^{-6}	R_7	471 Ω	475 Ω
γ_3	24.86×10^{-6}	7.11×10^{-6}	R	1 kΩ	1 kΩ
γ_4	160.22×10^{-6}	45.51×10^{-6}	C_1	0.022 μF	0.063 μF
			C_2	0.142 μF	0.404 μF
			C_3	0.25 μF	0.712 μF
			C_4	1.602 μF	4.55 μF

Note from Table 5 that the value of constants A_1 and A_3 for both conversion modes are very small and can be neglected. For this reason, the top part of electrical circuit for the controller in Figure 9b can be also omitted in the implementation with no effect in the final result, since its contribution is in the range of ηV.

Using PSIM 9.0 (Powersim Inc., 2001-2010), the proposed arrangement was tested through the electrical simulation of the complete system. In Figure 10, the output voltage V_o and inductor current i_L for converter in Figure 1 are shown. Synthesized controllers effectively regulated output voltage in both buck $V_o = 15$ V (Figure 10a) and boost $V_o = 35$ V (Figure 10b) conversion modes, while operating the converter in continuous conduction mode, as can be corroborated through inductor current.

On the other hand, implementation results confirmed the viability and effectiveness of the proposed approach. The components for the experiment are all commercial and were obtained from Mouser Electronics, Mexico. The experiment technical characteristics are the following: a very high current capacity inductor 1140-103K-RC of 10mH with ±20%, a DC resistance (DCR)of 2.76 Ω and 10 A. A polypropylene metalized film capacitor of 30 μF with max_{DC} voltage of 500 V, tolerance of 5% and equivalent series resistance (ESR)of 3.5 mΩ. A high current capability power MOSFET NTP5864NG with maximum drain-to-source voltage of 60 V, continuous drain current of 63 A and $R_{\text{ON}} = 12.4$ mΩ. Lastly, a diode SR504 R0 with forward voltage of 0.55 V. The controller was implemented with the high speed, 4 MHz wide bandwidth quad junction field effect transistor (JFET) inputs operational amplifier LF347N and the pulse width modulation (PWM) signal was created with the traditional TL494. Capacitors and resistances of described values with tolerances of ±5% and ±1%, respectively.

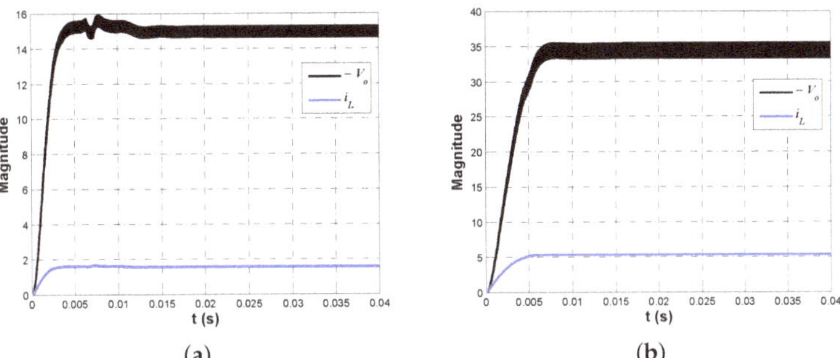

Figure 10. Regulated output voltage V_o and inductor current i_L of converter in Figure 1 for (**a**) buck and (**b**) boost conversion modes.

In Figure 11 evidence of the experiment table is shown. From left to right are the oscilloscope, the fractional-order controller approximation, the PWM generator, DC voltage sources and the buck–boost converter with the corresponding load.

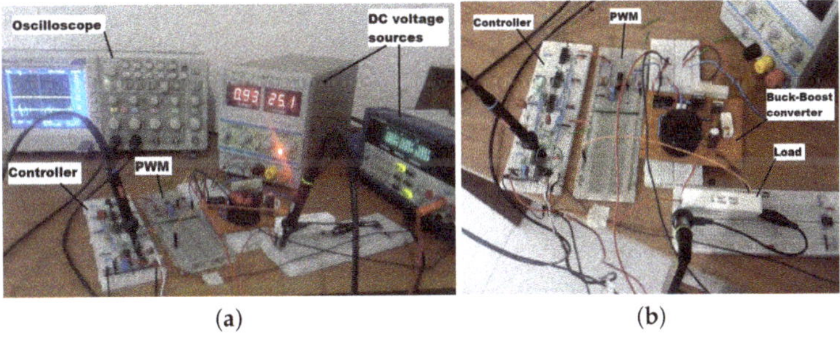

Figure 11. (**a**) Experiment table with oscilloscope, DC voltage sources and the electrical system. (**b**) Electrical system composed of fractional-order controller approximation, pulse width modulation (PWM) generator and buck–boost converter with the corresponding load.

In Figure 12 the voltage regulation of buck–boost converter can be corroborated in both conversion modes. In Figure 12a,b the output voltage (top signal) corroborates buck mode (V_o = 15 V) and boost mode (V_o = 35 V), respectively.

In Figure 13 the tracking characteristic of buck–boost converter is shown. As can be seen, the controller successfully regulates output voltage with a fast and stable tracking characteristic. It is important to mention the similarity of implementation results with those predicted through Figure 5 and Table 4, column 2, since boost mode regulation exhibits a faster response compared to the one produced in the buck mode.

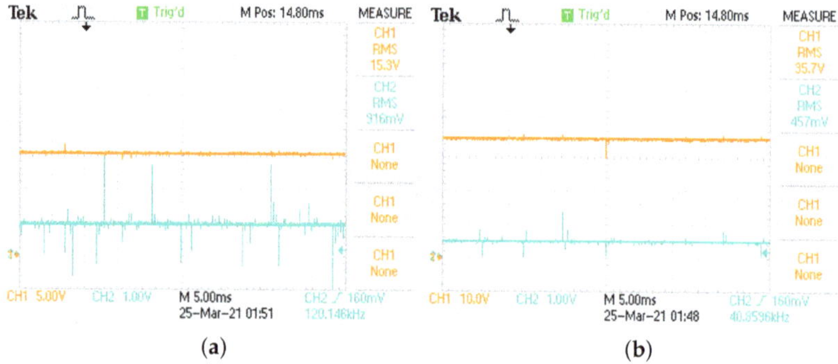

Figure 12. (**a**) Regulation to $V_o = 15$ V in buck mode. (**b**) Regulation to $V_o = 35$ V in boost mode.

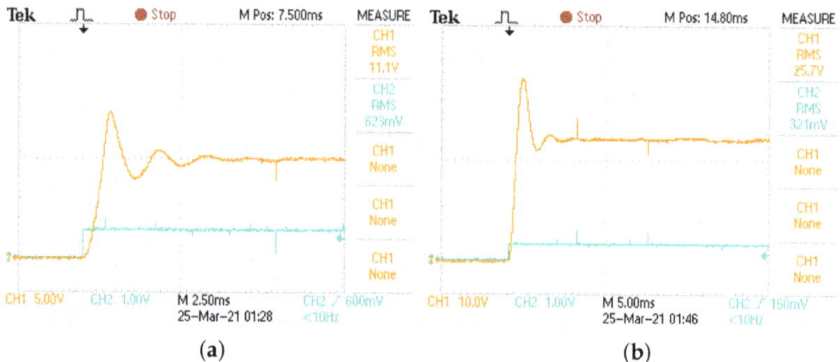

Figure 13. Tracking characteristic of buck–boost converter. (**a**) Buck mode. (**b**) Boost mode.

4. Discussion

In this paper, the effectiveness and viability of a fractional-order PID controller approximation to regulate voltage in a buck–boost converter were addressed. The importance of the converter rests in its variety of applications from which biomedicine ones take on special relevance, since the need for a stable and fast-response source of power is one of the main drawbacks of emerging cardiac technologies, which as in high-power systems, require an appropriate energy management/harvesting/storage strategy.

From the proposed approach, the controller design method considered both performance of closed-loop response and robustness. A biquadratic module that exhibits flat phase response was used to generate the controller structure. Fractional calculus is integrated due to its proven ability to describe systems with higher accuracy and robustness against parameter variations/uncertainties.

The proposed approach viability is investigated as an alternative for highly efficient converters such as Silicon-Carbide ones. Experimental results confirmed effectiveness of the controller to regulate output voltage in a buck–boost converter using a single control loop. These results open the possibility of applying this approach to a current control mode to determine if regulation velocity can be enhanced even more.

Although it could represent a disadvantage the required extra hardware for the implementation of the proposed controller, this could be dismissed considering the promising results of the proposed approach as well as the availability in commercial values of the extra components.

5. Conclusions

In this paper, a fractional-order PID controller approximation was suggested to regulate voltage in a buck–boost converted. The model of the system was linearized through the small-signal technique around the equilibrium point. The resulting linearized plant was divided into minimum and non-minimum phase parts, which simplified the controller design, allowing us to focus it on the minimum phase part only. The controller design considered both performance and robustness, while a simple tuning method allowed us to determine directly appropriate gains to achieve stability margin requirements. Synthesized controllers effectively regulated voltage in both conversion modes with a stable, fast, and good tracking characteristic. Superior disturbance/perturbation and noise rejection characteristic was also determined.

By comparing performance parameters of typical PID controller with those obtained by the fractional-order PID approximation allowed us to determine quantitatively the proposed approach superiority in time and frequency domains. To corroborate the experimental viability, a generalization for the controller structure was derived. The resulting electrical arrangement is implementable easily through RC circuits and OPAMPs. Experimental results confirmed the proposed controller effectiveness and its viability, since the parameter values are all commercial.

Lastly, it is worth mentioning that numerical simulations not only predicted effectiveness of the controller but its actual effect, since the obtained experimental results confirmed data of Figure 5 and Table 4, column 2 for both conversion modes.

Author Contributions: Conceptualization, A.G.S.S.; Formal analysis, A.G.S.S. and E.T.-C.; Investigation, A.G.S.S., J.S.-V. and E.T.-C.; Methodology, A.G.S.S., J.S.-V. and E.T.-C.; Resources, A.G.S.S., E.T.-C. and M.A.R.-L.; Validation, A.G.S.S. and M.A.R.-L.; Visualization, A.G.S.S., E.T.-C. and M.A.R.-L.; Writing—original draft, A.G.S.S.; Writing—review & editing, A.G.S.S., E.T.-C. and M.A.R.-L. All authors have read and agreed to the published version of the manuscript.

Funding: This research received no external funding.

Acknowledgments: The authors would like to thank CONACYT México for catedras 6782, 4155 and Josué Soto–Vega's M. Sc. scholarship.

Conflicts of Interest: The authors declare no conflict of interest.

References

1. Kuwahara, K.; Asadi, F.; Ishibashi, T.; Eguchi, K. The Development of an LED Lighting Circuit Using High Gain Buck-Boost Converters. *Int. J. Electr. Electron. Eng. Telecommun.* **2019**, *8*, 262–267. [CrossRef]
2. Ado, M.; BinArif, M.S.; Jusoh, A.; Mutawakkil, A.U.; Danmallam, I.M. Buck-boost converter with simple gate control for renewable energy applications. *Energy Sources Part A Recover. Util. Environ. Eff.* **2020**, 1–13. [CrossRef]
3. Rajavel, A.; Rathina Prabha, N. Fuzzy logic controller-based boost and buck-boost converter for maximum power point tracking in solar system. *Trans. Inst. Meas. Control.* **2020**, *43*, 945–957. [CrossRef]
4. Gheisarnejad, M.; Farsizadeh, H.; Tavana, M.R.; Khooban, M.H. A Novel Deep Learning Controller for DC/DC Buck-Boost Converters in Wireless Power Transfer Feeding CPLs. *IEEE Trans. Ind. Electron.* **2020**. [CrossRef]
5. Sarikhani, A.; Allahverdinejad, B.; Hamzeh, M. A Non-Isolated Buck-Boost DC-DC Converter with Continuous Input Current for Photovoltaic Applications. *IEEE J. Emerg. Sel. Top. Power Electron.* **2020**. [CrossRef]
6. Park, J.; Jeong, M.G.; Kang, J.G.; Yoo, C. Solar Energy-Harvesting Buck–Boost Converter With Battery-Charging and Battery-Assisted Modes. *IEEE Trans. Ind. Electron.* **2020**, *68*, 2163–2172. [CrossRef]
7. Co, M.L.; Khouzam, J.P.; Pour-Ghaz, I.; Minhas, S.; Ray, I.B. Emerging Technologies in Cardiac Pacing From Leadless Pacers to Stem Cells. *Curr. Probl. Cardiol.* **2021**, *46*, 100797. [CrossRef]
8. Azimi, S.; Golabchi, A.; Nekookar, A.; Rabbani, S.; Amiri, M.H.; Asadi, K.; Abolhasani, M.M. Self-powered cardiac pacemaker by piezoelectric polymer nanogenerator implant. *Nano Energy* **2021**, *83*, 105781. [CrossRef]
9. Adelstein, E.; Zhang, L.; Nazeer, H.; Loka, A.; Steckman, D. Increased incidence of electrical abnormalities in a pacemaker lead family. *J. Cardiovasc. Electrophysiol.* **2021**, *32*, 1111–1121. [CrossRef]
10. Hajihosseini, M.; Andalibi, M.; Gheisarnejad, M.; Farsizadeh, H.; Khooban, M.H. DC/DC power converter control-based deep machine learning techniques: Real-time implementation. *IEEE Trans. Power Electron.* **2020**, *35*, 9971–9977. [CrossRef]
11. Chanjira, P.; Tunyasrirut, S. Intelligent Control Using Metaheuristic Optimization for Buck-Boost Converter. *J. Eng.* **2020**, *2020*. [CrossRef]

12. Noriyati, R.; Musyafa, A.; Rahmadiansah, A.; Utama, A.; Asy'ari, M.; Abdillah, M. Design and Implemented Buck-Boost Converter Based Fuzzy Logic Control on Wind Power Plant. *Int. J. Mech. Mech. Eng.* **2020**, *20*, 115–122.
13. Soriano-Rangel, C.A.; He, W.; Mancilla-David, F.; Ortega, R. Voltage regulation in buck-boost converters feeding an unknown constant power load: An adaptive passivity-based control. *IEEE Trans. Control. Syst. Technol.* **2020**, *29*, 395–402. [CrossRef]
14. Linares-Flores, J.; Barahona-Avalos, J.L.; Sira-Ramirez, H.; Contreras-Ordaz, M.A. Robust passivity-based control of a buck–boost-converter/DC-motor system: An active disturbance rejection approach. *IEEE Trans. Ind. Appl.* **2012**, *48*, 2362–2371. [CrossRef]
15. Ghanghro, A.; Sahito, A.; Memon, M.; Soomro, A. Nonlinear Controller for Buck Boost Converter for Photovoltaic System. *Int. J. Electr. Eng. Emerg. Technol.* **2020**, *3*, 59–63.
16. Albira, M.E.; Alzahrani, A.; Zohdy, M. Model Predictive Control of DC-DC Buck-Boost Converter with Various Resistive Load Values. In Proceedings of the 2020 IEEE Global Congress on Electrical Engineering (GC-ElecEng), Valencia, Spain, 4–6 September 2020; pp. 124–128.
17. Arora, R.; Tayal, V.K.; Singh, H.; Singh, S. PSO optimized PID controller design for performance enhancement of hybrid renewable energy system. In Proceedings of the 2020 IEEE 9th Power India International Conference (PIICON), Sonepat, India, 28 February–1 March 2020; pp. 1–5.
18. Ardhenta, L.; Ansyari, M.R.; Subroto, R.K.; Hasanah, R.N. DC Voltage Regulator using Buck-Boost Converter Based PID-Fuzzy Control. In Proceedings of the 2020 IEEE 10th Electrical Power, Electronics, Communications, Controls and Informatics Seminar (EECCIS), Malang, Indonesia, 26–28 August 2020; pp. 117–121.
19. Almawlawe, M.D.; Kovandzic, M. A modified method for tuning PID controller for buck-boost converter. *Int. J. Adv. Eng. Res. Sci.* **2016**, *3*, 236938. [CrossRef]
20. Erickson, R.W.; Maksimovic, D. *Fundamentals of Power Electronics*; Springer Science & Business Media: Berlin, Germany, 2007.
21. Ang, S.; Oliva, A.; Griffiths, G.; Harrison, R. Continuous—Time modeling of switching converters. In *Power–Switching Converters*, 3rd ed.; CRC Press: Boca Raton, FL, USA, 2010.
22. El-Khazali, R. Fractional-order $PI^{\lambda}D^{\mu}$ controller design. *Comput. Math. Appl.* **2013**, *66*, 639–646. [CrossRef]
23. El-Khazali, R. On the biquadratic approximation of fractional–order Laplacian operators. *Analog Integr. Circuits Signal Process.* **2015**, *82*, 503–517. [CrossRef]
24. Podlubny, I.; Petráš, I.; Vinagre, B.M.; O'leary, P.; Dorčák, L. Analogue realizations of fractional-order controllers. *Nonlinear Dyn.* **2002**, *29*, 281–296. [CrossRef]
25. Wallén, A.; ÅRström, K.; Hägglun, T. Loop-Shaping Design Of PID Controllers With Constant Ti/Td RATIO. *Asian J. Control* **2002**, *4*, 403–409. [CrossRef]
26. Monje, C.A.; Vinagre, B.M.; Feliu, V.; Chen, Y. Tuning and auto-tuning of fractional order controllers for industry applications. *Control Eng. Pract.* **2008**, *16*, 798–812. [CrossRef]
27. Ogata, K. *Modern Control Engineering*; Prentice Hall: Upper Saddle River, NJ, USA, 2009.
28. Dorf, R.C.; Bishop, R.H. *Modern Control Systems*; Pearson: London, UK, 2011.

Review

Power Losses Models for Magnetic Cores: A Review

Daniela Rodriguez-Sotelo [1,†], Martin A. Rodriguez-Licea [2,†], Ismael Araujo-Vargas [3], Juan Prado-Olivarez [1], Alejandro-Israel Barranco-Gutiérrez [1] and Francisco J. Perez-Pinal [1,*,†]

1 Tecnológico Nacional de México, Instituto Tecnológico de Celaya, Antonio García Cubas Pte. 600, Celaya 38010, Mexico; d1903002@itcelaya.edu.mx (D.R.-S.); juan.prado@itcelaya.edu.mx (J.P.-O.); israel.barranco@itcelaya.edu.mx (A.-I.B.-G.)
2 CONACYT, Tecnológico Nacional de México, Instituto Tecnológico de Celaya, Antonio García Cubas Pte. 600, Celaya 38010, Mexico; martin.rodriguez@itcelaya.edu.mx
3 Escuela Superior de Ingeniería Mecánica y Eléctrica (ESIME), Unidad Culhuacán, Instituto Politécnico Nacional, Mexico City 04260, Mexico; iaraujo@ipn.mx
* Correspondence: francisco.perez@itcelaya.edu.mx
† These authors contributed equally to this work.

Abstract: In power electronics, magnetic components are fundamental, and, unfortunately, represent one of the greatest challenges for designers because they are some of the components that lead the opposition to miniaturization and the main source of losses (both electrical and thermal). The use of ferromagnetic materials as substitutes for ferrite, in the core of magnetic components, has been proposed as a solution to this problem, and with them, a new perspective and methodology in the calculation of power losses open the way to new design proposals and challenges to overcome. Achieving a core losses model that combines all the parameters (electric, magnetic, thermal) needed in power electronic applications is a challenge. The main objective of this work is to position the reader in state-of-the-art for core losses models. This last provides, in one source, tools and techniques to develop magnetic solutions towards miniaturization applications. Details about new proposals, materials used, design steps, software tools, and miniaturization examples are provided.

Keywords: core losses methods; power losses; ferromagnetic material; inductors; transformers

Citation: Rodriguez-Sotelo, D.; Rodriguez-Licea, M.A.; Araujo-Vargas, I.; Prado-Olivarez, J.; Barranco-Gutierrez, A.-I.; Perez-Pinal, F.J. Power Losses Models for Magnetic Cores: A Review. *Micromachines* **2022**, *13*, 418. https://doi.org/10.3390/mi13030418

Academic Editor: Jürgen J. Brandner

Received: 3 February 2022
Accepted: 25 February 2022
Published: 7 March 2022

1. Introduction

It is a fact that magnetic components are an essential part of our lives. We can find them in almost anything of quotidian use, from simple things such as cell phone charges [1] to TVs and home appliances. However, they have become more relevant in the development of electric and hybrid vehicles, electric machines [2], renewable energy systems [3–6], and recently in implanted electronics [7,8] to open the possibility to micro-scale neural interfaces [9]. Inside these devices and systems, a power stage is conformed by magnetic and electronic components.

A few years ago, these power stages used to be bigger and heavier, and they had considerable energy losses making them less efficient. Nowadays, silicon carbide (SiC) and gallium nitride (GaN) switching devices have improved power electronics, making them smaller and faster [10,11]. Notwithstanding, magnetic components oppose miniaturization; these remain stubbornly large and lossy [12]. In a modern power converter, magnetics are approximately half of the volume and weight , and they are the main source of power losses [13,14].

The trend in power electronics has two explicit purposes for the design of magnetic components. The former is to make maximum use of magnetics capabilities, achieving multiple functions from a single component [12]. The latter consists of minimizing the size of magnetic components substituting ferrite (the leading material in their fabrication) with ferromagnetic materials [15]. Those materials have a higher saturation point than ferrite, high permeability and are based on iron (Fe) and metallic elements (Si, Ni, Cr, and

Co). Some examples of these kinds of material are Fe-Si alloys, powder cores, amorphous materials, and nanocrystal material [16–18].

The design of magnetic components is the key to achieving two purposes. The design parameters are intimately dependent on geometric structure, excitation conditions, and magnetic properties such as power losses that determine if a core magnetic is suitable to be part of a magnetic component [19]. Power losses in magnetic components are important design parameters; which limit many high-frequency designs.

Power losses in a magnetic component are divided into two core power losses and winding power losses. Core power losses are related to the selection of the material to make the core of inductors and transformers; in this case, parameters such as frequency operating range, geometric shape, volume, weight, temperature operating range, magnetic saturation point, and relative permeability must be considered [20]. On the other hand, winding losses are related to phenomena such as skin effect, direct current (DC) and alternating current(AC) resistance in a conductor, and proximity effect [19,21,22]. Those phenomena increase the volume of a winding structure [23]. One way to overcome this kind of loss is by using Litz wire [24].

Nowadays, there are several models for studying, predicting, and analyzing the power losses in the ferromagnetic cores of magnetic components. However, many of them have been developed to work in a limited frequency range , temperature and low magnetic flux [25]. Usually, the ferromagnetic cores' manufacturers provide the graphs of the core losses in their datasheets, which it is commonly obtained with sinusoidal signal tests and for specific values of frequency (f) and magnetic flux (B) [26].

There are many ways to calculate the power losses in magnetic materials. The original Steinmetz equation is one of them; however, to the best of the author's knowledge, it is not always the best option. In the author's opinion, selecting a method to calculate power core losses must be based on its versatility to vary their measurement's parameters [27].

This document aims to provide the reader with a general panorama about the core power losses in ferromagnetic materials, emphasising the diverse models found in the literature, announcing their characteristics, advantages, and limitations. Besides, it will present a relationship about magnetic materials and tested models. Any magnetic component designer must know the basic methods, core losses' models, the conditions at which they are valid, and their mathematical fundamentals. Magnetic and thermal losses, calculus and modelling are open research areas due to the non-linear features of ferromagnetic materials, the complexity of developing a unique power core losses model for the overall ferromagnetic materials, and their respective validations.

The organization of this work consists of the following sections. In Section 2, the reader will find information about the characteristics of each ferromagnetic material. Losses in magnetic components are reviewed in Section 3. In Section 4, general core losses models are mentioned as empirical core losses proposals. Features of Finite Element Method software are given in Section 5. The design process of a magnetic component is discussed in Section 6; on the other hand, the importance of the core losses methods in miniaturization of magnetic devices is summarized in Section 7. The discussion of this work is presented in Section 8, and finally, in Section 9, conclusions are provided.

2. Ferromagnetic Alloys

Ferromagnetic materials are exciting materials, where several of their physical properties and chemical micro-structure allow being controlled. Susceptibilities, permeabilities, the shape of the hysteresis loop, power loss, coercivity, remanence, and magnetic induction are some examples of no intrinsic properties [18]. The saturation magnetization and the Curie temperature are the only intrinsic properties [17,18].

The magnetic behavior is ruled by the dipole moments' interaction of their atoms within an external magnetic field [28]. Ferromagnetic materials have strong magnetic properties due to their magnetic moments that tend to line up easily along an outward magnetic field direction [29]. These materials also have the property of remaining partially magnetized even when the external magnetic field is removed; this means that they can quickly change their magnetic polarization by applying a small field. Ferromagnetic materials are also profitable materials due to being composed by Fe, one of the eight more abundant elements on Earth [30].

Ferromagnetic materials are used in the core of magnetic components; they are classified in Fe-Si alloys, powder cores, amorphous material, and nanocrystalline material; on the other hand, materials as silver, gold, copper, aluminium, iron, steel, among others, are used in windings Figure 1. Except for powder cores, the rest of them are rolled materials [31]; in the following sections, the importance of this fact in the core losses calculation will be detailed. Each one has specific magnetic properties, manufacturing processes, and physical features that determine its feasibility [32].

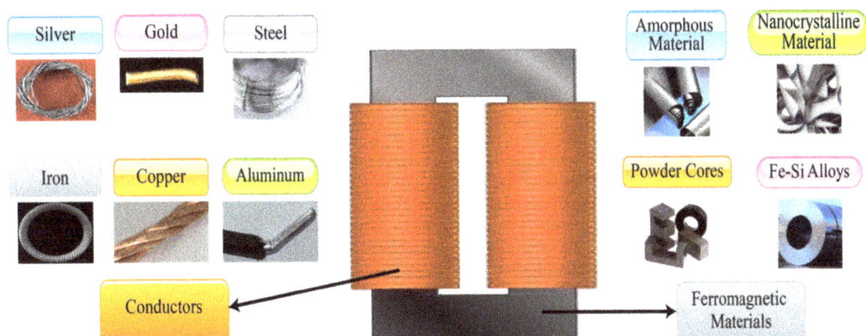

Figure 1. Ferromagnetic and conductor materials types. Source: Adapted from [33–43].

Fe-Si alloys are alloys based on Fe with small quantities of Si (not more than 4.5%). These alloys are attractive due to their reasonable cost and magnetic properties [44]. There are two kinds: grain non-oriented sheets (GNO) and grain-oriented sheets (GO) [30]. At present, GO material represents 80% of the electrical devices market [17,45].

On the other hand, powder cores are fabricated from metallic powders, typically iron; however, those can be composed of alloys with P, Si and Co [46]. The manufacturing process of this kind of material allows special fabrication geometry cores. Usually, powder cores are mixed with a binder or insulating material to reduce magnetic losses at high frequencies [47]. According to the material used in their fabrication, those can be classified into four groups: iron powder core, molybdenum permalloy powder cores (MPP), high flux powder cores, and sendust cores, also called Kool Mμ [48]. These materials present the more recent advances in magnetic elements [49].

Amorphous alloys or metallic glasses are materials without crystalline order [50]. Those alloys are very strong and hard, but also ductile. They contain approximately 80% of particles of Fe, Ni, Co, and their combinations; and 20% of metalloids particles or glass formed elements (C, Al, B, Si, and P) [51,52].

Nanocrystalline alloys consist of a Fe-Si ultra-thin grain alloy with a few quantities of Cu and Nb. Their manufacturing process is very similar to amorphous alloys [53]. Nanocrystals are materials with high mechanical hardness and extremely fragile. The four main kinds are: finemet, nanoperm, hitperm, and nanomet [54,55].

Nanocrystalline, amorphous and Fe-Si alloys are the materials more used in power electronics, so they have been widely studied, especially in terms of core power losses. In rolled materials as Fe-Si alloys and amorphous materials, their magnetic properties depend on the sheet's thickness (around a few mm, and 5–50 µm, respectively). Instead, in nanocrystalline alloys, magnetic properties depend on the diameter of their grains (of the order of 10–15 nm). Rolled materials are susceptible to Eddy currents and skin losses. In contrast, powder cores' magnetic properties can be manipulated during their manufacturing process , which includes the relative permeability variation according to the magnetic field intensity, high saturation point, fringing flux elimination, soft saturation, among others.

3. Losses in Magnetic Components

To any magnetic component designer, a real challenge to overcome is getting a magnetic component with high efficiency, small size, low cost, convenience, and low losses [56]. Usually, losses are the common factor in all requirements announced before; losses are the most difficult challenges to beat in a magnetic component.

Losses in a magnetic component are divided in two groups: core losses and winding losses (also called copper losses), [57]. Figure 2 shows each one of them, as well as their causing phenomena, methods, models, techniques and elements associated with them; nonetheless, phenomena such as the fringing effect and the flux linkage many times are not considered in the losses model. Still, they are necessary to achieve a complete losses model in any magnetic component.

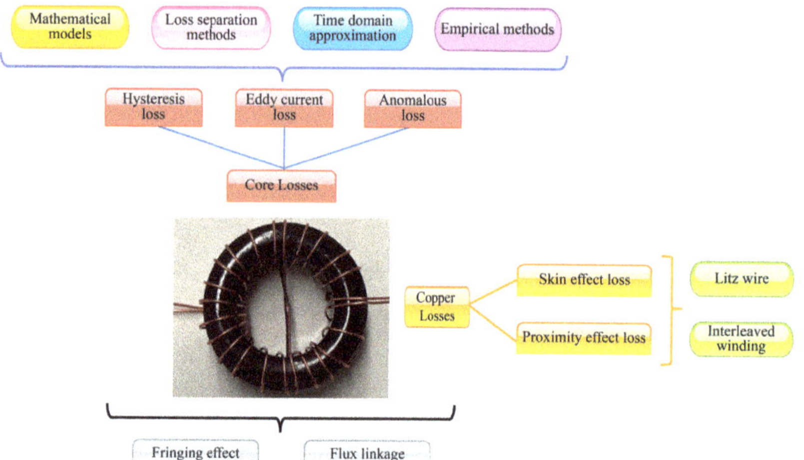

Figure 2. Types and classification of losses presented on magnetic components. Source: Adapted from [58].

Current windings generate the flux linkage density; therefore, it is the sum of the flux enclosed of each one of the turns wound around the core, the flux linkage linked them [50,59]. Flux linkage depends on the conductor's geometry and the quantity of flux contained in it. In the gap between windings is the maximum flux linkage [60–62].

Otherwise, the fringing effect is the counterpart of flux linkage; this is, the fringing effect is presented around the air gap instead of the windings of the magnetic component. This phenomenon depends on core geometry and core permeability, to higher permeability the fringing effect is low [50].

The copper losses are caused by the flow of DC and AC through the windings of a magnetic component, where losses always are more significant for AC than DC [63,64]. The circulating current in the windings generates Eddy phenomena as Eddy currents, skin effect loss and, proximity effect loss.

The skin effect and the proximity effect loss are linked to the conductor size, frequency, permeability and distance between winding wires. For the first of them, the current distribution trough the cross-area of the wire will define the current density in it. The skin effect and the proximity effect loss are linked to the winding wires. For the first of them, the current distribution through the cross-area of the wire will define the current density in it. It means that the conductor will have a uniform distribution for DC, higher on its surface and lower in its center for AC distribution [65].

The proximity effect is similar to the skin effect, but in this case, it is generated by the current carried nearby conductors [50,65]. Eddy currents are induced in a wire in this kind of loss due to a variant magnetic field in the vicinity of conductors at high frequency [46,66].

The Litz wire and interleaved windings help minimize winding losses in magnetic components, both are widely used currently. Indeed, interleaved windings are very efficient in high-frequency planar magnetics [67,68].

On the other hand, core losses directly depend on the intrinsic and extrinsic core materials' characteristics. Core losses are related to hysteresis loop, Eddy currents, and anomalous or residual losses. Without matter, the core losses model chosen by the magnetic component designer will be based on those three primary losses; core loss models will be detailed in the next section. Besides, core losses depend on the core's geometry and the core intrinsic properties' such as permeability, flux density, Curie temperature, among others [69–71]. Many methods and models have been developed; all of them have the predominant interest in studying, analyzing and understanding those kinds of losses to improve the magnetic components' performance.

4. General Core Losses Models

In a magnetic component, the core is the key to determine its magnetic properties and performance [72]. So to achieve an optimal magnetic component's performance, the core losses effects must be characterized [73].

Core loss depends on many aspects that must be considered [74,75]:

- Relative permeability.
- Magnetic saturation point.
- Temperature operation range.
- AC excitation frequency and amplitude
- Voltages' waveform.
- DC bias.
- Magnetization process.
- Peak-to-peak value of magnetic flux density.

In the case of the magnetization process, some factors are instantaneous values and time variation values. While to waveform's topic, the duty ratio of the excitation waveform also influence the core loss [76,77].

Generally, the core loss is provided at a specific frequency and a maximum flux density [78]. The variation frequency effect in ferromagnetic materials is related to Eddy currents and the wall-domain displacement [72].

Core losses in ferrimagnetic and ferromagnetic materials are similar. Both have losses due to Eddy currents, hysteresis, and anomalous; however, there are differences concerning flux density, magnetization process, and hysteresis loop shape that define the magnetic behavior of each one.

The hysteresis is one of the principal features of ferromagnetic materials; it describes the internal magnetization of magnetic components as a function of external magnetizing force and magnetization history [79]. The source of hysteresis loss is the domain wall movement and the magnetic domains' reorientations [80,81].

The hysteresis loss is defined as power loss in each cycle of magnetization and demagnetization into a ferromagnetic material [82]. If a magnetic sample is excited from zero to the maximum field value and later comes back at the initial field's value, it will be observed that the power returned is lower than the supplied it [83,84].

The loss is proportional to the area surrounded by the upper and lower traces of the hysteresis curve; it represents the per cycle loss and it is proportional to $f \cdot B^2$ [82]. However, if the curve's shape remains equal for each successive excitation, the loss power will be the product of the core's area and the applied frequency [80,82].

The hysteresis loops give a lot of information about the magnetic properties [72]. An accurate way to calculate core loss is by measuring the full hysteresis curve [85].

There are many methods to calculate core losses; Figure 3 shows a general classification. All of them are based on one, two, or three main effects (hysteresis, Eddy current, and anomalous); depending on the method's focus they are analyzed as macro or micro phenomena. Each one of the methods shown in Figure 3 will be detailed in the following paragraphs.

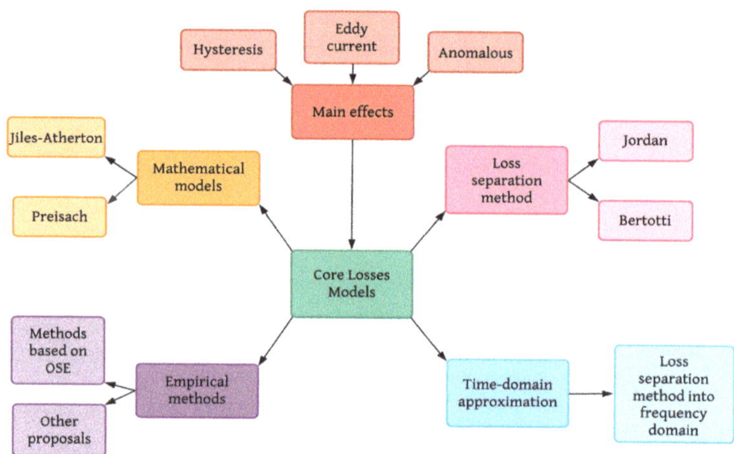

Figure 3. Classification of core losses methods.

4.1. Mathematical Models

Mathematical hysteresis modelling is divided into approaches based on the theory of micro-magnetics and the methods based on curve-fitting. The hysteresis models require complex computation to calculate the model parameters, the material parameter that manufacturers do not provide, or mathematical approximations where the accuracy depends on the number of data points to fit hysteresis loops [86,87]. These drawbacks remain if they are analyzed from a purely mathematical or a physical point of view [78].

The Preisach model and the Jiles-Atherton (J-A) model have been widely used in practical problems to calculate core loss, they are in continuous improvement, and they are considered a valuable and convenient tool to the hysteresis modelling [87–89].

The Preisach is a scalar-static model that considers several quantities of basic's domain-walls [72]. It is an accurate-phenomenological model, such that it could describe any system that shows a hysterical behavior [87,90]. This model can be a link between theory and experimentation to describe a microscopical system by measuring macroscopic behavior [78].

Given a typical hysteresis loop for separate domains as shown in Figure 4, H_d and H_u are the switching magnetic fields "down" and "up", respectively. The magnetization $M(t)$ of a particle having the hysteresis loop $\hat{m}(H_u, H_d)$ is described as the magnetic moment m_s while the particle is switched up $\hat{m}(H_u, H_d)H(t) = +m_s$ or down $\hat{m}(H_u, H_d)H(t) = -m_s$ [87]. It is assumed that all domains have a distribution of reversal fields H_u and H_d that can be characterized by a distribution function $\phi(H_u, H_d)$, so the Preisach model is usually defined as:

$$M(t) = \iint_{H_u \geq H_d} \phi(H_u, H_d) * \hat{m}(H_u, H_d) * H(t) dH_u dH_d. \tag{1}$$

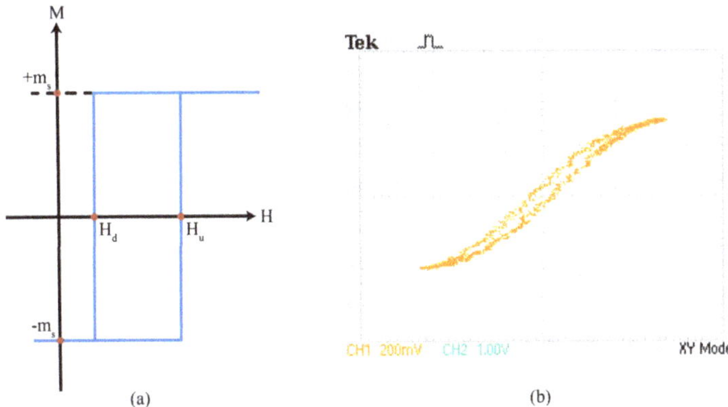

(a) (b)

Figure 4. Hysteresis loop of (**a**) an oriented magnetic domain, (**b**) a multidomain particle of a 3C90 Epcos ferrite core at 500 Hz and 28.5 °C. Source: Adapted from [91].

A drawback of the Preisach model is the several measurement data to adapt it to the B-H environment's changes [72]. Nonetheless, it can track complex magnetization processes and minor loops [78].

On the other hand, the J-A model requires solving a strongly non-linear system of equations, with five scalar parameters, which are determined in an experimental way; this is, measuring the hysteresis loop [92]. This model predicts the major hysteresis loops under quasi-static conditions, besides there is a dynamic conditions version in which eddy current loss, DC-bias fields, anisotropy, and minor loops are included [86,93].

The J-A model is described as the sum of reversible M_{rev} and irreversible M_{irr} magnetizations, hence, the total magnetization M is:

$$M = M_{irr} + M_{rev} \tag{2}$$

$$\frac{dM_{rev}}{dH} = c\left(\frac{dM_{an}}{dH} - \frac{dM_{irr}}{dH}\right) \tag{3}$$

$$\frac{dM_{irr}}{dH} = \frac{M_{an} - M_{irr}}{\delta x/\mu_0 - \alpha_{ic}(M_{an} - M_{irr})} \tag{4}$$

$$M_{an} = M_s\left(\coth\frac{H_{eff}}{a_s} - \frac{a_s}{H_{eff}}\right) \tag{5}$$

$$H_{eff} = H + \alpha_{ic}M. \tag{6}$$

In these equations, c is the reversibility coefficient, α_{ic} is the inter-domain coupling, δ indicates the direction for magnetizing field ($\delta = 1$ for the increasing field, and $\delta = -1$ for decreasing field), x is the Steinmetz loss coefficient, M_s is the saturation magnetization, M_{an} is anhysteretic magnetization, H_{eff} is the effective field, H is the magnetic field, μ_0 is the permeability in free-space, and a_s is the shape parameter for anhysteretic magnetization [73]. A drawback of the J-A model is the computing time and the computing resources to solve a strongly non-linear system.

On the other hand, Eddy's losses are produced due to changing magnetic fields inside the analyzed core. These fast changes generate circulating peak currents into it, in the form of loops, generating losses measured in joules [94].

These losses depend on the actual signal shape instead of only the maximum value of the density flux. The power loss is proportional to the area of the measured loops and inversely proportional to the resistivity of the core material [50,94].

Eddy currents are perpendicular to the magnetic field axis and domain at high-frequency, as well, they are proportional to the frequency. This phenomenon can be treated as a three-dimensional character or in a simplified way [95]. In a magnetic circuit, Eddy currents cause flux density changes in specific points of its cross-section [96].

One way to describe the Eddy currents distribution is based on Maxwell's equations [95,97]. In practice, especially on higher frequencies, there is a difference in the measured total loss between the sum of the hysteresis loss and the Eddy current loss; this difference is known as anomalous or residual losses [13]. The effect of temperature and the relaxation phenomena are enclosed in this category [82].

4.2. Time-Domain Approximation Models

For the time-domain model, the loss separation method is used in the frequency domain, using Fast Fourier Transform (FFT) [13].

Time-domain approximation (TDA) can be used for sinusoidal and non-sinusoidal magnetic flux density; however, it is only valid for linear systems [98]. The losses are calculated considering each frequency, separately, and adding them later [85,98].

$$P_c = \sum_{n=1}^{\infty} \pi(fn) B_n H_n \sin \phi_n. \tag{7}$$

According to Equation (7), time-domain approximation is a function of the fundamental frequency f, the peak values of the nth harmonic of the magnetic field H and magnetic flux density B, and ϕ_n the angle between B_n and H_n.

When a pulse width modulation (PWM) is induced in a magnetic component made with non-linear material, a non-sinusoidal ripple is generated [93]. Fourier's core loss decomposition accuracy is not acceptable for non-sinusoidal flux densities and it is limited for frequencies > 400 Hz, as it was reported in [99].

4.3. Loss Separation Models

The loss separation method (LSM) is well-known, accurate and straightforward in many applications [94]. It defines the core loss (P_C) under a dynamic magnetic excitation as the sum of three components: the hysteresis loss (P_h), the Eddy current loss (P_{eddy}) and the residual or anomalous loss (P_{anom}) [82] is given by:

$$P_c = P_h + P_{eddy} + P_{anom}. \tag{8}$$

According to Equation (8), the power loss per unit volume is the sum of a hysteresis and dynamic contribution; Eddy currents and residual loss are part of the latest one [100].

The loss separation method was proposed in 1924 by Jordan, which describes the core losses as the sum of the hysteresis loss (or static loss) and the Eddy loss (or dynamic loss) [98]

$$P_c = k_h f B_s + k_{eddy} f^2 B_s^2. \tag{9}$$

Years later, Bertotti extended Jordan's proposal, adding an extra term to calculate residual losses. G. Bertotti is the maximum exponent in the loss separation methods; his theory provides a solid physical background. The total power loss can be calculated at any magnetizing frequency as the sum of three components [44]. This is show in the following equation

$$P_c = k_h f B_s^x + k_{eddy} f^2 B_s^2 + k_{anom} f^{1.5} B_s^{1.5}, \tag{10}$$

where k_h, k_c, k_{anom} are the hysteresis, Eddy currents, and anomalous coefficients, respectively. The hysteresis coefficient decreases if the magnetic permeability increase. The frequency is represented by f, while x is the Steinmetz coefficient and has values between 1.5 and 2.5, according to the permeability of the material. Finally, B_s is the peak value of the flux density amplitude [23,100].

From Equation (10), Bertotti defined eddy currents and anomalous losses in Equations (11) and (12), respectively, for lamination materials.

$$P_{eddy} = \frac{\pi^2 d^2 f^2 B_s^2}{\rho\beta} \tag{11}$$

$$P_{anom} = 8\sqrt{\frac{GAV_0}{\rho}} B_s^{1.5}\sqrt{f}, \tag{12}$$

where G is about 0.2, ρ is the electrical resistivity, d is the lamination thickness, A is the cross sectional area of the lamination, β is the magnetic induction exponent, G is a dimensionless coefficient of Eddy current, and V_0 is a parameter that characterizes the statistical distribution of the magnetic objects responsible for the anomalous eddy currents [101]. However, this method only provides average information, so it is not able to calculate core loss under harmonic excitation [94]. Still, it can calculate core loss under square waveforms considering DC bias [102].

4.4. Empirical Models

A big group of empirical models is based on a simple power-law equation, which was proposed in 1982 by Charles Steinmetz to calculate hysteresis loss without including the frequency relation [103]. This equation is known as the Steinmetz equation or original Steinmetz equation (OSE),

$$P_{\text{OSE}} = kf^{\alpha}B_s^{\beta}, \tag{13}$$

where P_{OSE} is the time-average core loss per unit volume, B_s the peak induction of the sinusoidal excitation, k is a material parameter, α, and β are the frequency and magnetic induction exponents, respectively, often referred to as Steinmetz parameters [76]. Typically α is a number between 1 and 2, and β is typically between 1.5 and 3 [104].

The Steinmetz parameters can be determined from a double logarithm plot by linear curve fitting of the measured core loss data. Therefore Equation (13) assumes only sinusoidal flux densities with no DC bias [76,77,98,105].

In modern power electronics applications, sinusoidal-wave voltage excitation is not practical because many of them, like power converters, require square-wave voltage excitation [106]. Therefore, a square waveform's core loss can be lower than the sinusoidal-wave's losses for the same peak flux density and the same frequency [76,107].

To overcome the aforementioned situation, modifications to OSE have been made. The result was the developing Modified Steinmetz Equation(MSE) [108], General Steinmetz Equation (GSE) [109], Doubly Improved Steinmetz Equation (Improved-Improved Steinmetz Equation, i^2GSE) [110], Natural Steinmetz Equation (NSE) [85], and Waveform-Coefficient Steinmetz Equation (WcSE) [111]. It is important to mention that the OSEs's constants α, β, and k remain in those expressions.

Modified Steinmetz Equation was the first modification proposed to calculate core loss with non-sinusoidal excitation, incorporating into OSE the influence of a magnetization rate. MSE is proposed in [112] as follows

$$P_{\text{MSE}} = (kf_{eq}^{\alpha-1}B_s^{\beta})f_r \tag{14}$$

$$f_{eq} = \frac{2}{\Delta B^2\pi^2}\int_0^T \left(\frac{dB(t)}{dt}\right)^2 dt,$$

where f_r is the periodic waveform fundamental frequency, f_{eq} is the frequency equivalent, ΔB is the magnetic induction peak to peak, and $dB(t)/dt$ is the core loss magnetization rate.

A drawback of Equation (14) is that its accuracy decreases with the increasing of waveform harmonics, and for waveforms with a small fundamental frequency part [77,85,113].

Another modification of OSE is the Generalized Steinmetz Equation to overcome the drawbacks of OSE and MSE for sinusoidal excitations, whose expression is:

$$P_{GSE} = \frac{k_1}{T}\int_0^T \left|\frac{dB(t)}{dt}\right|^{\alpha} |B(t)|^{\beta-\alpha} dt \tag{15}$$

$$k_1 = \frac{k}{2^{\beta-\alpha}(2\pi)^{\alpha-1}\int_0^{2\pi}|\cos\theta|^{\alpha}|\sin\theta|^{(\beta-\alpha)}d\theta}.$$

In this equation, the current value of flux is considered additional to the instantaneous value of $dB(t)/dt$, $B(t)$, T is the waveform period, and θ is the sinusoidal waveform phase angle. The main GSE's advantage is it DC-bias sensitivity. However, its accuracy is limited if a higher harmonic part of the flux density becomes significant [85,98].

The Improved Generalized Steinmetz Equation is considered one of the best methods because it is a practical and accurate [114]. It is defined as follows:

$$P_{i\text{GSE}} = \frac{k_i}{T}\int_0^T \left|\frac{dB(t)}{dt}\right|^{\alpha} |\Delta B|^{\beta-\alpha} dt \tag{16}$$

$$k_i = \frac{k}{2^{\beta-\alpha}(2\pi)^{\alpha-1}\int_0^{2\pi}|\cos\theta|^{\alpha}d\theta}.$$

Unlike GSE, which uses the instantaneous value $B(t)$, *i*GSE considers its peak-to-peak value ΔB. Any excitation waveform can be calculated by it; therefore, it has better accuracy with waveforms that contained strong harmonics [114–117].

The Natural Steinmetz Extension is similar approach to the *i*GSE; this means that ΔB was also taken into account [85,118]. It is written as:

$$P_{NSE} = \left(\frac{\Delta B}{2}\right)^{\beta-\alpha}\frac{k_N}{T}\int_0^T \left|\frac{dB(t)}{dt}\right|^{\alpha} dt \tag{17}$$

$$k_N = \frac{k}{(2\pi)^{\alpha-1}\int_0^{2\pi}|\cos\theta|^{\alpha}d\theta}.$$

The NSE focuses on the impact of rectangular switching waveform like PWM, so Equation (17) can be modelled for a square waveform with duty ratio D by:

$$P_{NSE} = k_N(2f)^{\alpha}B_s\left[D^{1-\alpha} + (1-D)^{1-\alpha}\right]. \tag{18}$$

The Improved–Improved Generalized Steinmetz Equation (i^2GSE) considers the relaxation phenomena effect in the magnetic material used due to a transition to zero voltage [115]. The i^2GSE was developed to work with any waveform but its main application is with trapezoidal magnetic flux waveform [110]. For a trapezoidal waveform, i^2GSE is described as follows:

$$P_{i^2\text{GSE}} = \frac{1}{T}\int_0^T k_i\left|\frac{dB(t)}{dt}\right|^{\alpha}(\Delta B)^{(\beta-\alpha)}dt + \sum_{l=1}^{n} Q_{rl}P_{rl} \tag{19}$$

$$P_{rl} = \frac{k_r}{T}\left|\frac{d}{dt}B(t-)\right|^{\alpha_r}(\Delta B)^{\beta_r}(1-e^{-\frac{t_1}{\tau}})$$

$$Q_{rl} = e^{-q_r\left|\frac{dB(t+)/dt}{dB(t-)/dt}\right|}.$$

Note that P_{rl} and Q_{rl} calculate the variation of each voltage change and the voltage change, respectively. However, to use the i^2GSE requires additional coefficients as can be seen in Equation (19) where k_r, α_r, β_r, τ and q_r are material parameters and they have to be measured experimentally; given that, they are not provided by manufacturers and the steps to extract the model parameters are detailed in [119]. The expression given by Equation (19) can also be rewritten by a triangular waveform.

Finally, the Waveform-Coefficient Steinmetz Equation (WcSE) was proposed in [111] to include the resonant phenomena , and it applies only at situations with certain loss characteristics. This empirical equation is used in high-power, and high-frequency applications, where resonant operations are adopted to reduce switching losses. The WcSE is a simple method that correlates a non-sinusoidal wave with a sinusoidal one with the same peak flux density; the waveform coefficient (*FWC*) is the ratio between the average value of both types of signals [119,120]. WcSE can be written as follows:

$$P_{\text{WcSE}} = FWCkf^{\alpha}B_s^{\beta}. \quad (20)$$

The most important characteristics of each approximation based on the Steinmetz equation are listed in Table 1.

Table 1. Details of Steinmetz's equations.

Steinmetz's Equation	Characteristics
OSE	Only for sinusoidal signals. Hysteresis losses proportional to f. Eddy current losses proportional to f^2. The values for α and β are between 1 and 3.
MSE	Equivalent f calculus. Considers B_s in the core losses. Its accuracy decreases with harmonics increment.
GSE	Considers B_s variation and its instantaneous value. Compensates the mathematical error between OSE and GSE for sinusoidal excitations. Considers the DC-level in the signal.
*i*GSE	Takes the peak to peak value of B_s. Accurate with a high number of harmonics. Core losses calculus with frequencies and duty cycles variables.
NSE	It can be used for rectangular signals. The second and third harmonic are dominants at moderate values of D. For extreme values of D (~95%) a high alpha value will give a better adjustment.
i^2GSE	Characteristics similar to *i*GSE. Applications with trapezoidal B_s. Takes into account the material relaxation effect. Require parameters not provided by the materials manufacturers.
WcSE	Proposed to correlate a not sinusoidal signal with a sinusoidal with the same measured value of B_s.

In addition to the methods listed before, in [110] the authors provided several graphs for different materials at different operating temperatures making use of the Steinmetz Premagnetization Graph (SPG), which is a simple form to show the dependency of Steinmetz's parameters on premagnetization to calculate the core losses under bias conditions.

Using the SPG, the changes of Steinmetz in *i*GSE can be considered; it is also possible to calculate the core loss under any density flux waveform [19,74].

Empirical Core Losses Proposals

The primary purpose of this section is to show a few empirical core losses models proposals to emphasize the diversity of parameters that are taken into account: voltage waveforms, temperature effect, and the versatility of using them to calculate core losses.

Generally, square waveforms are used in power electronic applications resulting in triangular waveform induction and flux density; therefore, core losses models must contemplate this waveform. However, a sinusoidal approximation for a duty rate of 50% is valid [121].

Curve fitting is used to approximate and characterize the model's core losses parameters. Nonetheless, curve fitting by polynomials are unstable with minor changes at entry, resulting in a big variation on coefficients, so the designer must be meticulous in selecting the model as the curve fitting method and logarithm curve [121].

The Composite Waveform Hypothesis (CHW) was proposed in 2010; it is based on a hypothesis that establishes that if a rectangular waveform given is decomposed in two pulses, total core losses are the sum of the losses generated by each one of the two pulses [26,122]. This method is described by

$$P_{\text{CHW}} = \frac{1}{T}\left[P_{sqr}\left(\frac{V_1}{N}, t\right)t_1 + P_{sqr}\left(\frac{V_2}{N}, t\right)t_2\right], \tag{21}$$

where P_{sqr} is the power losses of a rectangular waveform with the parameters given, V_1/N and V_2/N are the voltage by turn in time t_1 and t_2, respectively.

Villar's proposal [120,123] calculates core losses for three-level voltage profiles through a lineal model by parts. Villar takes the models based on Steinmetz equations and modifies them by a linear model by parts (22)–(24); this includes a duty ratio parameter in the original equations, getting expressions to calculate core losses for three-level voltage profiles.

$$P_{Villar\text{OSE}} = kf^{\alpha}B_s^{\beta}D^{\beta} \tag{22}$$

$$P_{Villar\text{IGSE}} = 2^{\alpha+\beta}k_i f^{\alpha}B_s^{\beta}D^{\beta-\alpha+1} \tag{23}$$

$$P_{Villar\text{WcSE}} = \frac{\pi}{4}\left(1+\frac{\omega}{\pi}\right)kf^{\alpha}B_s^{\beta}D^{\beta}. \tag{24}$$

In Equation (24) the parameter ω is the duration of zero-voltage period this is, power electronics converters' switching devices take a short time to start their conduction mode; therefore, Villar includes this effect in its model by the flux density's equation as follows:

$$B_s = \frac{1}{2}\frac{U}{NA_{eff}}\left(\frac{T}{2} - T_{\omega}\right), T_{\omega} = \omega\frac{T}{2\pi}, \tag{25}$$

where U is a DC constant voltage, A_{eff} is the core's area effective, and T_{ω} is the length of zero-voltage period.

Villar's proposal also is applied to the equivalent elliptical loop (EEL); however, the proposal does not include thermal parameters.

Another alternative is the model proposed by Gorécki given by Equation (26) in [124,125], which includes 20 parameters. Those are divided into three groups: electrical parameters, magnetic parameters, and thermal parameters. At the same time, the magnetic parameters' are subdivided into three groups: core material parameters, geometrical parameters, and ferromagnetic material parameters corresponding to power losses. The model parameters can be obtained by the datasheets of the cores and experimental way when the inductor is tested under specific conditions, which is called local estimation.

$$P_{Gorecki} = P_{v0}f^{\alpha}B_s^{\beta}(2\pi)^{\alpha}\left[1+\alpha_p(T_R - T_M)^2\right][0.6336 - 0.1892\ln(\alpha)] \tag{26}$$

$$P_{v0} = \alpha exp\left(-\frac{f+f_0}{f_3}\right) + a_1(T_R - T_M) + a_2 exp\left(\frac{f-f_2}{f_1}\right)$$

$$\beta\begin{cases} 2\left[1-exp\left(-\frac{T_R}{\alpha_T}\right)\right]+1.5 & \text{if} \quad exp\left(\frac{-T_R}{\alpha_T}\right) > 0 \\ 1.5 & \text{if} \quad exp\left(\frac{-T_R}{\alpha_T}\right) < 0 \end{cases}.$$

In these equations, T_R is the core temperature, T_M is the core temperature at which the core has a minor loss, α_p is the losses' temperature coefficient in ferromagnetic material, a, a_1, a_2, f_0, f_1, f_2, f_3, and α_T are material parameters.

Gorecki's model is also valid for the triangular waveform following Equation (27), where D is the duty ratio of the waveform.

$$P_{Gorecki} = P_{v0}f^{\alpha}B_s^{\beta}(2\pi)^{\alpha}\left[1+\alpha_p(T_R - T_M)^2\right]\left[D^{(1-\alpha)} + (1+D)^{(1-\alpha)}\right]. \tag{27}$$

The main drawback of Gorecki's model is to find the overall parameters mentioned before, and solving the equations related to the power core losses.

5. Simulation Software

Several core loss models have been reviewed; they are the base for developing a series of numerical and theoretical models that are useful for designing magnetic components. Nonetheless, they fail to predict dynamic magnetic behaviors. Usually, the calculation of the core parameters' losses in a dynamic situation is complicated (specially with complex geometries), and it requires a rigorous numerical treatment [126].

Several methods are used in the simulation and calculation of core and wire windings. However, the finite element method (FEM) is the most widely used for designs in 2-D and 3-D [22].

In 1960 Clough introduced the name of Finite Element Method (FEM), which continues to these days [127]. FEM is a computational method whose basic idea consists on finding a complicated problem and replacing it with a simpler one. It is always possible to improve the approximation solution spending more computational effort [127,128].

To find the solution of a region, the FEM considers that it is built of many small, interconnected subregions called elements [128], and the global solution is obtained from the union of individual solutions on these regions [129].

The FEM is a tool for solving problems with partial differential equations that are part of physics problems [130]. FEM has quite benefits of using it; some of them is the freedom that it offers in the discretization's selection, and its well developed theoretical base that allows valid error estimates for the numerical model equations, and its flexibility to be adapted to a wide range of numerical problems [129,131].

The FEM was originally developed to solve problems in solid-state mechanics. Still, its versatility, excellent simulation technique, availability to optimize the mesh size, and accuracy have been implemented in a wide variety of applied science and engineering [127,129].

Mathematical models are discretized by FEM, resulting in numerical models. To solve the discretized equations, Finite Element Analysis (FEA) is used [131,132].

There are many FEM software developers in the market; however, Ansys® is the leader so far. Ansys, Inc. (Canonsburg, PA, USA) was founded in 1970s, and since then, it has developed, commercialized and brought support at several range of physics through engineering simulation software. The Ansys® catalogue includes simulation software for semiconductors, structures, materials, fluids, and electronics.

An exciting tool that Ansys has developed in recent years is Twin Builder®, a multi-technology platform to create digital representations simulations, recollected real-time data information through sensor inputs asset with real-world [133]. Twin Builder® is a powerful and robust multi-domain system modelling compatible with a series of standard languages and formats as SPICE, Python, C/C++, simplorer modelling language (SML), among others. It can be used to develop basic simulation experiments and advanced simulation studies, from 2-D and 3-D physics simulations . Inclusive the functional mock-up interface (FMI standard) can import and export models as available mockup units (FMU). An example of an electromagnetic digital twin is shown in Figure 5. Twin Builder® only is compatible with Ansys software and its main application is in the industry as virtual laboratory to test any kind of system.

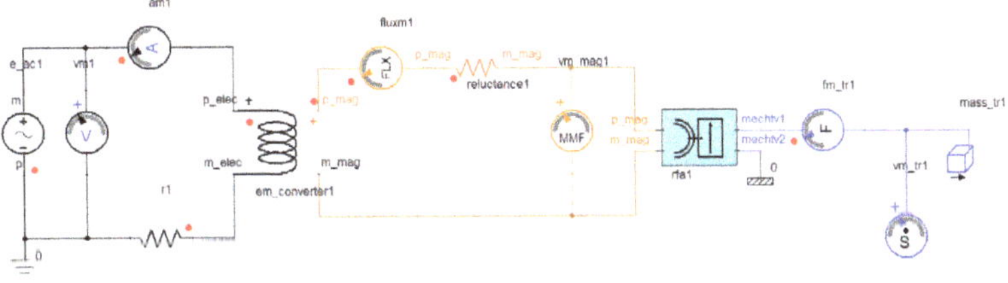

Figure 5. Twin Builder example.

Added to these FEM software developers, there is Comsol Multiphysics®, which is a powerful software tool for developing and simulating modelling designs in all fields of engineering, manufacturing, and scientific research [134]. This software was founded in 1986. Its main characteristic is its friendly graphical user interface. The Comsol Multiphysics® community is more significant than the other software. In Figure 6, two simulations are provided. Comsol Multiphysics® can import designs in CAD and export final designs to Simulink®, given that both belong to the MathWorks® family. Comsol Multiphysics® is oriented to students and academic researchers, it is not complex to learn and it is an excellent option to start with for FEM software.

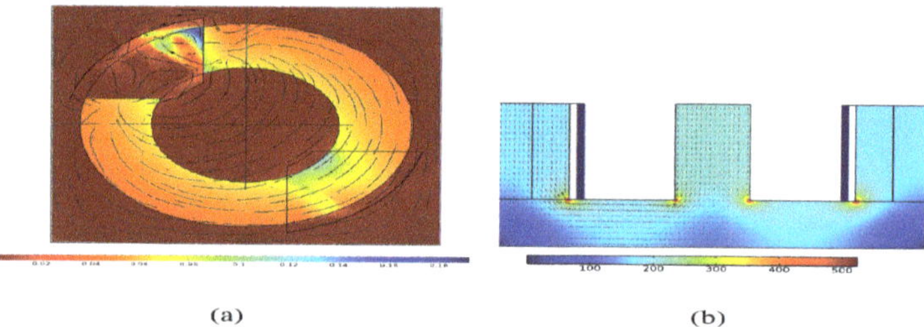

Figure 6. Examples of magnetic flux density simulations' in Comsol Multiphysics for a powder core with shape: (**a**) toroidal , and (**b**) "E".

Another big company of FEM software is JMAG®, a simulation software specialized for electric device design and development, including the accurate model of complex magnetic phenomena. It was founded in 1983, and currently, its most attractive feature is the capability to link various systems through its interface [135].

The JMAG® interface allows the data exchange at high speed without loss of precision with other software as SPEED, PSIM and MATLAB/Simulink; one example of this software is shown in Figure 7. Additionally, JMAG® allows to the user import and export multipurpose files and to run VB Script and other scripting languages [135].

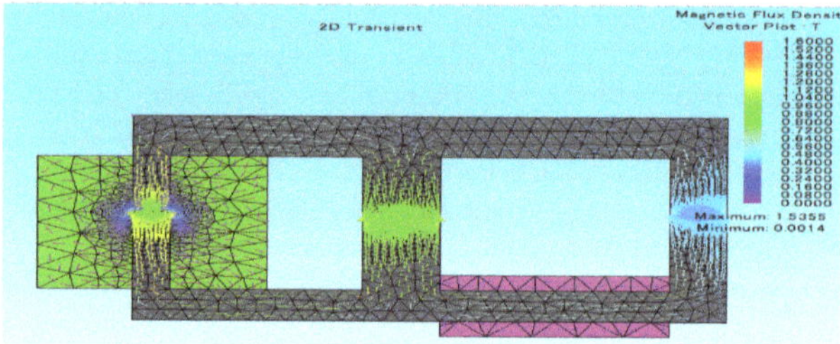

Figure 7. JMAG example of magnetic flux density simulation.

Another feature of JMAG® is its CAD interface to link and import files with software as SOLIDWORKS, CATIA V5, among others to test the model developed using software in the loop (SIL), model in the loop (MIL) and hardware in the loop (HIL) systems, which are applied for system-level and real-time simulations [135,136]. JMAG® is a specialized software to model inductors and magnetic motors adding vibration and thermal analysis; however, the information about it is limited, and it is complex to learn.

The software mentioned before are the most commonly used; however, there are many other finite element software packages available to any platform and they have different features [137].

6. Magnetic Components Design Process

In general, the design process of a magnetic component consists of four steps: design, simulate, implement, and evaluate Figure 8.

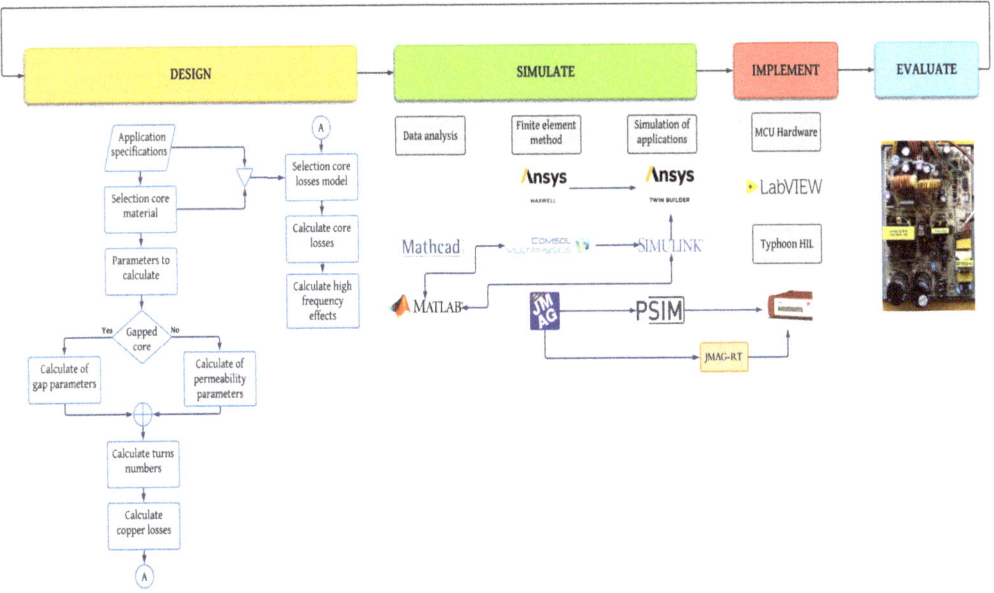

Figure 8. Steps to design a magnetic component. Source: Adapted from [138].

The design step can completely follow the diagram shown in Figure 8, where the application specifications will determine the selection core material up to the copper losses, the core losses model, and the high-frequency effects according to [139,140].

It is worth to mention, that there are some intermediate steps, which are related to core physical and magnetic parameters such as the cross section of the core, section of the core, length of it, effective relative permeability, peak to peak density ripple, among others [141]. In the same way, choosing a gapped core (a core with a concentrated air gap) instead of a distributed air gap core will impact its behavior and parameters to calculate; for the first one, gap parameters as length are a priority. For the second one, permeability parameters will be taken into account [142].

As applications specifications as selection core material are fundamental to select or design a core losses model, both will establish the minimum parameters to calculate power losses.

The simulate step is the masterpiece to validate the magnetic component designed, but at the same time, it will be a problematic step if the designer is not careful; FEM software always will give a solution but does not mean that it is correct. In a general way, simulation steps can be subdivided into three parts: data analysis, finite element method, and simulation of applications.

From the design steps depending on the models selected to calculate power losses, core physical parameters, and data related to the windings, many formulas are involved, and sometimes some depend on others; software such as Mathcad® and Matlab® are of great help in these cases. FEM is a big world, as it was explained in the section before. Once the magnetic component's design is complete (windings and core parameters) and validated in FEM; the next step consists in exporting the final FEM design to software like

Simulink® or Twin Builder®, PSIM®, to simulate it in a specific application and corroborate its correct performance [59,143–146].

The implementation step tests the simulated application through the interconnection of different sorts of software and platforms (MCU hardware, LabVIEW®, Typhoon HIL, among others) to simulate it with high fidelity, in similar conditions to real [147]. Typhoon HIL is a feasible example of this, as it can see in Figure 8. The reader interested in magnetic component testing is referred to [148–153] and the references therein.

The last one is to evaluate, that is, physically build the magnetic component designed and tested in the steps before and implement it in a circuit. Technically speaking, the tests obtained in the implementation step must coincide with the physical circuit measurements with a minimal error percentage. If the magnetic component behavior is not satisfactory, the steps must be repeated, modifying the necessary parameters until the desired results are obtained.

7. Magnetic Devices and Miniaturization

Transportation, electrification, medical applications, micromachines, wireless communication and wireless charging are some areas demanding new technology to develop compact and high performance applications; and the manufacturing of low losses magnetic components is key [154].

Magnetic components exhibit design difficulties when power conversion systems require low power levels and high frequencies due to the nonlinear behavior of magnetic materials, thermal limits, and the exponential increase of losses under these conditions.

Core data and core losses models are critical to design magnetic components properly and select the size of magnetic devices. According to core selection material different procedures may be applied [155].

Miniaturization of magnetic components and micromachines are an interesting couple to develop and implement in biomedical devices and applications. In biomedical applications magnetism-based systems are widely utilized, for example, to study and measure human tissue and organs (such as magnetocardiography, magnetoneurography, magnetoencephalography, and magnetomyography) [156–159]. By 2021, the market of magnetic sensors will worth around $7.6 billion due to their attractive commercial purposes and new application areas [160]. However, the development of efficient magnetic technologies that are sensitive, inexpensive, biocompatible and miniaturized is still far away [161].

The miniaturization, feasibility and integration of magnetic sensors for biomagnetic signal detection have been in constant evolution in terms of size, sensing signals magnitude (pico-Tesla scale), and environment conditions. In magnetic sensors, the magnetic field is converted into measurable quantities as voltage and current [156]. Some examples of them are the thin-film magnetoelectric (ME) sensors, optically pumped magnetometers (OPM), superconducting quantum interference device (SQUID), flux gate sensors, giant magnetoimpedance (GMI) sensor, magnetic sensors based on the thin-film magnetoresistive (MR), and conventional superconducting coils. Those are compared in Table 2 [156,160].

Table 2. Magnetic sensing technologies comparison in biomedical applications.

Magnetic Sensing Technology	Sensitivity	Frequency	Miniaturization Achieving	Portability	Cost
OPM	Acceptable	0 Hz	Unacceptable	Unacceptable	Marginally Acceptable
Coils	Acceptable	60 Hz	Unacceptable	Acceptable	Excellent
ME	Acceptable	0–1 kHz	Acceptable	Excellent	Excellent
Fluxgate	Acceptable	0–5 kHz	Excellent	Excellent	Acceptable
GMI	Marginally Acceptable	0–10 kHz	Excellent	Excellent	Excellent
SQUID	Excellent	0–100 kHz	Unacceptable	Unacceptable	Unacceptable
MR	Acceptable	0–GHz	Excellent	Excellent	Excellent

In [162], a GMI miniaturized magnetic sensor fabricated with a Co-based amorphous by Micro-Electro-Mechanical System (MEMS) technology wire is described, achieving a size of 5.6 × 1.5 × 1.1 mm^3. Co-based amorphous wire is selected for its high impedance change rate, increasing the sensitive of the GMI sensor, it is used to fabricated a pick-up coil of 200 turns and diameter of 200 μm.

A common problem in biomagnetic sensors is the noise at low frequencies, specifically between 10 and 100 Hz due to the small magnitude of measured signals. A set of bi-planar electromagnetic coils is a recent technique to cancelling the Earth's noise nearby the magnetic sensors and improving their sensitivity [156].

Magnetic signal detection including portable and handled devices in the point-of-care testing (POCT) including mini magnetic induction coils and electromagnets to do the devices more portable, flexible and with high detection capabilities. In [163] several examples of these applications are described, where magnetic devices have dimensions in centimeters scale and those consider other POCT techniques for their favorable accuracy, high reliability, innovation and novelty, although their cost might increase.

Portable and small size, or even single handled devices that provide rapid and accurate detection have potential application of POCT [163]. Portable and wearable biosensors are the future of healthcare sensor technologies. Those have demonstrated their utility in disease diagnosis with accurate prediction. Their incorporation with mobile phones is known as digital health or mobile health, which promises reduce the frequency of clinical visits, prevent health problems and revolutionary the demand of micro and wearable sensors technology [164–166].

The emerge of microrobots and the use of a magnetic resonance imaging help to improve diagnostic capabilities by minimally invasive procedures. A microrobot is a robot on a microscale that can perform high-precision operations. Microrobots' actuation system require input energy to act it, which usually requires special materials (soft and hard magnetic materials) or structural design (arrangement of coils) [167,168]. While the propulsion of microrobots, ferromagnetic cores and magnetic structures can be used to generate an image of many sites in the human body [169–171].

In [172], microcoils with a diameter <1 mm are used to harvest electromagnetic energy wirelessly by inductive coupling in the on-board energy robots, achieving several milliwatts of power, capable of controlled motion and actuation with a maximum efficiency of 40%. This has been a recent development to controlled and powered remotely a system-engineered miniaturized robot (SEMER).

To increase the adaptability and robustness of robotic systems in challenging environments, and inter disciplinary research is needed, including material science, biology, control, among others [172]. Magnetic materials and magnetoelectric concepts of micro and nanorobots in magnetic applications are detailed in [173].

In this authors opinion the versatility, scalability, and flexibility of developing a power core loss model to analyze and allowing a better comprehension of the magnetic phenomena in ferromagnetic materials, will be the first step to the miniaturization of magnetic devices and power conversion systems at any scale [174,175].

8. Discussion

Nowadays , there are several models for studying, predicting, and analyzing the power losses in the ferromagnetic cores of magnetic components. Usually, these models consider a series of parameters based on magnetic units and frequency to predict and calculate power losses. Figure 9 illustrates a comparison between core losses models depending several features.

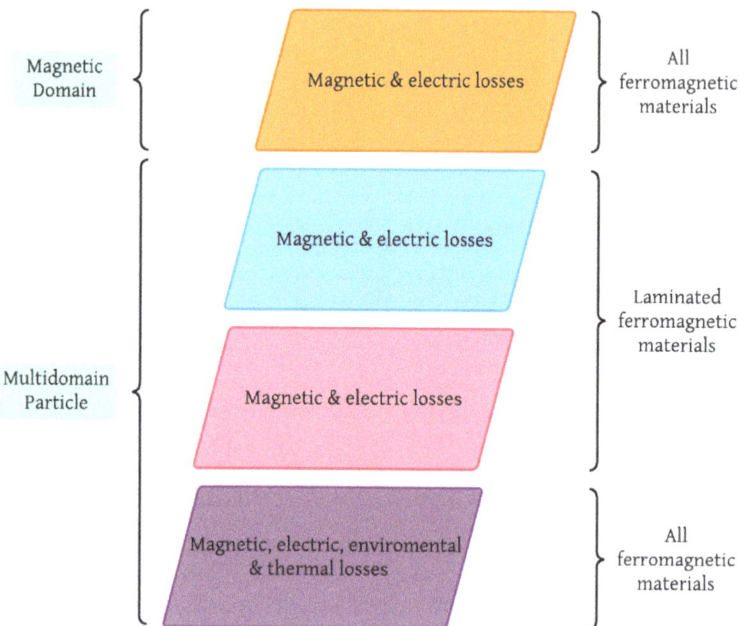

Figure 9. Power core losses models' comparison depending on the size of the phenomena analyzed, kind of losses to calculate and ferromagnetic materials. Features of mathematical models, time-domain approximations, loss separation, and empirical methods are in color yellow, blue, pink, and purple, respectively.

Nevertheless, one main application of magnetic components is in power electronics, where the measurement units are electric, which in many cases cause inexactness while designing a magnetic element. A common idea in the literature is that in a magnetic component with high operating frequency, its losses and size will be lower; however, this is not always true. There are many factors involved (kind of material, core losses, copper losses, among others) to determine the integrity of this expression. Although the last years' power electronic researchers have been working hard to find and develop numerous techniques to achieve better performance from magnetic components, the way is still long. For instance, the future design of magnetic components should take advantage of all capabilities of the final application (support multiple input/outputs, multiple voltage/current domains, characterization of the material, losses' reduction, among others), with the option to work in parallel with switches at high frequency [12,176].

On the other hand, the industry of FEM software is more versatile each year; several applications and tools are added to these, improving accuracy, functionality, scenarios to apply it, and user friendly environments. Nonetheless, a common point is a lack of compatibility among those, limiting the use of platforms to validate and complement the final design. This lack of compatibility motivates users to know different packages to solve a unique problem.

This is translated into a monetary cost to purchase different licenses and increase computer resources. Therefore, to motivate and improve the FEM designs, in this authors opinion, a standard format type should be implemented in the following years between those kinds of software to take advantage of each characteristic and unify all FEM files.

In the literature, static or dynamic core losses models can be found, depending on the designer choice, the loses calculus could be as complex as the model chosen [177]. So, the model selection in function of the material and the requirements' application of a magnetic component's design process are fundamental to achieving an accurate core losses calculus.

Table 3 illustrates a comparison between power core losses models, ferromagnetic materials, and accuracy. It is essential to mention that ferrite is the start point to validate any power core losses model, so this is not listed. Nonetheless, to the best authors knowledge, not all models had been tested in all ferromagnetic materials, or those are not suitable to calculate their power core losses.

Table 3. Comparison between empiric core losses models. The sinusoidal and non-sinusoidal waveform are indicated with light yellow and light green, respectively.

Loss Model	Steinmetz's Parameters	Additional Data	Ferromagnetic Materials	Characteristics	Accuracy
Preisach	No	Yes	All unless powder cores	Based on domains movements and *B-H* loop Many additional data.	Computing cost, approximations
J-A	No	Yes	All unless powder cores	Non-linear equations systems. Based on magnetization core process. Many additional data.	Computing cost, approximations
LSM	No	Yes	All unless powder core	Suitable only for lineal systems. Very accuracy for laminar materials.	Good
TDA	No	No	All unless powder cores	Suitable only for lineal systems. Valid for PWM signals <400 Hz.	Good
OSE	Yes	No	All	Base of empiric loss models.	Low
MSE	Yes	No	All	Not suitable for signals with many harmonics.	$M^{\alpha-2}$
GSE	Yes	No	All	Consider CD-level in the signal.	<*i*GSE
*i*GSE	Yes	No	All	Signals with strong harmonics.	Same as i^2GSE
NSE	Yes	No	No reported	A similar version of GSE.	No reported
i^2GSE	Yes	Yes	All	Consider the relaxation effect. Signals with strong harmonics.	Same as *i*GSE
WcSE	Yes	Yes	Nanocrystal, amorphous and powder core	Physical base not verified. A practical and direct method.	$D < 0.5$
CHW	No	Yes	Nanocrystal, amorphous and powder core	Square waveform as the sum of its components.	Good
Villar	Yes	No	Amorphous material	Core loss calculus implemented a piecewise linear model (PWL).	Good
Górecki	Yes	No	Nanocrystal, amorphous and powder core	Considers thermal, electrical and magnetic effects	Same as manufacturers

Designers can use the information mentioned in Table 3 and Figure 9 to improve some of the power core loss models listed before or propose their model, always keeping in mind that any design parameter will be sacrificed to improve another one; magnetic components size will be reduced with frequency increment.

Due to the high energy consumption worldwide, renewable energy sources and energy efficiency have accelerated research in energy applications, conversion, transportation, and telecommunication [178]. Magnetic materials are the key piece of those research fields, mainly for their potential for energy efficiency and their impact on consumption power [179]. Materials' demand as electric steel, iron, and cobalt will be exponential growth for 2026. Studies on the effect of magnetic materials in renewable energy have to be a priority to guarantee the supply chain, options for recycling, the footprint impact, the blueprint impact, and the socio-economical impact of their extraction [180,181].

9. Conclusions

In the literature, many core-losses models are available to select. Researchers continue developing new core losses models with specific features (core material, waveform, density flux, application, frequency range, among others), but the majority are not easy to replicate.

To develop a power core losses model that involves electrical, magnetic, thermal effects, suitable for all kinds of ferromagnetic materials, and also competent to power electronic is fundamental. Not only for advantages to calculate power core losses accurately but not to reach optimal magnetic components. The challenge is to understand the comportment of the magnetic materials and model them differently than ferrite magnetic elements because, until now, both are very similar, although their behaviour is quite different.

Each of the core losses models in this document is a point of start for researchers. Despite promising results using Steinmetz's equations, those are not enough to calculate the core losses accuracy according to the need for the power electronic community. FEM software is indispensable nowadays to validate the power losses model as the magnetic component design. At the same time, virtual laboratories are promising to be the trend in a few years, allowing a reduction of the cost of a physical test bench to that of a virtual one and getting good approximations of the magnetic component behavior in a real environment. In addition, the option to interconnect different kinds of software to implement complete power electronics systems in platforms such as HIL, SIL, and MIL, makes the capability of design more versatile.

Author Contributions: Conceptualization, D.R.-S. and F.J.P.-P.; methodology, M.A.R.-L.; software, I.A.-V.; validation, J.P.-O., A.-I.B.-G. and F.J.P.-P.; formal analysis, M.A.R.-L.; investigation, D.R.-S.; resources, F.J.P.-P. and A.-I.B.-G.; data curation, J.P.-O.; writing original draft preparation, D.R.-S.; writing—review and editing, D.R.-S., M.A.R.-L., J.P.-O. and F.J.P.-P.; visualization, D.R.-S.; supervision, I.A.-V.; project administration, A.-I.B.-G.; funding acquisition, F.J.P.-P. and I.A.-V. All authors have read and agreed to the published version of the manuscript.

Funding: The authors thank Tecnológico Nacional de México, Instituto Tecnológico de Celaya. The work of Daniela Rodriguez-Sotelo was supported by CONACYT through a Ph.D. grant.

Acknowledgments: The authors thank IPN, Tecnológico Nacional de México, Instituto Tecnológico de Celaya and CONACYT.

Conflicts of Interest: The authors declare no conflict of interest.

Abbreviations

The following abbreviations are used in this manuscript:

P_c	Core loss	P_h	Hysteresis or static loss
P_{eddy}	Eddy current loss	P_{anom}	Anomalous or residual loss
k_h	Hysteresis coefficient	k_{eddy}	Eddy currents coefficient
k_{anom}	Anomalous coefficient	f	Fundamental frequency
B	Magnetic flux density	ϕ_n	Angle between B_n and H_n
B_n	Magnetic flux density of the nth harmonic	H_n	Magnetic field of the nth harmonic
B_s	Magnetic induction peak	x	Steinmetz coefficient
d	Lamination thickness	ρ	Electrical resistivity
A	Cross-sectional area lamination	f_{eq}	Equivalent frequency
M_{an}	Anhysteretic magnetization	α_{ic}	Inter-domain coupling
ΔB	Magnetic induction peak to peak	$dB(t)/dt$	Core loss magnetization rate
β	Magnetic induction exponent	α	Frequency exponent
δ	Magnetizing field direction	f_r	Fundamental frequency
θ	Sinusoidal waveform phase angle	T	Waveform period
D	Duty ratio	N	Number of turns
FWC	Waveform coefficient	V_1, V_2	Voltage
t	Time	ω	Zero-voltage period duration
U	DC constant voltage	T_ω	Length of zero-voltage period
T_M	Core temperature with smallest losses	T_R	Core temperature
M	Total magnetization	H	Magnetic field
H_{eff}	Effective magnetic field	μ_0	Free-space permeability
M_{irr}	Irreversible magnetization	M_{rev}	Reversible magnetization
A_{eff}	Effective core area	α_p	Losses temperature coefficient
$B(t)$	Instantaneous value of core loss magnetization rate	V_0	Magnetic objects statistical distribution
G	Eddy current dimensionless coefficient	a_s	Anhysteretic magnetization shape parameter

$M(t)$	Instantaneous magnetization	M_s	Saturation magnetization
m_s	Magnetic moment	$H(t)$	Instantaneous magnetic field
P_{sqr}	Rectangular waveform power losses	c	Reversibility coefficient
$k_r, \alpha_r, \beta_r, \tau, q_r$ $a, a_1, a_2, f_0, f_1,$ f_2, f_3, α_T, k	Material parameters		

References

1. Lee, H.; Jung, S.; Huh, Y.; Lee, J.; Bae, C.; Kim, S.J. An implantable wireless charger system with ×8.91 Increased charging power using smartphone and relay coil. In Proceedings of the IEEE Wireless Power Transfer Conference (WPTC), San Diego, CA, USA, 1–4 June 2021; pp. 1–4. [CrossRef]
2. Yu, W.; Hua, W.; Zhang, Z. High-frequency core loss analysis of high-speed flux-switching permanent magnet machines. *Electronics* **2021**, *10*, 1076. [CrossRef]
3. Pérez, E.; Espiñeira, P.; Ferreiro, A. *Instrumentación Electrónica*; Acceso Rápido; Marcombo: Barcelona, Spain, 1995.
4. Chitode, J.; Bakshi, U. *Power Devices and Machines*; Technical Publications: Maharashtra, India, 2009.
5. López-Fernández, X.; Ertan, H.; Turowski, J. *Transformers: Analysis, Design, and Measurement*; CRC Press: Boca Raton, FL, USA, 2017.
6. Muttaqi, K.M.; Islam, M.R.; Sutanto, D. Future power distribution grids: Integration of renewable energy, energy storage, electric vehicles, superconductor, and magnetic bus. *IEEE Trans. Appl. Supercond.* **2019**, *29*, 3800305. [CrossRef]
7. Phan, H.P. Implanted flexible electronics: Set device lifetime with smart nanomaterials. *Micromachines* **2021**, *12*, 157. [CrossRef] [PubMed]
8. Ivanov, A.; Lahiri, A.; Baldzhiev, V.; Trych-Wildner, A. Suggested research trends in the area of micro-EDM—Study of some parameters affecting micro-EDM. *Micromachines* **2021**, *12*, 1184. [CrossRef]
9. Kozai, T.D.Y. The history and horizons of microscale neural interfaces. *Micromachines* **2018**, *9*, 445. [CrossRef]
10. Rashid, M.H. *Power Electronics Handbook*; Elsevier: Amsterdam, The Netherlands, 2018; p. 1510. [CrossRef]
11. Matallana, A.; Ibarra, E.; López, I.; Andreu, J.; Garate, J.; Jordà, X.; Rebollo, J. Power module electronics in HEV/EV applications: New trends in wide-bandgap semiconductor technologies and design aspects. *Renew. Sustain. Energy Rev.* **2019**, *113*, 109264. [CrossRef]
12. Hanson, A.J.; Perreault, D.J. Modeling the magnetic behavior of n-winding components: Approaches for unshackling switching superheroes. *IEEE Power Electron. Mag* **2020**, *7*, 35–45. [CrossRef]
13. Bahmani, A. Core loss evaluation of high-frequency transformers in high-power DC-DC converters. In Proceedings of the Thirteenth International Conference on Ecological Vehicles and Renewable Energies (EVER), Monte Carlo, Monaco, 10–12 April 2018; pp. 1–7. [CrossRef]
14. Hossain, M.Z.; Rahim, N.A.; Selvaraj, J. Recent progress and development on power DC-DC converter topology, control, design and applications: A review. *Renew. Sustain. Energy Rev.* **2018**, *81*, 205–230. [CrossRef]
15. Silveyra, J.M.; Ferrara, E.; Huber, D.L.; Monson, T.C. Soft magnetic materials for a sustainable and electrified world. *Science* **2018**, *362*, eaao0195. [CrossRef]
16. Zurek, S. *Characterisation of Soft Magnetic Materials under Rotational Magnetisation*; CRC Press: Boca Raton, FL, USA, 2017; p. 568.
17. Cullity, B.; Graham, C. Soft magnetic materials. In *Introduction to Magnetic Materials*; John Wiley & Sons, Ltd.: Hoboken, NJ, USA, 2008; Chapter 13, pp. 439–476. [CrossRef]
18. Krishnan, K.M. *Fundamentals and Applications of Magnetic Materials*; Oxford University Press: Oxford, UK, 2016; p. 816. [CrossRef]
19. Islam, M.R.; Rahman, M.A.; Sarker, P.C.; Muttaqi, K.M.; Sutanto, D. Investigation of the magnetic response of a nanocrystalline high-frequency magnetic link with multi-input excitations. *IEEE Trans. Appl. Supercond.* **2019**, *29*, 0602205. [CrossRef]
20. Dobrzanski, L.A.; Drak, M.; Zieboqicz, B. Materials with specific magnetic properties. *J. Achiev. Mater. Manuf. Eng.* **2006**, *17*, 37.
21. Duppalli, V.S. Design Methodology for a High-Frequency Transformer in an Isolating DC-DC Converter. Ph.D. Thesis, Purdue University, West Lafayette, IN, USA, 2018.
22. Liu, X.; Wang, Y.; Zhu, J.; Guo, Y.; Lei, G.; Liu, C. Calculation of core loss and copper loss in amorphous/nanocrystalline core-based high-frequency transformer. *AIP Adv.* **2016**, *6*, 055927. [CrossRef]
23. Balci, S.; Sefa, I.; Bayram, M.B. Core material investigation of medium-frequency power transformers. In Proceedings of the 2014 16th International Power Electronics and Motion Control Conference and Exposition, Antalya, Turkey, 21–24 September 2014; pp. 861–866.
24. de Sevilla Galán, J.F. Mejora del Cálculo de las Pérdidas por el Efecto Proximidad en alta Frecuencia Para Devanados de Hilos de Litz. Available online: https://oa.upm.es/56807/1/TFG_JAVIER_FERNANDEZ_DE_SEVILLA_GALAN.pdf (accessed on 24 February 2022).
25. Herbert, E. User-Friendly Data for Magnetic Core Loss Calculations. Technical Report 1. 2008. Available online: https://www.psma.com/coreloss/eh1.pdf (accessed on 24 February 2022).
26. Sullivan, C.R.; Harris, J.H.; Herbert, E. Testing Core Loss for Rectangular Waveforms the Power Sources Manufacturers Association. Technical Report 973. 2010. Available online: https://www.psma.com/coreloss/phase2.pdf (accessed on 24 February 2022).

27. Tacca, H.; Sullivan, C. *Extended Steinmetz Equation*; Report of Posdoctoral Research; Thayer School of Engineering: Hanover, NH, USA, 2002.
28. Ulaby, F.T. *Fundamentos de Aplicaciones en Electromagnetismo*; Pearson: London, UK, 2007.
29. Cheng, D. *Fundamentos de Electromagnetismo para Ingeniería*; Mexicana, A., Ed.; 1997. Available online: https://www.academia.edu/36682331/Fundamentos_de_Electromagnetismo_para_Ingenieria_David_K_Cheng (accessed on 24 February 2022).
30. Coey, J.M.D. *Magnetism and Magnetic Materials*; Cambridge University Press: Cambridge, UK, 2010. [CrossRef]
31. Goldman, A. *Handbook of Modern Ferromagnetic Materials*; Springer: Boston, MA, USA, 1999. [CrossRef]
32. Gupta, K.; Gupta, N. Magnetic Materials: Types and Applications. In *Advanced Electrical and Electronics Materials*; John Wiley & Sons, Ltd.: Hoboken, NJ, USA, 2015; Chapter 12, pp. 423–448. [CrossRef]
33. Hitachi. *Nanocrystalline Material Properties*; Hitachi: Tokyo, Japan, 2019.
34. County Council, K.; Richardson, A. Silver Wire Ring. 2017. [image/JPEG]. Available online: https://www.academia.edu/2306387/With_Dickinson_T_M_and_Richardson_A_Early_Anglo_Saxon_Eastry_Archaeological_Evidence_for_the_Beginnings_of_a_District_Centre_in_the_Kingdom_of_Kent_Anglo_Saxon_Studies_in_Archaeology_and_History_17 (accessed on 24 February 2022).
35. Alisdojo. Detail of an Enamelled Litz Wire. 2011. [image/JPEG]. Available online: https://commons.wikimedia.org/wiki/File:Enamelled_litz_copper_wire.JPG (accessed on 24 February 2022).
36. Magnetics. *Powder Cores Properties*; Magnetics: Pittsburgh, PA, USA, 2019.
37. Laboratory, N.E.T. FINEMET Properties. 2018. Available online: https://www.hitachi-metals.co.jp/e/products/elec/tel/p02_21.html#:~:text=FINEMET%C2%AE%20has%20high%20saturation,and%20electronics%20devices%20as%20well (accessed on 24 February 2022).
38. CATECH®. *Amorphous and Nanocrystalline Core*; CATECH: Singapore, 2022.
39. Soon, H. Wire. 2017. [image/JPEG]. Available online: https://iopscience.iop.org/article/10.1088/1742-6596/871/1/012098/pdf (accessed on 24 February 2022).
40. Aluminium Wire, 16 mm^2. 2012. [image/JPEG]. Available online: https://ceb.lk/front_img/specifications/1540801489P.V_.C_._INSULATED_ALUMINIUM_SERVICE_MAIN_WIRE-FINISHED_PRODUCT_.pdf (accessed on 24 February 2022).
41. County Council, K.; Richardson, A. Length of Gold Wire, Ringlemere. 2017. [image/JPEG]. Available online: https://www.sachsensymposion.org/wp-content/uploads/2012/04/68th-International-Sachsensymposion-Canterbury.pdf (accessed on 24 February 2022).
42. SpinningSpark. Transformer Core. 2012. [image/JPEG]. Available online: https://commons.wikimedia.org/wiki/File:Transformer_winding_formats.jpg (accessed on 24 February 2022).
43. TheDigitalArtist. coil-gd137ed32f-1920. 2015. [image/JPEG]. Available online: https://plato.stanford.edu/entries/digital-art/ (accessed on 24 February 2022).
44. Fiorillo, F.; Bertotti, G.; Appino, C.; Pasquale, M., Soft magnetic materials. In *Wiley Encyclopedia of Electrical and Electronics Engineering*; American Cancer Society: Atlanta, GA, USA, 2016; pp. 1–42. [CrossRef]
45. Aguglia, D.; Neuhaus, M. Laminated magnetic materials losses analysis under non-sinusoidal flux waveforms in power electronics systems. In Proceedings of the 15th European Conference on Power Electronics and Applications (EPE), Lille, France, 2–6 September 2013; pp. 1–8. [CrossRef]
46. McLyman, C. *Transformer and Inductor Design Handbook*, 3rd ed.; Taylor & Francis: Abingdon-on-Thames, UK, 2004.
47. Tsepelev, V.; Starodubtsev, Y.; Konashkov, V.; Belozerov, V. Thermomagnetic analysis of soft magnetic nanocrystalline alloys. *J. Alloys Compd.* **2017**, *707*, 210–213. [CrossRef]
48. Ouyang, G.; Chen, X.; Liang, Y.; Macziewski, C.; Cui, J. Review of Fe-6.5 wt% silicon steel-a promising soft magnetic material for sub-kHz application. *J. Magn. Magn. Mater.* **2019**, *481*, 234–250. [CrossRef]
49. Yuan, W.; Wang, Y.; Liu, D.; Deng, F.; Chen, Z. Impacts of inductor nonlinear characteristic in multi-converter microgrids: Modelling, analysis and mitigation. *IEEE J. Emerg. Sel. Top. Power Electron.* **2020**, *8*, 3333–3347. [CrossRef]
50. Kazimierczuk, M. *High-Frequency Magnetic Components*, 2nd ed.; John Wiley & Sons, Ltd.: Hoboken, NJ, USA, 2013. [CrossRef]
51. Tumanski, S. *Handbook of Magnetic Measurements*; Series in Sensors; CRC Press: Boca Raton, FL, USA, 2016.
52. Goldman, A. *Magnetic Components for Power Electronics*; Springer: Berlin/Heidelberg, Germany, 2012.
53. Wang, Y.; Calderon-Lopez, G.; Forsyth, A. Thermal management of compact nanocrystalline inductors for power dense converters. In Proceedings of the IEEE Applied Power Electronics Conference and Exposition (APEC), San Antonio, TX, USA, 4–8 March 2018; pp. 2696–2703. [CrossRef]
54. Conde, C.; Blázquez, J.; Conde, A. Nanocrystallization process of the Hitperm Fe-Co-Nb-B alloys. In *Properties and Applications of Nanocrystalline Alloys from Amorphous Precursors*; Idzikowski, B., Švec, P., Miglierini, M., Eds.; Springer: Dordrecht, The Netherlands, 2005; pp. 111–121.
55. Tsepelev, V.S.; Starodubtsev, Y.N. Nanocrystalline soft magnetic iron-based materials from liquid state to ready product. *Nanomaterials* **2021**, *11*, 108. [CrossRef]
56. Lidow, A. Accelerating adoption of magnetic resonant wireless power based on the AirFuel, 2021. *IEEE Wirel. Power Week* **2021**. Available online: https://airfuel.org/wireless-power-week-2021-highlights/ (accessed on 24 February 2022).
57. Jafari, M.; Malekjamshidi, Z.; Zhu, J. Copper loss analysis of a multiwinding high-frequency transformer for a magnetically-coupled residential microgrid. *IEEE Trans. Ind. Appl.* **2019**, *55*, 283–297. [CrossRef]

58. Slionnnnnn. The Prototype of the Balanced Twisted Winding CM Choke. 2017. [image/JPEG]. Available online: https://en.wikipedia.org/wiki/File:4ffffss123.jpg (accessed on 24 February 2022).
59. Corti, F.; Reatti, A.; Lozito, G.M.; Cardelli, E.; Laudani, A. Influence of non-linearity in losses estimation of magnetic components for DC-DC converters. *Energies* **2021**, *14*, 6498. [CrossRef]
60. Zhu, F.; Yang, B. *Power Transformer Design Practices*; CRC Press: Boca Raton, FL, USA, 2021.
61. Glisson, T. *Introduction to Circuit Analysis and Design*; Springer: Berlin/Heidelberg, Germany, 2011.
62. Wang, W.; Nysveen, A.; Magnusson, N. The influence of multidirectional leakage flux on transformer core losses. *J. Magn. Magn. Mater.* **2021**, *539*, 168370. [CrossRef]
63. Kulkarni, S.; Khaparde, S. *Transformer Engineering: Design, Technology, and Diagnostics*, 2nd ed.; Taylor & Francis: Abingdon-on-Thames, UK, 2012.
64. Tian, H.; Wei, Z.; Vaisambhayana, S.; Thevar, M.P.; Tripathi, A.; Kjær, P.C. Calculation and experimental validation on leakage inductance of a medium frequency transformer. In Proceedings of the IEEE 4th Southern Power Electronics Conference (SPEC), Singapore, 10–13 December 2018; pp. 1–6. [CrossRef]
65. Kothari, D.; Nagrath, I. *Modern Power System Analysis*; Tata McGraw-Hill Publishing Company: New York, NK, USA, 2003.
66. Barg, S.; Alam, M.F.; Bertilsson, K. Optimization of high frequency magnetic devices with consideration of the effects of the magnetic material, the core geometry and the switching frequency. In Proceedings of the 22nd European Conference on Power Electronics and Applications (EPE'20 ECCE Europe), Lyon, France, 7–11 September 2020; pp. 1–8. [CrossRef]
67. Ouyang, Z. *High Frequency Planar Magnetics for Power Converters*; Presentation; ECPE Online Tutorial: Nuremberg, Germany, 2021.
68. Barrios, E.L.; Ursúa, A.; Marroyo, L.; Sanchis, P. Analytical design methodology for Litz-wired high-frequency power transformers. *IEEE Trans. Ind. Electron.* **2015**, *62*, 2103–2113. [CrossRef]
69. Brighenti, L.L.; Martins, D.C.; Dos Santos, W.M. Study of magnetic core geometries for coupling systems through a magnetic bus. In Proceedings of the IEEE 10th International Symposium on Power Electronics for Distributed Generation Systems (PEDG), Xi'an, China, 3–6 June 2019; pp. 29–36. [CrossRef]
70. Rodriguez-Sotelo, D.; Rodriguez-Licea, M.A.; Soriano-Sanchez, A.G.; Espinosa-Calderon, A.; Perez-Pinal, F.J. Advanced ferromagnetic materials in power electronic converters: A state of the art. *IEEE Access* **2020**, *8*, 56238–56252. [CrossRef]
71. Detka, K.; Górecki, K. Influence of the size and shape of magnetic core on thermal parameters of the inductor. *Energies* **2020**, *13*, 3842. [CrossRef]
72. Jiandong, D.; Yang, L.; Hao, L. Research on ferromagnetic components J-A model—A review. In Proceedings of the International Conference on Power System Technology (POWERCON), Guangzhou, China, 6–8 November 2018; pp. 3288–3294.
73. Saeed, S.; Georgious, R.; Garcia, J. Modeling of magnetic elements including losses-application to variable inductor. *Energies* **2020**, *13*, 1865. [CrossRef]
74. Ishikura, Y.; Imaoka, J.; Noah, M.; Yamamoto, M. Core loss evaluation in powder cores: A comparative comparison between electrical and calorimetric methods. In Proceedings of the International Power Electronics Conference (IPEC-Niigata 2018-ECCE Asia), Niigata, Japan, 20–24 May 2018; pp. 1087–1094.
75. Hanif, A. Measurement of Core Losses in Toroidal Inductors with Different Magnetic Materials. Master's Thesis, Tampere University of Technology, Tampere, Finland, 2017.
76. Wang, Y. Modelling and Characterisation of Losses in Nanocrystalline Cores. Ph.D. Thesis, The University of Manchester, Manchester, UK, 2015.
77. Yue, S.; Li, Y.; Yang, Q.; Yu, X., Zhang, C. Comparative analysis of core loss calculation methods for magnetic materials under nonsinusoidal excitations. *IEEE Trans. Magn.* **2018**, *54*, 6300605. [CrossRef]
78. Bi, S. Charakterisieren und Modellieren der Ferromagnetischen Hysterese. Ph.D. Thesis, Friedrich-Alexander-Universität Erlangen-Nürnberg (FAU), Erlangen, Germany, 2014.
79. Iniewski, K. *Advanced Circuits for Emerging Technologies*; Wiley: Hoboken, NJ, USA, 2012.
80. Sudhoff, S. *Power Magnetic Devices: A Multi-Objective Design Approach*; IEEE Press Series on Power Engineering; Wiley: Hoboken, NJ, USA, 2014.
81. Spaldin, N. *Magnetic Materials: Fundamentals and Applications*; Cambridge University Press: Cambridge, UK, 2010.
82. Dong Tan, F.; Vollin, J.L.; Cuk, S.M. A practical approach for magnetic core-loss characterization. *IEEE Trans. Power Electron.* **1995**, *10*, 124–130. [CrossRef]
83. Fiorillo, F.; Mayergoyz, I. *Characterization and Measurement of Magnetic Materials*; Elsevier: Amsterdam, The Netherlands, 2004.
84. O'Handley, R. *Modern Magnetic Materials: Principles and Applications*; Wiley: Hoboken, NJ, USA, 1999.
85. Krings, A.; Soulard, J. Overview and comparison of iron loss models for electrical machines. *J. Electr. Eng. Elektrotechnicky Cas.* **2010**, *10*, 162–169.
86. Sai Ram, B.; Kulkarni, S. An isoparametric approach to model ferromagnetic hysteresis including anisotropy and symmetric minor loops. *J. Magn. Magn. Mater.* **2019**, *474*, 574–584. [CrossRef]
87. Mayergoyz, I. *Mathematical Models of Hysteresis and Their Applications*, 2nd ed.; Elsevier: Amsterdam, The Netherlands, 2003.
88. Zhao, X.; Liu, X.; Zhao, Z.; Zou, X.; Xiao, Y.; Li, G. Measurement and modeling of hysteresis characteristics in ferromagnetic materials under DC magnetizations. *AIP Adv.* **2019**, *9*, 025111. [CrossRef]
89. Jiles, D.; Atherton, D. Theory of ferromagnetic hysteresis. *J. Magn. Magn. Mater.* **1986**, *61*, 48–60. [CrossRef]

90. Bernard, Y.; Mendes, E.; Bouillault, F. Dynamic hysteresis modeling based on Preisach model. *IEEE Trans. Magn.* **2002**, *38*, 885–888. [CrossRef]
91. Magnetic, M. Jiles-Atherton Model of Magnetic Hysteresis Loops of Mn-Zn Ferrite. 2016. [image/PNG]. Available online: https://commons.wikimedia.org/wiki/File:Jiles-Atherton_model_of_magnetic_hysteresis_loops_of_Mn-Zn_ferrite.png (accessed on 24 February 2022) .
92. Aboura, F.; Touhami, O. Modeling and analyzing energetic hysteresis classical model. In Proceedings of the International Conference on Electrical Sciences and Technologies in Maghreb (CISTEM), Algiers, Algeria, 28–31 October 2018; pp. 1–5.
93. Chang, L.; Jahns, T.M.; Blissenbach, R. Characterization and modeling of soft magnetic materials for improved estimation of PWM-induced iron loss. *IEEE Trans. Ind. Appl.* **2020**, *56*, 287–300. [CrossRef]
94. Ducharne, B.; Tsafack, P.; Tene Deffo, Y.; Zhang, B.; Sebald, G. Anomalous fractional magnetic field diffusion through cross-section of a massive toroidal ferromagnetic core. *Commun. Nonlinear Sci. Numer. Simul.* **2021**, *92*, 105450. [CrossRef]
95. Szular, Z.; Mazgaj, W. Calculations of eddy currents in electrical steel sheets taking into account their magnetic hysteresis. *COMPEL—Int. J. Comput. Math. Electr. Electron. Eng.* **2019**, *38*, 1263–1273. [CrossRef]
96. Yu, Q.; Chu, S.; Li, W.; Tian, L.; Wang, X.; Cheng, Y. Electromagnetic shielding analysis of a canned permanent magnet motor. *IEEE Trans. Ind. Electron.* **2020**, *67*, 8123–8130. [CrossRef]
97. Marchenkov, Y.; Khvostov, A.; Slavinskaya, A.; Zhgut, A.; Chernov, V. The eddy current diagnostics method for the plastically deformed area sizes evaluation in non-magnetic metals. *J. Appl. Eng. Sci.* **2020**, *18*, 92–97. [CrossRef]
98. Müller, S.; Keller, M.; Maier, M.; Parspour, N. Comparison of iron loss calculation methods for soft magnetic composite. In Proceedings of the Brazilian Power Electronics Conference (COBEP), Juiz de Fora, Brazil, 19–22 November 2017; pp. 1–6.
99. Ionel, D.M.; Popescu, M.; McGilp, M.I.; Miller, T.J.E.; Dellinger, S.J.; Heideman, R.J. Computation of Core Losses in Electrical Machines Using Improved Models for Laminated Steel. *IEEE Trans. Ind. Appl.* **2007**, *43*, 1554–1564. [CrossRef]
100. Bertotti, G. General properties of power losses in soft ferromagnetic materials. *IEEE Trans. Magn.* **1988**, *24*, 621–630. [CrossRef]
101. Mehboob, N. Hysteresis Properties of Soft Magnetic Materials. Ph.D. Thesis, University of Vienna, Wien, Austria, 2012.
102. Sun, H.; Li, Y.; Lin, Z.; Zhang, C.; Yue, S. Core loss separation model under square voltage considering DC bias excitation. *AIP Adv.* **2020**, *10*, 015229. [CrossRef]
103. Steinmetz, C.P. On the law of hysteresis. *Trans. Am. Inst. Electr. Eng.* **1892**, *IX*, 1–64. [CrossRef]
104. Mu, M. High Frequency Magnetic Core Loss Study. 2013. Available online: https://vtechworks.lib.vt.edu/bitstream/handle/10919/19296/Mu_M_D_2013.pdf (accessed on 24 February 2022).
105. Zhou, T.; Zhou, G.; Ombach, G.; Gong, X.; Wang, Y.; Shen, J. Improvement of Steinmetz's parameters fitting formula for ferrite soft magnetic materials. In Proceedings of the IEEE Student Conference on Electric Machines and Systems, Huzhou, China, 14–16 December 2018; pp. 1–4.
106. Erickson, R.; Maksimović, D. *Fundamentals of Power Electronics*; Springer International Publishing AG: Berlin/Heidelberg, Germany, 2020.
107. Chen, D.Y. Comparisons of high frequency magnetic core losses under two different driving conditions: A sinusoidal voltage and a square-wave voltage. In Proceedings of the IEEE Power Electronics Specialists Conference, Syracuse, NY, USA, 13–15 June 1978; pp. 237–241.
108. Mulder, S.A. Fit formula for power loss in ferrites and their use in transformer design. In Proceedings of the 26th International Conference on Power Conversion, PCIM, Orlando, FL, USA, 1983; pp. 345–359.
109. Venkatachalam, K.; Sullivan, C.R.; Abdallah, T.; Tacca, H. Accurate prediction of ferrite core loss with nonsinusoidal waveforms using only Steinmetz parameters. In Proceedings of the 2002 IEEE Workshop on Computers in Power Electronics, Mayaguez, PR, USA, 3–4 June 2002; pp. 36–41. [CrossRef]
110. Muhlethaler, J.; Biela, J.; Kolar, J.W.; Ecklebe, A. Improved core-loss calculation for magnetic components employed in power electronic systems. *IEEE Trans. Power Electron.* **2012**, *27*, 964–973. [CrossRef]
111. Shen, W.; Wang, F.; Boroyevich, D.; Tipton, C.W. Loss Characterization and calculation of nanocrystalline cores for high-frequency magnetics applications. In Proceedings of the APEC 07—Twenty-Second Annual IEEE Applied Power Electronics Conference and Exposition, Anaheim, CA, USA, 25 February–1 March 2007; pp. 90–96.
112. Reinert, J.; Brockmeyer, A.; De Doncker, R.W.A.A. Calculation of losses in ferro- and ferrimagnetic materials based on the modified Steinmetz equation. *IEEE Trans. Ind. Appl.* **2001**, *37*, 1055–1061. [CrossRef]
113. Yu, X.; Li, Y.; Yang, Q.; Yue, S.; Zhang, C. Loss characteristics and model verification of soft magnetic composites under non-sinusoidal excitation. *IEEE Trans. Magn.* **2019**, *55*, 6100204. [CrossRef]
114. Yue, S.; Yang, Q.; Li, Y.; Zhang, C. Core loss calculation for magnetic materials employed in SMPS under rectangular voltage excitations. *AIP Adv.* **2018**, *8*, 056121. [CrossRef]
115. Yue, S.; Yang, Q.; Li, Y.; Zhang, C.; Xu, G. Core loss calculation of the soft ferrite cores in high frequency transformer under non-sinusoidal excitations. In Proceedings of the 2017 20th International Conference on Electrical Machines and Systems (ICEMS), Sydney, Australia, 11–14 August 2017; pp. 1–5.
116. Karthikeyan, V.; Rajasekar, S.; Pragaspathy, S.; Blaabjerg, F. Core loss estimation of magnetic links in DAB converter operated in high-frequency non-sinusoidal flux waveforms. In Proceedings of the IEEE International Conference on Power Electronics, Drives and Energy Systems (PEDES), Chennai, India, 18–21 December 2018; pp. 1–5.

117. Marin-Hurtado, A.J.; Rave-Restrepo, S.; Escobar-Mejía, A. Calculation of core losses in magnetic materials under nonsinusoidal excitation. In Proceedings of the 2016 13th International Conference on Power Electronics (CIEP), Guanajuato, Mexico, 20–23 June 2016; pp. 87–91.
118. Valchev, V.; Van den Bossche, A. *Inductors and Transformers for Power Electronics*; CRC Press; Boca Raton, FL, USA, 2018.
119. Barg, S.; Ammous, K.; Mejbri, H.; Ammous, A. An improved empirical formulation for magnetic core losses estimation under nonsinusoidal induction. *IEEE Trans. Power Electron.* **2017**, *32*, 2146–2154. [CrossRef]
120. Villar, I.; Rufer, A.; Viscarret, U.; Zurkinden, F.; Etxeberria-Otadui, I. Analysis of empirical core loss evaluation methods for non-sinusoidally fed medium frequency power transformers. In Proceedings of the IEEE International Symposium on Industrial Electronics, Cambridge, UK, 30 June–2 July 2008; pp. 208–213. [CrossRef]
121. Ridley, R.; Nace, A. Modeling ferrite core losses. *Switch. Power Mag.* **2006**, *3* , 1–7.
122. Bar, S.; Tabrikian, J. Adaptive waveform design for target detection with sequential composite hypothesis testing. In Proceedings of the IEEE Statistical Signal Processing Workshop (SSP), Palma de Mallorca, Spain, 26–29 June 2016; pp. 1–5. [CrossRef]
123. Villar, I.; Viscarret, U.; Etxeberria-Otadui, I.; Rufer, A. Global loss evaluation methods for nonsinusoidally fed medium-frequency power transformers. *IEEE Trans. Ind. Electron.* **2009**, *56*, 4132–4140. [CrossRef]
124. Górecki, K.; Detka, K. The Parameter Estimation of the Electrothermal Model of Inductors. 2015. Available online: http://www.midem-drustvo.si/Journal%20papers/MIDEM_45(2015)1p29.pdf (accessed on 24 February 2022).
125. Górecki, K.; Detka, K. Influence of power losses in the inductor core on characteristics of selected DC–DC converters. *Energies* **2019**, *12*, 1991. [CrossRef]
126. Dlala, E. Comparison of models for estimating magnetic core losses in electrical machines using the finite-element method. *IEEE Trans. Magn.* **2009**, *45*, 716–725. [CrossRef]
127. Pradhan, K.K.; Chakraverty, S. Chapter four—Finite element method. In *Computational Structural Mechanics*; Pradhan, K.K., Chakraverty, S., Eds.; Academic Press: Cambridge, MA, USA, 2019; pp. 25–28.
128. Peyton, A. 3—Electromagnetic induction tomography. In *Industrial Tomography*; Wang, M., Ed.; Woodhead Publishing Series in Electronic and Optical Materials; Woodhead Publishing: Sawston, UK, 2015; pp. 61–107. [CrossRef]
129. Rapp, B.E. Chapter 32—Finite Element Method. In *Microfluidics: Modelling, Mechanics and Mathematics*; Rapp, B.E., Ed.; Micro and Nano Technologies; Elsevier: Oxford, UK, 2017; pp. 655—678. [CrossRef]
130. Meunier, G. *The Finite Element Method for Electromagnetic Modeling*; ISTE, Wiley: Hoboken, NJ, USA, 2010.
131. Multiphysics, C. The Finite Element Method (FEM). 2017. Available online: https://www.comsol.com/multiphysics/finite-element-method (accessed on 24 February 2022).
132. Calderon-Lopez, G.; Wang, Y.; Forsyth, A.J. Mitigation of gap losses in nanocrystalline tape-wound cores. *IEEE Trans. Power Electron.* **2019**, *34*, 4656–4664. [CrossRef]
133. Ansys, I. Ansys Twin Builder Digital Twin: Simulation-Based & Hybrid Analytics. 2021. Available online: https://www.ansys.com/products/digital-twin/ansys-twin-builder (accessed on 24 February 2022).
134. COMSOL Inc. *Comsol Multiphysics*; COMSOL Inc.: Burlington, MA, USA, 2021.
135. JDJ Co. *JMAG Products Catalog*; JDJ Co.: Singapore, 2020.
136. Hansen, N.; Wiechowski, N.; Kugler, A.; Kowalewski, S.; Rambow, T.; Busch, R. Model-in-the-loop and software-in-the-loop testing of closed-loop automotive software with arttest. In *Informatik 2017*; Eibl, M., Gaedke, M., Eds.; Gesellschaft für Informatik: Bonn, Germany, 2017; pp. 1537–1549. [CrossRef]
137. Wikipedia. List of Finite Element Software Packages. 2021. Available online: https://en.wikipedia.org/wiki/List_of_finite_element_software_packages (accessed on 24 February 2022).
138. Haase, H. Forward Converter ATX PC Power Supply. 2013. [image/JPEG]. Available online: https://commons.wikimedia.org/wiki/File:Forward_Converter_ATX_PC_Power_Supply_IMG_1092.jpg (accessed on 24 February 2022).
139. Hurley, W.G. *Passives in Power Electroics: Magnetic Component Design and Simulation*; Presentation; ECPE Online Tutorial: Nuremberg, Germany, 2021.
140. Ionita, V.; Cazacu, E.; Petrescu, L. Effect of voltage harmonics on iron losses in magnetic cores with hysteresis. In Proceedings of the 2018 18th International Conference on Harmonics and Quality of Power (ICHQP), Ljubljana, Slovenia, 13–16 May 2018; pp. 1–5. [CrossRef]
141. Ruiz-Robles, D.; Figueroa-Barrera, C.; Moreno-Goytia, E.L.; Venegas-Rebollar, V. An experimental comparison of the effects of nanocrystalline core geometry on the performance and dispersion inductance of the MFTs applied in DC-DC converters. *Electronics* **2020**, *9*, 453. [CrossRef]
142. Jiang, C.; Li, X.; Ghosh, S.; Zhao, H.; Shen, Y.; Long, T. Nanocrystalline powder cores for high-power high-frequency applications. *IEEE Trans. Power Electron.* **2020**, *35*, 10821–10830. [CrossRef]
143. Islam, M.R.; Farrok, O.; Rahman, M.A.; Kiran, M.R.; Muttaqi, K.M.; Sutanto, D. Design and characterisation of advanced magnetic material-based core for isolated power converters used in wave energy generation systems. *IET Electr. Power Appl.* **2020**, *14*, 733–741. [CrossRef]
144. Bolsi, P.C.; Sartori, H.C.; Pinheiro, J.R. Comparison of core technologies applied to power inductors. In Proceedings of the 13th IEEE International Conference on Industry Applications (INDUSCON), Sao Paulo, Brazil, 12–14 November 2018; pp. 1100–1106. [CrossRef]

145. Imaoka, J.; Yu-Hsin, W.; Shigematsu, K.; Aoki, T.; Noah, M.; Yamamoto, M. Effects of high-frequency operation on magnetic components in power converters. In Proceedings of the IEEE 12th Energy Conversion Congress Exposition—Asia (ECCE-Asia), Singapore, 24–27 May 2021; pp. 978–984. [CrossRef]
146. Delgado, A.; Oliver, J.A.; Cobos, J.A.; Rodriguez-Moreno, J. Macroscopic modeling of magnetic microwires for finite element simulations of inductive components. *IEEE Trans. Power Electron.* **2020**, *35*, 8452–8459. [CrossRef]
147. Abourida, S.; Dufour, C.; Belanger, J.; Yamada, T.; Arasawa, T. Hardware-in-the-loop simulation of finite-element based motor drives with RT-LAB and JMAG. In Proceedings of the IEEE International Symposium on Industrial Electronics, Montreal, QC, Canada, 9–13 July 2006; Volume 3, pp. 2462–2466. [CrossRef]
148. Bjørheim, F.; Siriwardane, S.C.; Pavlou, D. A review of fatigue damage detection and measurement techniques. *Int. J. Fatigue* **2022**, *154*, 106556. [CrossRef]
149. Faba, A.; Quondam Antonio, S. An Overview of Non-Destructive Testing of Goss Texture in Grain-Oriented Magnetic Steels. *Mathematics* **2021**, *9*, 1539. [CrossRef]
150. Ibrahim, M.; Singh, S.; Barman, D.; Bernier, F.; Lamarre, J.M.; Grenier, S.; Pillay, P. Selection of Soft Magnetic Composite Material for Electrical Machines using 3D FEA Simulations. In Proceedings of the 2021 IEEE Energy Conversion Congress and Exposition (ECCE), Vancouver, BC, Canada, 10–14 October 2021; pp. 3860–3865. [CrossRef]
151. Xie, S.; Zhang, L.; Zhao, Y.; Wang, X.; Kong, Y.; Ma, Q.; Chen, Z.; Uchimoto, T.; Takagi, T. Features extraction and discussion in a novel frequency-band-selecting pulsed eddy current testing method for the detection of a certain depth range of defects. *NDT E Int.* **2020**, *111*, 102211. [CrossRef]
152. Li, E.; Chen, Y.; Chen, X.; Wu, J. Defect Width Assessment Based on the Near-Field Magnetic Flux Leakage Method. *Sensors* **2021**, *21*, 5424. [CrossRef]
153. Galluzzi, R.; Amati, N.; Tonoli, A. Modeling, Design, and Validation of Magnetic Hysteresis Motors. *IEEE Trans. Ind. Electron.* **2020**, *67*, 1171–1179. [CrossRef]
154. Kang, S.G.; Song, M.S.; Kim, J.W.; Lee, J.W.; Kim, J. Near-Field Communication in Biomedical Applications. *Sensors* **2021**, *21*, 703. [CrossRef]
155. Calderon-Lopez, G.; Todd, R.; Forsyth, A.J.; Wang, J.; Wang, W.; Yuan, X.; Aldhaher, S.; Kwan, C.; Yates, D.; Mitcheson, P.D. Towards Lightweight Magnetic Components for Converters with Wide-bandgap Devices. In Proceedings of the 2020 IEEE 9th International Power Electronics and Motion Control Conference (IPEMC2020-ECCE Asia), Nanjing, China, 29 November–2 December 2020; pp. 3149–3155. [CrossRef]
156. Zuo, S.; Heidari, H.; Farina, D.; Nazarpour, K. Miniaturized Magnetic Sensors for Implantable Magnetomyography. *Adv. Mater. Technol.* **2020**, *5*, 2000185. [CrossRef]
157. Trohman, R.G.; Huang, H.D.; Sharma, P.S. The Miniaturization of Cardiac Implantable Electronic Devices: Advances in Diagnostic and Therapeutic Modalities. *Micromachines* **2019**, *10*, 633. [CrossRef] [PubMed]
158. Murzin, D.; Mapps, D.J.; Levada, K.; Belyaev, V.; Omelyanchik, A.; Panina, L.; Rodionova, V. Ultrasensitive Magnetic Field Sensors for Biomedical Applications. *Sensors* **2020**, *20*, 1569. [CrossRef] [PubMed]
159. Jeon, S.; Park, S.H.; Kim, E.; Kim, J.y.; Kim, S.W.; Choi, H. A Magnetically Powered Stem Cell-Based Microrobot for Minimally Invasive Stem Cell Delivery via the Intranasal Pathway in a Mouse Brain. *Adv. Healthc. Mater.* **2021**, *10*, 2100801. [CrossRef]
160. Khan, M.A.; Sun, J.; Li, B.; Przybysz, A.; Kosel, J. Magnetic sensors-A review and recent technologies. *Eng. Res. Express* **2021**, *3*, 022005. [CrossRef]
161. Sharma, A.; Jain, V.; Gupta, D.; Babbar, A. A Review Study on Miniaturization : A Boon or Curse. In *Advanced Manufacturing and Processing Technology*; CRC Press: Boca Raton, FL, USA, 2020; pp. 111–131. [CrossRef]
162. Chen, J.; Li, J.; Li, Y.; Chen, Y.; Xu, L. Design and Fabrication of a Miniaturized GMI Magnetic Sensor Based on Amorphous Wire by MEMS Technology. *Sensors* **2018**, *18*, 732. [CrossRef]
163. Yang, J.; Wang, K.; Xu, H.; Yan, W.; Jin, Q.; Cui, D. Detection platforms for point-of-care testing based on colorimetric, luminescent and magnetic assays: A review. *Talanta* **2019**, *202*, 96–110. [CrossRef] [PubMed]
164. Tu, J.; Torrente-Rodríguez, R.M.; Wang, M.; Gao, W. The Era of Digital Health: A Review of Portable and Wearable Affinity Biosensors. *Adv. Funct. Mater.* **2020**, *30*, 1906713. [CrossRef]
165. Purohit, B.; Kumar, A.; Mahato, K.; Chandra, P. Smartphone-assisted personalized diagnostic devices and wearable sensors. *Curr. Opin. Biomed. Eng.* **2020**, *13*, 42–50. [CrossRef]
166. Dinis, H.; Mendes, P. A comprehensive review of powering methods used in state-of-the-art miniaturized implantable electronic devices. *Biosens. Bioelectron.* **2021**, *172*, 112781. [CrossRef]
167. Sitti, M.; Wiersma, D.S. Pros and Cons: Magnetic versus Optical Microrobots. *Adv. Mater.* **2020**, *32*, 1906766 . [CrossRef]
168. Zhou, H.; Mayorga-Martinez, C.C.; Pané, S.; Zhang, L.; Pumera, M. Magnetically Driven Micro and Nanorobots. *Chem. Rev.* **2021**, *121*, 4999–5041. [CrossRef] [PubMed]
169. Mathieu, J.B.; Beaudoin, G.; Martel, S. Method of propulsion of a ferromagnetic core in the cardiovascular system through magnetic gradients generated by an MRI system. *IEEE Trans. Biomed. Eng.* **2006**, *53*, 292–299. [CrossRef] [PubMed]
170. Vitol, E.A.; Novosad, V.; Rozhkova, E.A. Microfabricated magnetic structures for future medicine: From sensors to cell actuators. *Nanomedicine* **2012**, *7*, 1611–1624 . [CrossRef] [PubMed]
171. Li, Z.; Li, C.; Dong, L.; Zhao, J. A Review of Microrobot's System: Towards System Integration for Autonomous Actuation In Vivo. *Micromachines* **2021**, *12*, 1249. [CrossRef]

172. Bandari, V.K.; Schmidt, O.G. System-Engineered Miniaturized Robots: From Structure to Intelligence. *Adv. Intell. Syst.* **2021**, *3*, 2000284 . [CrossRef]
173. Xu, K.; Xu, S.; Wei, F. Recent progress in magnetic applications for micro- and nanorobots. *Beilstein J. Nanotechnol.* **2021**, *12*, 744–755. [CrossRef]
174. Hein, H.; Li, Y.; Yue, S.; Sun, H. Core Losses Analysis for Soft Magnetic Materials under SPWM Excitations. *Int. J. Electromagn. Appl.* **2020**, *10*, 1–6.
175. Sundaria, R.; Nair, D.G.; Lehikoinen, A.; Arkkio, A.; Belahcen, A. Effect of Laser Cutting on Core Losses in Electrical Machines—Measurements and Modeling. *IEEE Trans. Ind. Electron.* **2020**, *67*, 7354–7363. [CrossRef]
176. Lu, K.; Liu, X.; Wang, J.; Yang, T.; Xu, J. Simultaneous improvements of effective magnetic permeability, core losses and temperature characteristics of Fe-Si soft magnetic composites induced by annealing treatment. *J. Alloys Compd.* **2022**, *892*, 162100. [CrossRef]
177. Corti, F.; Reatti, A.; Cardelli, E.; Faba, A.; Rimal, H. Improved Spice simulation of dynamic core losses for ferrites with nonuniform field and its experimental validation. *IEEE Trans. Ind. Electron.* **2020**, *68*, 12069–12078. [CrossRef]
178. Gutfleisch, O. Magnetic Materials in Sustainable Energy. 2012. Available online: https://www.jst.go.jp/sicp/ws2011_eu/presentation/presentation_04.pdf (accessed on 24 February 2022).
179. Group, H.M. Examples of Products That Help Realize a Sustainable Society. 2020. Available online: https://www.hitachi-metals.co.jp/e/ir/pdf/ar/2020/2020_17.pdf (accessed on 24 February 2022).
180. Research, A.M. Soft Magnetic Materials Market Outlook—2026. 2019. Available online: https://www.alliedmarketresearch.com/soft-magnetic-materials-market#:~:text=Soft (accessed on 24 February 2022).
181. Matizamhuka, W. The impact of magnetic materials in renewable energy-related technologies in the 21st century industrial revolution: The case of South Africa. *Adv. Mater. Sci. Eng.* **2018**, *2018*, 3149412. [CrossRef]

MDPI
St. Alban-Anlage 66
4052 Basel
Switzerland
Tel. +41 61 683 77 34
Fax +41 61 302 89 18
www.mdpi.com

Micromachines Editorial Office
E-mail: micromachines@mdpi.com
www.mdpi.com/journal/micromachines

www.ingramcontent.com/pod-product-compliance
Lightning Source LLC
LaVergne TN
LVHW071014170726
843515LV00006B/1132

9783036563060